# Resistance to Targeted Anti-Cancer Therapeutics

## Volume 1

*Series Editor*

Benjamin Bonavida

For further volumes:
http://www.springer.com/series/11727

Benjamin Bonavida
Editor

# Molecular Mechanisms of Tumor Cell Resistance to Chemotherapy

## Targeted Therapies to Reverse Resistance

Springer

*Editor*
Benjamin Bonavida, Ph.D.
Department of Microbiology
Immunology & Molecular Genetics
University of California, Los Angeles
Los Angeles, CA
USA

ISBN 978-1-4614-7069-4 ISBN 978-1-4614-7070-0 (eBook)
DOI 10.1007/978-1-4614-7070-0
Springer New York Heidelberg Dordrecht London

Library of Congress Control Number: 2013937199



Printed on acid-free paper

Springer is part of Springer Science+Business Media (www.springer.com)

# Preface

Patients with cancer are treated with conventional chemotherapeutic drugs and/or radiation. Initially, the majority of patients respond well to such therapies. However, there is a subset of patients who does not respond initially and another subset who no longer responds to further treatments. Clearly, in those two subsets of patients, the cancer cells exhibit mechanisms of resistance to apoptotic stimuli, resulting in cancer progression and metastasis and ultimately death of the patients. One of the main challenges facing us to date is to develop new therapies to treat patients with the resistant tumors in an effort to prolong their survival and to preferentially achieve long-term cures. The development of new effective therapies will be dependent on delineating the biochemical, molecular, and genetic mechanisms that regulate tumor cell resistance to cytotoxic drugs-induced apoptosis. Such mechanisms have revealed gene products that directly regulate resistance and, thus, one may develop new drugs that target these resistance factors. Such new drugs may be either selective or common to various cancers. If successful, doses used by such new drugs may not be toxic and may be used effectively in combination with subtoxic conventional drugs to achieve synergy and to reverse the resistance. Clearly, the development of sensitizing drugs in combination should lead to better clinical responses in patients with the resistant tumors and poor prognosis. Above efforts and approaches have resulted in several FDA-approved drugs that were able to overcome drug resistance and are successfully used clinically. They have been used as monotherapy or synergized with other therapies for the treatment of resistant tumors.

This volume assembles several selective reviews by scientist leaders in the field of cancer drug resistance. Specific mechanisms in drug resistance are reviewed and novel approaches for each mechanism are being proposed for intervention based on the targets that regulate resistance in different cancers. This volume consists of a total of 12 chapters each of which has been selected to cover a different topic in the field of drug resistance. Clearly, these chapters only review selected topics and it was not the intent of this volume to review the whole field of drug resistance. The highlights of each chapter are briefly summarized below.

*Doctors Ambudkar and associates* describe the multiple resistance phenotypes mediated by the multiple drug resistance (MDR) gene products that are exhibited by many cancer types. They discuss the mechanisms by which the three

ATP-binding cassettes (multiple drug transporters, namely, ABCB1, ABCC1, and ABCG2) prevent cytotoxic drugs to mediate their cytotoxic effects. How these gene products regulate the drug transport and how a large number of chemically unrelated drugs are regulated by the same ABC drug transporter have been thoroughly discussed. Of interest, the authors report single nucleotide polymorphisms (SNPs) in the ABC transporters in certain human populations. These findings open the door to possibly individualize specific treatments of such populations. Moreover, this SNP polymorphism and its relationship to the ABC transporter functions for drugs may also be used as biomarkers and screening to predict a priori the patients' response to certain drugs. The authors also suggest various means of intervention to reverse drug resistance by the use, for example, of tyrosine kinase inhibitors and various novel products. *Doctors Daniel and Rauch* extend the review by Dr. Ambudkar and colleagues on the biochemical and molecular mechanisms upon which the MDR functions in cancer cells. They describe how the physical properties of the cell membrane affect MDR functions and, primarily, discuss the pH changes responsible for altering the cell membrane physical properties and MDR activities. They also proposed an integrated approach unifying all reported studies and mechanisms of drug resistance. *Doctors Freeman and Morrison* discuss the role of the constitutively activated Raf/MEK/ERK pathways in cancer and how inhibitors of these pathways initially resulted in the reversal of resistance. However, tumor cells also develop resistance to such inhibitors. The constitutive activation of the above pathways is due, in part, in the acquisition of oncogenic mutations in components of this pathway. For instance, it has been reported that Ras mutation is found in approximately 20 % of cancers and in melanoma, Ras mutation is found in over 50 % of cases. The development of resistance to inhibitors of the Raf/MEK/ERK pathway, however, is a critical clinical problem for the treatment of patients with such resistance. The authors provide convincing evidence of how the resistance to inhibitors may develop and how interventions may obviate therapy in otherwise resistant patients. While there are currently many active clinical trials assessing the therapeutic drugs targeted MEK and Raf kinases in cancer, it is clear that a large body of data will be generated assessing clinical responses in relationship with the genotype of the tumor. Therefore, it may be possible to select the proper inhibitor for a particular cancer and prevent the development of resistance to the inhibitor. (2) Doctors Tantravahi, Hoffman, and Reddy extend the above studies by Doctors Freeman and Morrison and review published work by others and theirs on the mechanism of resistance of cancer cells to B-Raf therapies. They discuss the FDA-approved B-Raf inhibitor Vemurafinib (Plexxikon) in melanoma and the development of resistance to this drug. In this review, they describe how the B-Raf signaling pathway regulates the development of melanoma and how diverse mechanisms regulate resistance and suggest approaches to be considered for the reversal resistance. The above authors also suggested a different approach than the one used by conventional approaches whereby most of targeted therapies used ATP mimetic small molecules as competitive inhibitors of kinases. They proposed that cancer cells cytotoxicity takes precedence over affinity toward the targets. Hence,

ATP mimetics can be assayed in cultures of tumor cells for cytotoxicity and safety. *Doctors Mukherjee and colleagues* review the role of $\beta 1$ integrins in drug resistance in lung cancer. They discussed that the interaction between transmembrane integrin receptors and ligands in the extracellular milieu is of paramount importance in the pathogenesis of lung cancer. They review the primary role of $\beta 1$ integrins that are preferentially expressed in lung cancer and their involvement in cell proliferation, survival, invasion, and metastasis. $\beta$-integrins are also involved in drug resistance. They raise the issue whether targeting $\beta 1$ integrins can reverse drug resistance and may be of therapeutic importance.

*Doctor Hara and colleagues* reviewed the critical role of aldo-keto reductases (AKR) in colon cancer. In addition to the reported factors that regulate drug resistance such as P-glycoproteins, growth factors, proteasomes, Dr. Hara and colleagues emphasize the important role of aldo-keto-reductases as well. These are upregulated and are involved in both carcinogenesis and resistance. The AKR family consists of NAD(P)(H)-dependent enzymes that catalyzes the oxidoreductase of a variety of substrates. They report that three members only, AKR1B10, AKR1C1, and AKR1C3 are important in the regulation of drug resistance. The mechanism by which these enzymes regulate resistance are also discussed. They suggest that the above three enzymes are not only diagnostic/prognostic marker candidates, but are themselves therapeutic targets. *Doctor Maiti* reviews the role of elevated levels of ROS in cancer cells in relationship to drug resistance. Based on the observation in ovarian cancer that anti-cancer drugs initially increase ROS levels and induce cytotoxicity whereas long exposure of the same drug reduces ROS levels and leads to drug resistance; they suggest that the constant maintenance of high ROS levels is critical. They explain that since most drugs induce toxicity in the cells based on the induction of ROS, cancer cells that have low levels of ROS may be resistant to drugs-induced cytotoxicity. The underlying mechanism of ROS generation and maintenance of genes that regulate ROS in drug resistance may identify new targets for elevation of ROS levels and sensitization to drugs-induced cytotoxicity. *Doctors Nie and colleagues* review a clinically relevant question regarding the inherent resistance of cancer stem cells (CSCs) to cytotoxic drugs. They reviewed the putative underlying mechanisms of CSC resistance and various approaches to target CSCs to improve the clinical response to chemotherapy. *Doctors Maga and colleagues* review the potential use of DNA repair in the reversal of drug resistance and propose to target both the signal transduction pathways and DNA repair pathways. They review the differences that exist between normal cells and cancer cells with respect to underlying mechanisms with respect to cell proliferation and genetic stability in an effort to unravel novel targets for therapeutic intervention. Combination treatment targeting signal transduction and DNA repair pathways should lead to the cancer cells' demise by apoptosis. *Doctors Efferth and colleagues* review an important mechanism in drug resistance, namely, "collateral sensitivity." Collateral sensitivity is defined as the hypersensitivity of otherwise drug-resistant cells. Collateral sensitivity has not been examined in detail when compared to a large number of mechanisms of drug resistance. The mode of action of collateral sensitivity of currently used new drugs has not been studied in detail. They suggest the development of drugs with high degree of collateral sensitivity to treat drug-resistant cancer.

*Doctors Hao and Bellail* reviewed the resistance of cancer cells to TRAIL-mediated apoptosis. Phase I and phase II clinical trials have examined recombinant TRAIL and agonist monoclonal antibodies to TRAIL DR4 and DR5 receptors. These clinical trials have not yielded any encouraging results, although, preclinically in mice, the results were highly encouraging. They suggest that it is possible that the previous analyses were done by using primarily tumor cell lines and that freshly derived tumor tissues may have either the absence of TRAIL apoptotic pathways or they may be defective. They propose to analyze freshly-derived human tumor tissues for analysis of TRAIL-mediated cytotoxicity. *Doctor Shekhar* reviews the paradoxical role of apoptosis in cancer cells. They propose that apoptosis results in the accumulation of genomic instability and the promotion of malignant progression of tumors, findings that may have been observed clinically. They suggest that rather than using combination of drugs to induce apoptosis, on the contrary, we have to use drugs that inhibit apoptosis.

I wish to acknowledge the assistance of Daphne Liang, Melissa Cao, Kathy Nguyen, and Suzie Vardanyan for their valuable assistance in the preparation and editing the various contributions in this volume.

Benjamin Bonavida, Ph.D.

# Contents

# Contributors

**Suresh V. Ambudkar** Laboratory of Cell Biology, Center for Cancer Research, National Cancer Institute, NIH, Bethesda, MD 20892, USA

**Sunil kumar Arora** Department of Immunopathology, Research block A, Post Graduate Institute of Medical Education and Research, Chandigarh, India

**Anita C. Bellail** Department of Pathology and Laboratory Medicine, Emory University School of Medicine, Atlanta, GA, USA

**Federica Bertoletti** Institute of Molecular Genetics, IGM-CNR, via Abbiategrasso 207, I-27100 Pavia, Italy

**Debasish Boral** Department of Medical Microbiology, Immunology, and Cell Biology, Southern Illinois University School of Medicine and Simmons Cancer Institute, Springfield, IL 62794, USA

**Emmanuele Crespan** Institute of Molecular Genetics, IGM-CNR, via Abbiategrasso 207, I-27100 Pavia, Italy

**Chloe Daniel** School of Veterinary Medicine and Science, University of Nottingham, College Road, Sutton Bonington, LE12 5RD, UK

**Thomas Efferth** Department of Pharmaceutical Biology, Institute of Pharmacy and Biochemistry, University of Mainz, Staudinger Weg 5, Mainz 55128, Germany

**Ossama El-Kabbani** Monash Institute of Pharmaceutical Sciences, Monash University, Parkville, Victoria 3052, Australia

**Alyson K. Freeman** Laboratory of Cell and Developmental Signaling, Frederick National Laboratory for Cancer Research, National Cancer Institute, Frederick, MD 21702, USA

**Henry Johannes Greten** Heidelberg School of Traditional Chinese Medicine, Heidelberg, Germany; Biomedical Sciences Institute Abel Salazar, University of Porto, Portugal

**Chunhai Hao** Department of Pathology and Laboratory Medicine, Emory University School of Medicine, Atlanta, GA, USA

**Akira Hara** Laboratory of Biochemistry, Gifu Pharmaceutical University, Gifu 501-1196, Japan

**Benjamin Hoffman** Onconova Therapeutics, Inc., 375 Pheasant Run, Newtown, PA 18940, USA

**Hongmei Jiang** Department of Medical Microbiology, Immunology, and Cell Biology, Southern Illinois University School of Medicine and Simmons Cancer Institute, Springfield, IL 62794, USA

**Khyati Kapoor** Laboratory of Cell Biology, Center for Cancer Research, National Cancer Institute, NIH, Bethesda, MD 20892, USA

**Miroslava Kissova** Institute of Molecular Genetics, IGM-CNR, via Abbiategrasso 207, I-27100 Pavia, Italy

**Giovanni Maga** Institute of Molecular Genetics, IGM-CNR, via Abbiategrasso 207, I-27100 Pavia, Italy

**Amit K. Maiti** Department of Arthritis and Clinical Immunology, Oklahoma Medical Research Foundation, Oklahoma, OK 73104, USA

**Parth Malik** Department of Biotechnology, Maharishi Markandeshwar University, Mullana, Ambala, India

**Toshiyuki Matsunaga** Laboratory of Biochemistry, Gifu Pharmaceutical University, Gifu 501-1196, Japan

**Deborah K. Morrison** Laboratory of Cell and Developmental Signaling, Frederick National Laboratory for Cancer Research, National Cancer Institute, Frederick, MD 21702, USA

**Tapan K Mukherjee** Department of Biotechnology, Maharishi Markandeshwar University, Mullana, Ambala, India; Department of Biology, IISER Mohali, Sector-26, Chandigarh 160036, India

**Srirupa Mukhopadhyay** Department of Immunopathology, Research block A, Post Graduate Institute of Medical Education and Research, Chandigarh, India

**Daotai Nie** Department of Medical Microbiology, Immunology, and Cell Biology, Southern Illinois University School of Medicine and Simmons Cancer Institute, Springfield, IL 62794, USA

**Cyril Rauch** School of Veterinary Medicine and Science, University of Nottingham, College Road, Sutton Bonington, LE12 5RD, UK

**E. Premkumar Reddy** Department of Oncological Sciences, Mount Sinai School of Medicine, Room 15-20 Icahn Medical Institute, 1425 Madison Avenue, New York, NY 10029, USA

**Mohamed Saeed** Department of Pharmaceutical Biology, Institute of Pharmacy and Biochemistry, University of Mainz, Staudinger Weg 5, Mainz 55128, Germany

**Malathy P. V. Shekhar** Tumor and Microenvironment Program, Karmanos Cancer Institute, Department of Oncology, Wayne State University School of Medicine, 110 East Warren Avenue, Detroit, MI 48201, USA

**Hong May Sim** Laboratory of Cell Biology, Center for Cancer Research, National Cancer Institute, NIH, Bethesda, MD 20892, USA

**Ramana V. Tantravahi** Onconova Therapeutics, Inc., 375 Pheasant Run, Newtown, PA 18940, USA

**Man-Tzu Wang** Department of Medical Microbiology, Immunology, and Cell Biology, Southern Illinois University School of Medicine and Simmons Cancer Institute, Springfield, IL 62794, USA

**Elisa Zucca** Institute of Molecular Genetics, IGM-CNR, via Abbiategrasso 207, I-27100 Pavia, Italy

# Multidrug Resistance in Cancer: A Tale of ABC Drug Transporters

**Khyati Kapoor, Hong May Sim and Suresh V. Ambudkar**

**Abstract** ATP-binding Cassette (ABC) drug transporters are present throughout prokaryotes as well as eukaryotes and have an important physiological role in detoxification and protection of sensitive tissues from toxic compounds. The overexpression of these transporters contributes to the development of multidrug resistance (MDR) by decreasing the intracellular concentration of therapeutic drugs. MDR is characterized by the ability of drug-resistant tumors to exhibit simultaneous resistance to a number of structurally and functionally unrelated chemotherapeutic agents. This chapter discusses the phenomenon of MDR and the proposed ways to overcome its challenges. It particularly focuses on how the three major ABC drug transporters, ABCB1, ABCC1 and ABCG2 hamper chemotherapy in the treatment of cancer. We provide an overview of the structural and functional aspects of these three transporters, discuss the role of single nucleotide polymorphisms in these transporters in the overall pharmacokinetics and pharmacodynamics of chemotherapeutic agents, and summarize current knowledge on the reversal of MDR mediated by these transporters using tyrosine kinase inhibitors and natural products as modulators.

**Keywords** ABC transporter • ABCG2 • Catalytic cycle • Drug efflux • Modulators • Multidrug resistance • MRP1 • P-glycoprotein • Tyrosine kinase inhibitors

## Abbreviations

| | |
|---|---|
| ABC | ATP-binding cassette |
| BCRP | Breast cancer resistance protein |
| CS | Collateral sensitivity |
| EGFR | Epithelial growth factor receptor |
| EPR | Electron paramagnetic resonance |

K. Kapoor · H. M. Sim · S. V. Ambudkar (✉)
Laboratory of Cell Biology, Center for Cancer Research, National Cancer Institute, NIH, Bethesda, MD 20892, USA
e-mail: ambudkar@mail.nih.gov

B. Bonavida (ed.), *Molecular Mechanisms of Tumor Cell Resistance to Chemotherapy*, Resistance to Targeted Anti-Cancer Therapeutics 1, DOI: 10.1007/978-1-4614-7070-0_1,

| | |
|---|---|
| GSH | Glutathione |
| HDL | High-density lipoprotein |
| HER | Human epidermal growth factor receptor |
| LTC4 | Leukotriene C4 |
| MAP | Mitogen-activated protein |
| MDR | Multidrug resistance |
| MRP | Multidrug-resistance protein |
| NBD | Nucleotide-binding domain |
| PDGFR | Platelet-derived growth factor receptor |
| P-gp | P-glycoprotein |
| SNP | Single nucleotide polymorphism |
| TK | Tyrosine kinase |
| TKI | TK inhibitor |
| TM | Transmembrane |
| TMD | Transmembrane domain |
| VEGFR | Vascular endothelial growth factor receptor |
| Vi | Sodium ortho-vanadate |

# Introduction

## *Multidrug Resistance in Cancer*

According to recent statistics from the American Cancer Society (http://www.cancer.org), the top three cancers in men are prostate, lung and colorectal cancers. In women, the lung, breast, colon and rectum are the cancer sites most frequently leading to death. About 45 % of all cancers are resistant to chemotherapy and the main hurdle to treating resistant cancer cells is the development of multidrug resistance (MDR). By definition, MDR is a phenomenon associated with the resistance of tumor cells to the cytostatic or cytotoxic actions of structurally dissimilar and functionally divergent drugs commonly used in cancer chemotherapy [1]. The mechanisms of MDR have been discussed in numerous reviews and can be generalized to the following mechanisms: (1) Over-expression of ATP-binding cassette (ABC) efflux transporters (2) decrease in drug uptake by solute carriers (3) inhibition of apoptosis (4) increase in DNA repair and (5) inactivation of chemotherapeutic drugs by metabolic enzymes [2–4]. In this chapter, we will focus on the role of ABC drug transporters in the development of MDR.

## *Intrinsic Versus Acquired Resistance*

The clinical problem of MDR can be broadly classified as either intrinsic or acquired. Intrinsic MDR refers to pre-existing resistance present in the tumor cells at the time of diagnosis, hence giving rise to the MDR phenotype, while acquired MDR occurs when resistance develops during the course of chemotherapy. To differentiate the two mechanisms of MDR in clinical samples, it is usually necessary to identify the population of self-renewing cells (also referred to as cancer stem cells) present in the tumor. Cancer stem cells have the ability to initiate and sustain the growth of a heterogeneous cancer through self-renewal and differentiation [5]. Often, to isolate this population of cancer stem cells is a challenge, as the properties of these cells depend primarily on their niche environment and they may not be easily identified within the tumor population. Cancer stem cells are able to evade cancer chemotherapy through many avenues. A recent review includes a thorough discussion of the mechanisms of MDR in cancer stem cells. It ascribes the MDR phenotype of cancer stem cells to certain features of the tumor microenvironment and immunosuppression, in addition to the increased expression of ABC transporters [6]. An alternative way for tumor cells to acquire MDR is proposed by Levchenko et al. [7] who suggest that cancer cells can acquire the MDR phenotype by functional intercellular transfer of P-glycoprotein (P-gp). While this concept is interesting, the exact mechanism is not known and the physiological relevance of this mechanism remains to be determined [8].

MDR is multi-factorial and the over-expression of ABC transporters is one major contributor to MDR. Nonetheless, MDR can also develop with anticancer drugs that are not substrates of ABC transporters and clinical MDR can be mediated by other mechanisms besides the involvement of ABC transporters. An example is 'atypical MDR', a term which was applied to a structurally varied class of topoisomerase inhibitors including doxorubicin and etoposide. This type of MDR is mediated by the altered expression of topoisomerase II [9]. Other mechanisms of acquired MDR largely involve the suppression of signaling pathways that leads to the apoptosis of cancer cells.

Recently, the notion of cancer drug pan-resistance has emerged, referring to the state of a tumor when it has acquired resistance to all available chemotherapy treatments and often also to ionizing radiation [10]. According to a review by Piet Borst, pan-resistance should be differentiated from MDR, as its underlying mechanisms of resistance are more perplexing than MDR mediated by drug efflux pumps. This review also explains pan-resistance of cancer cells as mimicry, superior defense or compensation [10]. Cancer cells adopt the proliferation strategies of normal cells and thus mimic them. Also, cancer cells can set up a strong defense system by upregulation of expression of ABC efflux pumps (e.g., P-gp) or by compensation mechanisms by which the cancer cells adapt indirectly without affecting drug-target interaction.

To counteract MDR as well as pan-resistance of cancer cells mediated by the over-expression of efflux transporters such as P-gp, the use of agents that specifically target cells with upregulated P-gp can be employed through a strategy referred to as collateral sensitivity (CS). CS is the ability of small molecules to selectively kill multidrug-resistant cells and not their parental cells [11, 12]. Examples of compounds that promote CS include verapamil [13], desmosdumotin B analog [14], NSC73306 [15] and tiopronin [16]. Although the development of CS agents to overcome MDR is a novel strategy, the precise mechanisms for the action of these CS agents have yet to be elucidated and warrant more research.

## ABC Transporters and their Participation in Diseases

The ABC superfamily is one of the largest superfamilies of transporters, found in all living kingdoms from prokaryotes to eukaryotes. The transporters of this family perform a wide variety of functions. At present there are 48 known ABC members in the human genome and the over-expression of many of these results in various human diseases. These transporters can be subdivided into seven sub-families (A-G) based on structural and sequence similarities. Out of these seven, three subfamilies are commonly linked to MDR. They are the 'B' subfamily, which includes P-gp (MDR1 or ABCB1), the 'C' subfamily, which includes multidrug-resistance protein (MRP1 or ABCC1) and the 'G' subfamily, which includes breast cancer resistance protein (ABCG2 or MXR). P-gp, ABCG2 and MRP1 are the three most common ABC transporters reported to contribute to the development of MDR. Their contribution to resistance is still not well defined, largely because of the difficulties associated with the quantification of the expression level and the activity of these ABC transporters in tumor tissues compared to that in normal tissues in carefully designed trials [17]. Besides these three heavily investigated ABC transporters that are linked to MDR, there are at least 11 other members that can transport drugs [18]. They are ABCA2 [19], ABCA3 [20], ABCB4 [21], ABCB5 [22, 23], ABCB11 [24], ABCC2 [25], ABCC3 [26], ABCC4, ABCC5 [27], ABCC10 [28] and ABCC11 [29]. Defects in 14 ABC transporters cause 12 known genetic diseases, the most common of which are cystic fibrosis, Stargardt disease, adrenoleukodystrophy, Pseudoxanthoma elasticum and Tangier disease (Table 1).

Other than ABCC8 and ABCC9, which encode proteins not known to transport drugs, the ABCC family consists of members that are involved in the transport of various types of molecules. Some of the molecules are organic anions formed from conjugated drug metabolites. The MRPs also transport endogenous metabolites such as cysteinyl leukotrienes, prostaglandins, steroids and glucuronide conjugates. ABCC10 (MRP7) has a membrane topology similar to that of other MRPs but has the least sequence identity to them. The expression of ABCC10 has been seen in small-cell lung cancer cells after exposure to paclitaxel. *In vitro*, ABCC10 is known to confer resistance to taxols such as docetaxel and paclitaxel and a few other compounds such as vincristine, vinblastine and epothilone. Studies indicate MRP7 functions as a lipophilic anion transporter [30].

**Table 1** Disease-associated human ABC transporters

| ABC transporter | Associated disease[a] |
|---|---|
| ABCA1 (ABC1) | Tangier disease and familial HDL deficiency |
| ABCA4 (ABCR) | Stargardt disease, age-related macular degeneration, retinitis pigmentosa |
| ABCB1 (P-gp) | Cancer, inflammatory bowel disease |
| ABCB4 (MDR2) | Progressive familial intrahepatic cholestasis |
| ABCB7 (ABC7) | Sideroblastic anemia and ataxia |
| ABCB11 (SPGP) | Progressive familial intrahepatic cholestasis |
| ABCC1 (MRP1) | Cancer, chronic obstructive pulmonary disease, cystic fibrosis |
| ABCC2 (MRP2) | Dubin-Johnson syndrome |
| ABCC6 (MRP6) | Pseudoxanthoma elasticum |
| ABCC7 (CFTR) | Cystic fibrosis |
| ABCC8 (SUR1) | Persistent hypoglycemia of infancy, neonatal diabetes |
| ABCC9 (SUR2) | Persistent hypoglycemia of infancy, neonatal diabetes |
| ABCD1 (ALD) | Adrenoleukodystrophy |
| ABCG2 (BCRP) | Cancer, gout |
| ABCG5 | Sitosterolemia |
| ABCG8 | Sitosterolemia, gallstone disease |

[a] In some instances association with disease condition is due to non-synonymous single nucleotide polymorphisms [125]

The localization of ABCC2, also known as MRP2, is exclusively on the apical membrane of polarized epithelial and endothelial cells in the liver, kidney and intestines. A mutation in the *ABCC2* gene results in an autosomal recessive disorder known as Dubin-Johnson syndrome. Individuals harboring this mutation have increased serum-conjugated bilirubin levels, as the biliary elimination of bilirubin glucuronides is impaired [30]. Both ABCC1 and ABCC2 are able to confer resistance to a variety of natural product anticancer drugs. However, unlike ABCC1, ABCC2 is able to confer resistance to an important class of chemotherapeutics including platinum-containing drugs [17, 31–34]. Despite its role in conferring resistance to these drugs, the clinical significance of ABCC2 is limited, as there is no clear association between its expression and clinical outcome [17].

In 2003, ABCB5 was reported to regulate rhodamine transport in progenitor cells [35]. ABCB5 has also been identified in malignant melanoma-initiating cells (MMICs) and its expression was reported to be associated with the resistance of melanoma cells to a broad range of anticancer drugs and/or of melanocytes to toxic melanin intermediates and metabolites [36]. ABCB5-transfected HEK293 cells showed higher resistance to many anticancer drugs, which are also known substrates of P-gp. In addition, a reversal of this resistance in the transfectants was observed upon treatment with ABCB5 siRNA, thus indicating that the full-length ABCB5 may play a role in MDR in melanoma cells. However, additional work will be required to better understand the function of this transporter [22].

## Tissue Localization and Substrate Specificity of ABC Transporters

Out of the 48 ABC proteins known in humans [37], only 14 appear to transport chemotherapeutic drugs [38]. Further, only three have been demonstrated to have an impact on drug resistance observed in the clinic. There are differences observed in their functions, substrate specificities, molecular mechanisms, and in vivo localizations. We will limit ourselves here to the discussion of only P-gp, ABCG2 and MRP1, which are known to be the major ABC drug transporters linked to MDR in cancer [39]. In addition to their role in drug resistance, they are expressed in non-malignant tissues and are believed to be involved in protecting those tissues from xenobiotic accumulation and resulting toxicity [31]. The localization of these three transporters in various human tissues is listed in Table 2.

One of the well-studied and best-characterized ABC transporters is P-gp, which was first discovered by Juliano and Ling because of its involvement in the development of MDR in cancer cells [40]. P-gp is expressed at the apical membranes of epithelial cells in kidney proximal tubules, the lungs, intestines, liver, brain microvascular endothelia, blood-nerve barrier, placenta, adrenal cortex, blood-testis barrier, uterus, lymphocytes, and in hematopoietic cells [41]. Its primary pharmacological function is to prevent the uptake of toxic compounds from the gut into the body, to expel them in the bile, urine or feces and to protect some very sensitive organs, such as the brain, from them [42–44]. P-gp is also involved in other physiologic processes, such as the control and regulation of apoptosis, stress, hypoxia, stem cell differentiation, cellular immune response and plasma membrane dynamics [45–47].

The second transporter that impacts the pharmacology of chemotherapeutic drugs and the latest to be discovered is ABCG2. Since this gene was isolated from a breast cancer cell line, it was named BCRP [48–50]. It is a half transporter, but functions as a homodimer or oligomer. In addition to its expression in stem cells, ABCG2 localizes mainly in the placental syncytiotrophoblast plasma membrane, the bile canalicular membrane of liver hepatocytes, the luminal membrane of villous epithelial cells in the small and large intestine, the apical side of ducts and lobules in the breast, and in the venous and capillary endothelial cells of almost all tissues [51, 52]. The tissue distribution of ABCG2 overlaps with that of P-gp, which indicates that it might have similar pharmacological functions as P-gp. The

**Table 2** Tissue localization of three ABC drug transporters

| ABC transporter | Tissue localization |
|---|---|
| ABCB1 (P-gp) | Intestine, liver, kidney, placenta, adrenal gland, blood–brain barrier, stem cells and hematopoietic cells |
| ABCG2 (BCRP) | Placenta, intestine, breast, liver, blood–brain barrier, liver, kidney and stem cells |
| ABCC1 (MRP1) | Ubiquitous and present in all tissues |

localization of these three ABC drug transporters strongly suggests their importance in restricting the penetration of toxic substances in crucial tissues such as the central nervous system, intestines, and in the fetus and hematopoietic cells, thus playing a protective role.

Another ABC transporter, MRP1, is one of the best-characterized members of the MRP family and was identified by Cole et al. in 1992 [53]. The main physiological function of MRP1 is presumably leukotriene C4 transport. It is also known to maintain glutathione homeostasis in vivo. MRP1 is expressed in most tissues throughout the body, with relatively high levels found in the lung, testis, kidneys, skeletal muscle, and peripheral blood mononuclear cells. Unlike P-gp, it is mainly localized to the basolateral membrane of epithelial cells [54]. The endogenous expression of MRP1 can contribute to the basal resistance of cell lines to a wide range of chemotherapeutic drugs [55, 56].

Of the other ABCC subfamily members, ABCC4 (MRP4) has been found to be present at higher levels in the prostate and is implicated in the aggressiveness of prostate tumors and neuroblastomas [57, 58]. Depending on the tissue where MRP4 is found, it can be localized to either the basolateral or the apical membranes in polarized cells. In prostate cells and hepatocytes, MRP4 is localized to the basolateral membrane while in renal proximal tubules and brain capillaries, it is found on the apical side [59, 60]. MRP5 is detected at higher levels in brain capillary endothelial cells, pyramidal neurons and astrocytes and in smooth muscle cells of tissues in the genitourinary system [60, 61]. MRP5 can also be localized either on the apical side (in brain capillary endothelial cells) or the basolateral side (some polarized epithelial cells) [60, 62]. *ABCC11* (*MRP8*) is considered to be a non-essential gene, since no orthologous genes have been found in mammals, except in primates [17]. Full-length MRP8 is reported to localize to the apical membrane in stably-transfected polarized epithelial cells [17] and this protein has been detected in axons of neurons in human central and peripheral nervous systems [63].

## Topologies of ABC Transporters

The topology of ABC transporters typically consists of four domains, which include two transmembrane domains (TMDs) and two nucleotide-binding domains (NBDs), located in the lipid bilayer and the cytoplasm, respectively. These four domains of ABC transporters form the minimum functional unit needed for transport of substrates across the lipid bilayer, powered by the energy from ATP binding and hydrolysis at the NBDs. ABC transporters have varying domain organizations and can exist as a complex of four separate single-domain polypeptides of two NBDs and two TMDs (e.g. bacterial maltose or histidine permease) or they can be two half-transporters with each half of the transporter having one NBD and one TMD, such as members of the ABCG subfamily (ABCG2/BCRP). They can also exist as a single four-domain polypeptide chain, such as P-gp, which has interconnecting NBDs and TMDs. Interestingly, in some of the ABCC subfamily members including MRP1, besides the four critical domains; there is an additional N-terminal TMD (TMD0) that is smaller than the core TMDs. It has been shown that the deletion of TMD0 does not significantly affect function, but the linker region between TMD0 and TMD1 is important for activity [30, 64].

## Substrate Specificity of ABC Drug Transporters

The most striking property of the ABC drug transporters is their broad substrate specificity, which leads to the efflux of a large number of chemically unrelated hydrophobic or amphipathic compounds from the cells. The substrate specificities of the three ABC drug transporters are notably overlapping, but each transporter can handle its own unique set as well (http://www.genecards.org/). P-gp can interact with a vast array of compounds including natural products, chemotherapeutic drugs, steroids, linear and cyclic peptides, fluorescent dyes and ionophores. Amphipathicity and hydrophobicity are a few of the common structural characteristics of P-gp substrates. Because of their hydrophobic nature, these compounds can easily cross the plasma membrane and penetrate into tissues and other pharmacological compartments. The half transporter ABCG2 also has a remarkable range of substrate specificity, and confers resistance to mitoxantrone, topotecan, methotrexate and many other drugs [50, 65]. On the other hand, MRP1 functions as co-transporter of amphipathic organic anions. Further, it extrudes drugs conjugated to glutathione (GSH), glucuronate, or sulfate. MRP1 also mediates the co-transport of unconjugated amphiphilic anions, together with free GSH. The high affinity substrates of MRP1 are leukotriene C4 (LTC4), 17 $\beta$-glucuronosyl estradiol and bis-glucuronosyl bilirubin [66–68]. The overlapping substrate specificity of the three drug transporters P-gp, ABCG2 and MRP1, as listed in Table 3, is very clear.

## ABC Transporters: Polymorphisms and Pharmacogenomics

ABC transporters often have naturally occurring genetic polymorphisms, mainly single nucleotide polymorphisms (SNP), which result in changes in a single nucleotide (substitution, deletion or insertion) within the DNA sequence, some of which may or may not have an impact on the expression and function of the protein. Many SNPs, both synonymous (sSNP) as well as non-synonymous (nsSNP), have been identified in human ABC transporters. Although synonymous mutations are thought to have no effect on protein structure, there is a possibility that the secondary structure of the mRNA could be affected in such a way that its stability is altered, thus leading to a change in the protein expression levels as well as the functional properties [69]. Polymorphisms in P-gp and ABCG2 are widely

**Table 3** Substrate specificity of three ABC drug transporters

| Substrate | P-GP | ABCG2 | MRP1 |
|---|---|---|---|
| Apatinib | + | + | – |
| Bisantrene | + | + | – |
| Calcein | – | – | + |
| Calcein AM | + | – | + |
| Colchicine | + | – | – |
| Cyclosporine A | + | – | – |
| Daunorubicin | + | + | + |
| Dihydropyridines | + | + | – |
| Digoxin | + | – | – |
| Diltiazem | + | – | – |
| Docetaxel | + | – | – |
| Doxorubicin | + | + | + |
| Estrone-3 sulfate | – | + | + |
| Erythromycin | + | – | – |
| Etoposide | + | + | + |
| Flavopiridol | – | + | – |
| Fluo-3-AM | + | – | + |
| H33342 | + | + | – |
| Indinavir | + | – | – |
| $LTC_4$ | – | – | + |
| Losartan | + | – | – |
| Methotrexate | – | + | + |
| Mitoxantrone | + | + | + |
| Nelfinavir | + | – | – |
| NEM-GS | – | – | + |
| Nilotinib | + | + | – |
| Paclitaxel | + | – | – |
| Pheophorbide A | – | + | – |
| Prazosin | + | + | – |
| Sparfloxacin | + | – | – |
| Sulfasalazine | – | + | – |
| Topotecan | + | + | + |
| Uric acid | – | + | – |
| Verapamil | + | – | – |
| Vinblastine | + | – | + |
| Vincristine | + | – | + |

+ indicates a substrate, – indicates a non-substrate

discussed, as they can alter the pharmacokinetics and pharmacodynamics of chemotherapeutic drugs. These SNPs can result in changes in substrate specificity, thus altering drug responses in certain individuals. Details on most of these polymorphisms are available in several databases (www.ncbi.nlm.nih.gov/projects/SNP and www.genecards.org).

There are over 60 known SNPs in the coding region of human P-gp, which include both synonymous SNPs as well as non-synonymous SNPs. These SNPs have been discussed in detail in a recent review by Wolf et al. [70]. Genetic polymorphisms in P-gp have been reported to change the mRNA expression, protein expression and function of P-gp [71]. Since P-gp plays a role in cellular defense against toxic substances, it is not surprising that polymorphisms in this transporter have been linked to inflammatory bowel diseases such as Crohn's disease and ulcerative colitis [72, 73] (Table 1). Some polymorphisms are also known to increase the risk of renal epithelial tumors. Most of the SNPs in the coding region of P-gp can be mapped to the homology model of human P-gp based on the crystal structure of mouse Abcb1 (mdr1a). Notably, the occurrence of SNPs in the transmembrane region of the protein is much lower in comparison to the extracellular or intracellular regions of the protein [70]. Two synonymous SNPs and 12 non-synonymous SNPs have been associated with differences in function or expression of P-gp. Further, none of these SNPs is known to result in non-functional P-gp. Thus, the exact functional consequences of most of the polymorphisms are not yet clear. Naturally occurring null mutations in P-gp have been reported in mice and dogs but not yet in humans [74, 75]. Expressing human ABC proteins and their variants in yeast allows the function of individual variants to be assessed directly [76].

Analysis of the *ABCG2* gene has identified over 80 SNPs and other naturally occurring mutations in various ethnic groups. Individuals carrying these polymorphisms are predicted to show altered drug responses due to reduced activity of this efflux pump. Some polymorphisms even affect the stability of ABCG2 in cellular organelles, leading to increased degradation of the transporter [77, 78]. Most of the pharmacogenomic effects of ABCG2 polymorphisms are with respect to its substrates. One of the most characterized SNPs in ABCG2 is the Q141K mutation, which is located between the Walker A and signature motifs in the NBD. Similar to mutations in ABCC7 (CFTR), the Q141K mutation affects cell surface expression of ABCG2 [79]. This SNP has been shown to be associated with Gout disease [80] by genome-wide association studies [81]. The Q141K mutation is commonly found in Chinese and Japanese patients and affects their response to chemotherapy. ABCG2 mediates transport of urate and the Q141K mutation affects this function, which would lead to increased accumulation of uric acid in the kidneys of Gout patients. Another ABC transporter gene, *MRP1*, displays high haplotype diversity; a number of polymorphisms have been determined and many more are being investigated [82]. However, most of the polymorphisms in MRP1 do not appear to have any effect on drug disposition and only a very few have been associated with clinical disease or altered drug response.

Pharmacogenomic analysis is a way to predict the response of an individual to a certain drug, which can lead to improved treatment strategies. Drug therapy could be more effective if we were able to tailor drugs to an individual's genetic makeup. Polymorphisms in ABC drug transporters definitely affect the pharmacokinetics and pharmacodynamics of drugs. However, there are many other factors, such as age, race and diet, which will continue to influence the drug responses in different individuals.

## Structure and Mechanism of ABC Drug Transporters

Despite their substrate diversity, ABC transporters share a basic modular structure. It has been well established that the TMDs harbor the drug-binding sites, while the more conserved NBDs are involved in the ATP binding and hydrolysis. We further explain the structure and mechanism of a typical ABC transporter in the sections below.

### *Drug-Binding Sites of P-gp*

While the NBDs across the superfamily are highly conserved in primary sequence, the TMDs vary between the subfamilies of ABC transporters in both the primary sequence and protein folding. These differences in the TMDs lead to differences in the substrate specificities of the transporters. Some ABC transporters recognize very specific substrates, such as the *E. coli* vitamin B12 importer, BtuCD. Others, such as P-gp, are highly promiscuous and recognize a broad range of chemically and structurally diverse compounds, a property known as polyspecificity. Furthermore, the highly flexible nature of P-gp allows it to have a polyspecific nature, binding at least two chemically different drug molecules simultaneously. Identifying the residues in transmembrane (TM) helices that line the drug-binding pocket of P-gp has been intensively investigated by Loo and Clarke, who used cysteine scanning mutagenesis and cross-linking experiments, reporting that TM helices 4, 5, 6, 7, 10, 11 and 12 contribute to drug-binding [83–85]. Using methanethiosulfonate cross linkers and stipiamide derivatives of different lengths, the central putative drug-substrate binding region was proposed to be large enough to bind two drugs simultaneously [86–89]. With the determination of the structure of mouse P-gp, the structural basis for the polyspecificity of P-gp is explained by the large central cavity of 6,000 Å, involving $\pi$–$\pi$ and cation-$\pi$ stacking as well as hydrophobic and hydrogen bonding interactions between structurally diverse ligands and this transporter [90].

The crystal structures of some ABC exporters have been resolved, including those of mouse Abc1a (mdr1a) in the apo conformation [91] and bacterial Sav1866 in the ADP-bound conformation [92]. Most recently, a high resolution structure of *Caenorhabditis elegans* P-gp in the apo conformation was published [93]. In addition, the structure of the bacterial MsbA homodimer has been reported in various conformations [94]. The X-ray crystal structure of mouse P-gp was reported at resolutions of 3.8–4.4 Å [91], in the apo state in which the substrate is absent and in states in which stereoisomeric cyclic hexapeptide inhibitors (QZ59-RRR and QZ59-SSS) are bound to the transporter. The stereoselectivity of P-gp is based on differences in hydrophobic and aromatic interactions between the protein and the isomers. In the inward-facing apo conformation (in the shape of an inverted 'V'), the NBDs in the apo conformation are wide apart, separated by approximately 30 Å. To convert from the inward facing to the outward-facing conformation (in the shape of a regular 'V'), the protein has to undergo drastic conformational changes involving the TMDs, following a bellows-like mechanism of action that is commonly proposed for ABC transporters [95]. The evidence for a regular 'V' shape of the outward conformation stems from the crystal structure of Sav1866 (a bacterial homolog of P-gp) [92], in which the NBDs are in a close arrangement with ADP in both the nucleotide-binding pockets and the TMDs are open on the extracellular side, forming a central cavity. The homology models of human P-gp using the structures of mouse P-gp in apo conformation and Sav1866 in ADP-bound conformation is given in Fig. 1.

The validity of the crystal structure of mouse P-gp is controversial [95] with regards to the large separation of the NBDs and whether the solved apo conformation is physiologically relevant [96]. Nonetheless, there is also a body of evidence that supports the mouse P-gp model. The crystal structure of the lipid A exporter, MsbA, in the apo state reveals a large distance of ~50 Å between the

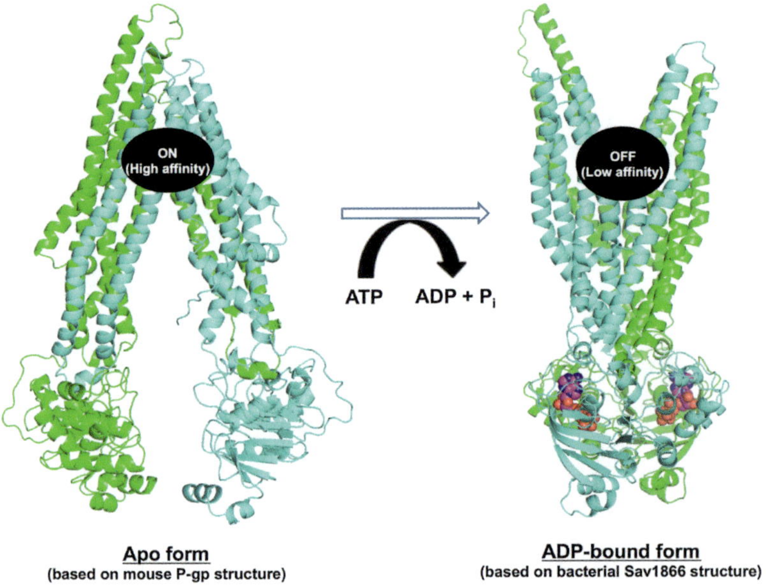


**Fig. 1** *Conformation of P-gp before and after ATP hydrolysis.* The homology model of human P-gp in apo (open) form based on mouse P-gp structure [91] and in ADP-bound (closed) form based on Sav1866 structure [92] is shown. Binding of ATP *per se* has no effect on the conformation of either the ATP site or the transport substrate site(s). However, formation of the P-gp•ATP (E•S) reaction intermediate is accompanied by conformation changes in the drug binding site. These changes result in the high affinity (or "ON") site being transformed to the low affinity (or "OFF") site in the drug-binding site. "ON" site refers to the drug-binding site that is exposed to the cytoplasm and affinity for the drug substrate is high. "OFF" site refers to the drug-binding site that is exposed extracellularly with decreased affinity for the drug substrate

NBDs, although the resolution is poor. Using another technique known as electron paramagnetic resonance (EPR), the distance between the NBDs of MsbA are reported as 55 Å in the apo state and 27 Å in the ADP-Vi trapped state.

The crystal structure of *C. elegans* in the apo state was solved recently at a resolution of 3.4 Å [93]. *C. elegans* P-gp, like mouse P-gp, confers a multidrug-resistant phenotype in the insect cells [93]. Compared to mouse P-gp, which has 87 % sequence identity to human P-gp, *C. elegans* P-gp is only 46 % identical. However, similar to the mouse crystal structure, which was determined at the slightly lower resolution of 3.8 Å, *C. elegans* P-gp also showed an inward-facing conformation and the NBDs were seen to have a larger degree of separation as compared to the NBDs of mouse P-gp. In the apo form of *C. elegans* P-gp, the distance between the two serine residues located at the ABC signature motif and the Walker A motif is 35 Å. However, in the mouse P-gp, the same pair of serine residues would be separated by a shorter distance of 19 Å. This inward-facing conformation of the *C. elegans* P-gp in an inverted 'V' shape is similar to the crystal structure of mouse P-gp in the apo conformation, which is not consistent with the in silico model proposed by Jones and George [97]. The in silico model involves constant contact of the NBDs throughout the transport cycle, oscillating around the vertical axis to allow alternating access to the two nucleotide-binding pockets.

Furthermore, in the crystal structure of *C. elegans* P-gp, the drug transport pathway is open to the cytoplasmic surface and is continuous with the membrane inner leaflet. This structure is consistent with the hypothesis that drugs enter ABC exporters through the membrane's inner leaflet according to "hydrophobic vacuum cleaner" model instead of entering directly from the cytoplasm (the "flippase" model). An interesting structural feature observed in *C. elegans* P-gp that is different from the mouse P-gp is the discontinuity of both helices TM10 and TM12 lining the lateral opening of *C. elegans* P-gp. Between the discontinuous TM helices are extended loops that can potentially mediate drug binding. These loops flanking the lateral opening are also highly flexible and can function as hinges to gate the pathway of drug transport. These loop structures may account for the polyspecificity of P-gp. Although the crystal structures of mouse and *C. elegans* P-gp may help reveal the molecular basis of polyspecificity, there is still a significant gap in our understanding of the molecular complexity of multiple drug-binding sites in human P-gp.

## *Structure of NBDs of ABC Transporters*

The transport of substrates by ABC transporters is driven by the binding and hydrolysis of ATP. Even within the ABC exporters, the catalytic and transport mechanisms are not fully conserved [98, 99]. However, the structure and function of NBDs is quite conserved among ABC transporters. Each NBD contains an A-loop, the Walker A and the Walker B motifs, the signature motif (also known as the C motif with consensus sequence LSGGQ), and the D- and H-loops. The two

NBDs are arranged in a head-to-tail manner with the two ATPs sandwiched between the Walker A and B motifs from one NBD and the ABC signature motif from the apposed NBD to form two nucleotide-binding sites. The signature motif as well as the Walker A motifs are involved in the binding of ATP, while the Walker B motif is involved in $Mg^{2+}$ and water co-ordination at the catalytic sites. Other conserved motifs in the NBDs include the Q-loop, which has a role similar to that of Walker B, the A loop (upstream of the Walker A motif), which is a conserved aromatic residue that interacts with the adenine ring of the bound ATP to form $\pi$-$\pi$ interactions [100], the D loop (downstream of Walker B motif), which is involved in indirect co-ordination of the $\gamma$-phosphate of ATP through a water molecule [101] and lastly the H-loop that participates in the hydrogen bonding of the $\gamma$-phosphate of ATP. An interesting feature to note is the presence of the X-loop (with consensus sequence TEVGERG), which is only present in ABC exporters. These are involved in interdomain communication and cross-talk between NBDs and TMDs.

## *Drug Transport Models for P-gp*

Two main models that describe drug transport by P-gp are the "hydrophobic vacuum cleaner" and the "flippase" models (Fig. 2). These two models involve the hydrophobic or amphipathic drug partitioned in the lipid bilayer, unlike the normal action of a transporter that expels water soluble substrates directly from the cytoplasm to the extracellular medium (pump model on the left in Fig. 2). P-gp substrates are relatively hydrophobic or amphipathic and are readily partitioned in the lipid bilayer by diffusion. P-gp has been suggested to function as a "hydrophobic vacuum cleaner", binding nonpolar substances that are partitioned in the membrane and then expelling them into the extracellular medium [102, 103]. This model is supported by the recent X-ray crystal structure of mouse P-gp where two stereoisomeric cyclic peptides (QZ59-RRR and QZ59-SSS) are shown bound deep within the transmembrane helices of mouse P-gp. This suggests that these cyclic peptides can gain access to the protein from within the lipid bilayer [91]. Alternatively, P-gp has been suggested to function as a "flippase" for physiological substrates such as lipid-like drugs and platelet-activating factor [103–105], moving these substrates from the cytoplasm to the extracellular medium by flipping from the inner leaflet to the outer leaflet of the lipid bilayer. At present, the experimental data are lacking to distinguish between the vacuum cleaner and the flippase model for P-gp.

## *Transport Cycle of P-gp*

As P-gp is the prototypical ABC exporter involved in MDR, there are extensive studies investigating the molecular features of its catalytic cycle and drug

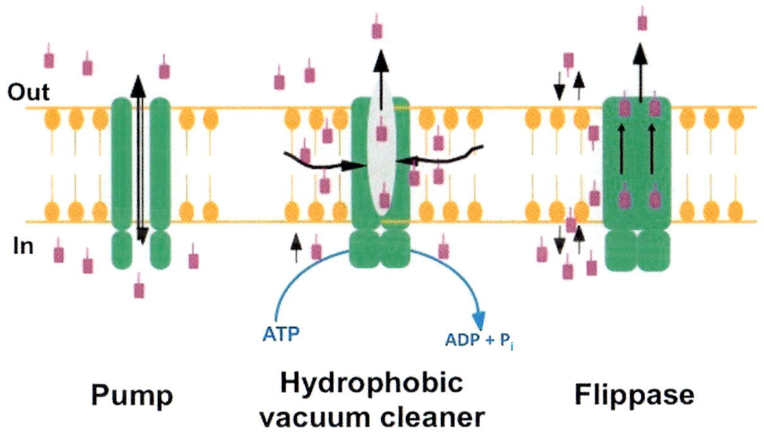


**Fig. 2** *Models for drug transport by P-gp.* Substrates of P-gp are proposed to be transported via two models, which involve the partitioning of the amphipathic or hydrophobic substrate molecule such as an anticancer drug (▮) into the lipid bilayer. In the conventional pump model (*left*), a water soluble substrate such as calcium, sodium, potassium or proton is directly effluxed out of the cell from the cytoplasm to the extracellular medium by P-type ATPases [195]. In the "hydrophobic vacuum cleaner" model (*middle*), the drug molecule from the extracellular medium is partitioned into the lipid bilayer and is effluxed out of the cell by P-gp, with only a few molecules entering the cytoplasm [196]. In the "flippase" model (*right*), the drug molecule is partitioned into the lipid bilayer either from the extracellular medium or from the cytoplasm, and then it is flipped from the inner leaflet to the outer leaflet and subsequently effluxed out of the cell by P-gp [103]. In all cases, the energy for transport is provided by hydrolysis of ATP by the NBDs. *Black arrows* indicate the direction of movement of a drug molecule

transport. The functional NBDs were shown to be essential for the hydrolysis of ATP and the inactivation of either one of the NBDs, for example, by the trapping of an inorganic phosphate analogue, orthovanadate ($V_i$), which inhibits this hydrolysis [106]. It is recognised that in order to hydrolyze ATP, the NBDs must dimerize to form an 'ATP sandwich dimer' with two ATP molecules bound along the interface [107]. However, there are many aspects of the catalytic and transport cycle that are still not well understood. Molecular details on how ATP hydrolysis is co-ordinated between the two NBDs, and how energy from this hydrolysis is transmitted to the TMDs to drive drug transport and whether one or two ATP molecules is/are hydrolyzed per drug molecule transported, are still unknown [100, 108, 109].

In 1966, in his report on membrane pumps, Jardetzky introduced the current consensus view of the mechanism of ABC exporters to be that of the alternating-access switch model [110]. As proposed, a transport event should involve at least three steps: (1) binding of the transport substrate to the TMDs in the 'high-affinity inward-facing orientation' (2) the binding of ATP to the NBDs to form the 'ATP sandwich dimer' and (3) ATP hydrolysis, which leads to transmission of conformational changes from the NBDs to the TMDs to effect a high-affinity to low-affinity switch at the substrate-binding site, as illustrated in Fig. 1. Notably, both nucleotide and the transport substrate can bind in the absence of each other. Different models have been proposed to identify the power stroke step which is used to drive the conformation from a high to a low affinity for drug-substrates. Is the power stroke obtained from ATP dimerization in the ATP sandwich or is it from ATP hydrolysis? Our work suggests that the power stroke for conformational change is provided only after formation of a pre-hydrolysis transition-like (E•S) state [111, 112].

## *Catalytic Cycle of P-gp*

P-gp is energized by the hydrolysis of ATP to efflux substrates out of the cell. Senior et al. [113] proposed that there are two independent events of ATP hydrolysis in a single catalytic cycle and at a given time, only one NBD hydrolyzes ATP. Therefore, the NBDs work via an "alternating catalytic sites" system with no preference for either the N- or the C-terminal ATP site during ATP hydrolysis [114]. A simplified reaction scheme for ATP hydrolysis by P-gp is given in Fig. 3. The end products of ATP hydrolysis are inorganic phosphate ($P_i$) and ADP. Preceding the hydrolysis, there is evidence for the formation of an asymmetric

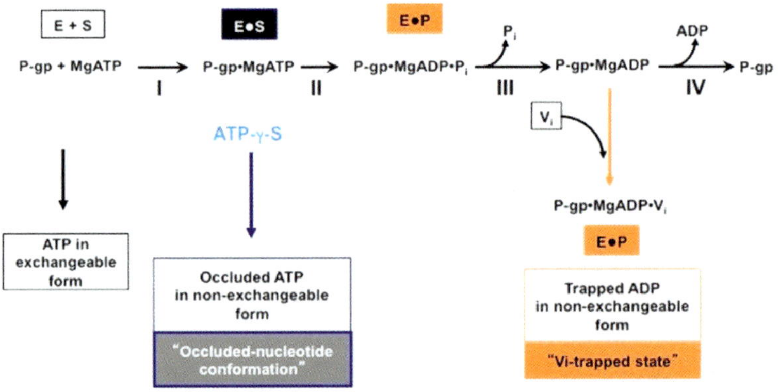


**Fig. 3** *Reaction intermediates in the catalytic cycle of ATP hydrolysis by human P-gp.* A simplified scheme for the ATPase reaction of P-gp is shown. The P-gp•MgATP intermediate is equivalent to the E•S complex, which represents the "occluded-nucleotide conformation" that can be obtained with the use of ATP-γ-S, a non-hydrolyzable analog of ATP or by using the Walker B E556Q/E1201Q double mutant [112]. One of two ATPs is occluded at the ATP site and only the ATP that is occluded is committed to hydrolysis. Similarly, the P-gp•MgADP•$P_i$ intermediate is equivalent to the E•P complex and can be captured using Vi, which is an analog of $P_i$, to form a "Vi-trapped state". The occluded pre-hydrolysis intermediate state at the NBD is proposed to be the "power stroke" for conformational change at the drug-binding site(s) in the TMDs [197]. The release of ADP at step IV (E + P) appears to be the rate-limiting step in the catalytic cycle [198]

occluded nucleotide as a reaction intermediate in the catalytic cycle of P-gp. This occluded reaction intermediate (E•S) is different from the nucleotide-bound state of P-gp. In the occluded nucleotide conformation, P-gp has reduced affinity for transport substrates. The role of this intermediate was characterized by Sauna et al. [112] using ATP-$\gamma$-S, which is a non-hydrolyzable analog of ATP. Those studies demonstrated that two ATP molecules initiate dimerization but only one is driven to an occluded pre-hydrolysis intermediate state at the NBD and this is the basis for the power stroke for conformational changes at the TMDs resulting in the movement of the drug substrate. Similarly, the ADP-Vi trapped post-hydrolysis conformation (E•P) also exhibits reduced affinity for drug-substrates [115].

## Modulation of ABC Drug Transporters

Compounds having the ability to reverse resistance against anticancer drugs are termed chemosensitizers, MDR modulators or inhibitors. Modulators targeting P-gp function belong to a number of chemical classes and have been classified as first, second or third generation MDR reversal agents on the basis of their affinity for the transporter proteins and relative toxicity towards normal cells. Drugs that were used for other pharmacological functions and were coincidently found to be effective in sensitizing drug-resistant tumors towards chemotherapy are referred to as first generation modulators. These include compounds like verapamil (calcium channel blocker) and cyclosporine A (immunosuppressant) [116]. Obviously, these had lower affinity and were required in high doses and thus failed to manipulate MDR. Additionally, the higher dosage resulted in toxicity and other adverse effects towards normal cells [117]. Thus, these drugs were modified, giving rise to second generation modulators such as R verapamil (R-enantiomer of verapamil, a weaker calcium channel blocker). However, these drugs also failed to deliver the desired efficacy due to their low affinity for their target transporter proteins. Thus, third generation inhibitors were designed specifically for high transport affinity and low pharmacokinetic interaction. These included drugs such as tariquidar and mitotane, which exhibit high affinity and selectivity for target MDR transporter(s) at a low nanomolar range and subsequently low toxicity towards normal cells [118]. These molecules, when co-administered with the anticancer drug, act as competitive inhibitors for the drug-binding site of the transporter proteins, allowing the drug to accumulate inside the cells to effective lethal concentrations.

There are many alternative approaches to handle the problem of MDR, among which targeting RNA is the most popular. This is done by using antisense oligonucleotides, hammerhead ribozymes or siRNA. Other treatments include transcriptional regulators, agents to alter the plasma membrane and also compounds that selectively target MDR cancer cells [119]. All of these approaches, although effective, have not been rigorously tested in clinical trials. MDR reversal agents are known to inhibit the efflux activity of ABC transporters, thus leading to cytotoxicity and are, therefore, important alternative treatments to overcome MDR. Most modulators act by binding to these ABC transporters in the drug-binding pocket, while others can influence gene expression; for example, the alkaloid piperine was found to lower the expression levels of *ABCB*1, *ABCC*1, and *ABCG*2 genes possibly by acting on the transcription factors [120].

## *Tyrosine Kinase Inhibitors (Small Molecule Targeted Therapies)*

The practice of targeted therapies has significantly changed the treatment of cancer over the past decade. These therapies involve the use of monoclonal antibodies and small molecule inhibitors. Keeping in mind the scope of this chapter, we will discuss only small molecule inhibitors and natural products as modulators of MDR. Cancer treatment has undergone revolutionary changes with the introduction of tyrosine kinase inhibitors (TKIs) that selectively inhibit growth factor pathways critical for tumor growth. TKs can be further classified as receptor TKs (RTKs) and non-receptor TKs (NRTKs) and TKIs block the phosphorylation mediated by these TKs.

TKs initiate a downstream signaling response leading to cell growth, proliferation, migration or angiogenesis in normal as well as malignant tissues. Small-molecule inhibitors target these kinases by directly affecting the tumor cells, rather than by causing immune responses, as monoclonal antibodies do. These small molecule inhibitors typically interrupt cellular processes by interfering with the intracellular signaling of tyrosine kinases. These can be broadly divided into three categories: (1) Inhibitors that block the binding of ATP to the active site of kinases, such as the BCR-ABL kinase inhibitors imatinib, nilotinib and dasatinib, the epidermal growth factor receptor inhibitors gefitinib and erlotinib, and the cyclin-dependent kinase inhibitor, roscovitine. (2) The allosteric inhibitors block sites other than the catalytic sites. These sites are usually required for the activation of kinases, such as the p38 inhibitor BIRB 796, the RAF inhibitor BAY43-9006, and the MAP kinase inhibitor PD184352. (3) Other TKIs include drugs that may inhibit the activation of fusion of TKs by blocking their dimerization, antibodies against RTKs or their ligands, which interrupt TK signaling through neutralization of ligands, blockade of ligand binding, or receptor internalization. Most small-molecule inhibitors of TKs are ATP mimetics. These either block the active site or allosteric site within the intracellular catalytic domain of TKs, thereby inhibiting the subsequent downstream signaling processes (Table 4).

Imatinib or Gleevec, which was approved in 2001 by the FDA, was the first small molecule inhibitor targeted to the BCR-ABL kinase. It is used for the treatment of chronic myeloid leukemia and acts by inhibiting the continuously active BCR-ABL that results from the reciprocal translocation between chromosomes 9 and 22 (the Philadelphia chromosome). Unlike monoclonal antibodies, most small molecule inhibitors are metabolized by cytochrome P450 enzymes.

**Table 4** Tyrosine kinase inhibitors, their kinase and ABC transporter targets

| TKIs | Target kinase | ABC transporter | References |
|---|---|---|---|
| AG1478 | EGFR | ABCB1, ABCG2 | [126] |
| Apatinib | VEGFR2 | ABCB1, ABCG2 | [127] |
| Bortezomib | 28s protease | ABCB1 | [128] |
| Canertinib | EGFR, HER2, 3, 4 | ABCG2 | [129] |
| Cediranib | VEGFR2, PDGFR, c-KIT | ABCB1, ABCC1 | [130] |
| Danusertib | BCR-ABL, Aurora kinases | ABCG2 | [131] |
| Dasatinib | BCR-ABL, Src, c-KIT, PDGFR | ABCB1, ABCG2 | [132] |
| Erlotinib | EGFR | ABCB1, ABCG2, ABCC10 | [133, 134] |
| Gefitinib | EGFR | ABCB1, ABCG2 | [135] |
| Imatinib | BCR-ABL, c-KIT, PDGFR | ABCB1, ABCG2, ABCC1, ABCC10 | [121, 136] |
| Lapatinib | EGFR, HER2 | ABCB1, ABCG2. ABCC10 | [134] |
| Neratinib | EGFR, HER2 | ABCB1 | [137] |
| Nilotinib | BCR-ABL, c-SRC | ABCB1, ABCG2, ABCC10 | [136, 138] |
| Ponatinib | PDGFR, VEGFR2, BCR-ABL and c-SRC | ABCG2 | [139] |
| Saracatinib | c-SRC | ABCB1 | [140] |
| SGI-1776 | FLT-3, Pim | ABCB1, ABCG2 | [141] |
| Sorafenib | EGFR, VEGFR 2, PDGFR, B-raf | ABCC2 | [142] |
| Sunitinib | PDGFR, VEGFR2, c-KIT, FLT-3 | ABCB1, ABCG2 | [143] |
| Vemurafenib | B-Raf, MAP | ABCB1, ABCG2 | [144] |

But these TKIs such as imatinib (Gleevec), dasatinib (Sprycel), sorafenib (Nexavar), and sunitinib (Sutent) are multitargeting in nature, so they are much less specific as compared to the monoclonal antibodies used for treament. Gefitinib (Iressa) and erlotinib (Tarceva) selectively inhibit the epidermal growth factor receptor (EGFR), and both are efficacious against EGFR-expressing cancers. Erlotinib in combination with an antimetabolite, gemcitabine, is also approved for treating advanced pancreatic cancer. More recently, small molecule inhibitors targeting the EGFR pathway have been used in the treatment of solid tumors, such as non-small-cell lung cancer.

Emerging resistance towards the TKIs is a cause of major concern. Some of the most reported and studied mechanisms of resistance to TKIs are point mutations within the kinase domain resulting in decreased sensitivity of the kinases towards the TKIs. These mutations block the inhibitors from accessing the catalytic site and/or activation loop of the kinase. Another mechanism is the efflux of TKIs mainly by P-gp and ABCG2, which has been reported as a contributing factor to the development of resistance towards these inhibitors. A majority of TKIs are transported by P-gp and ABCG2 [121]. It is important to note that TKIs which inhibit ATP-binding to the active site of kinases interact at the drug-binding site in the TMDs of the ABC transporters. Such TKIs do not bind to the ATP sites in these ABC transporters, suggesting that the ATP sites in kinases and those in the ABC transporters are structurally different. Taken together, an ideal TKI would be a molecule that can inhibit the activity of the target TK but would not be a substrate or modulator for ABC drug transporters. The development of such a molecule will lead to novel TKIs that will be effective and will be able to kill tumor cells at much lower concentrations.

## *Natural Products as Modulators of ABC Drug Transporters*

It is widely known that natural compounds found in vegetables, fruits, plant-derived beverages and herbal dietary supplements not only have anticancer properties, but may also modulate the activity of ABC transporters. Such compounds extracted from natural sources are now referred to as "fourth generation inhibitors". The secondary metabolites of plants, which include alkaloids, phenolics and terpinoids, can interfere with the activity of the well-known efflux pumps [122]. These may either interact directly with the protein, disturbing the tertiary structure, or might function as competitive inhibitors for therapeutic drugs. They can also be used in combination with a cytotoxic agent, resulting in reversal of MDR. Out of these numerous natural products, flavonoids are the most well-known and well-studied. Over 6,000 flavonoids have now been identified and are the most abundant polyphenols in the human diet. They are found in foods such as fruits, vegetables, food supplements, tea and wine. This class of secondary metabolites includes flavonols (quercetin, myricetin), flavones (luteolin, apigenin, and chrysin), isoflavones (genistein, diadzein), flavonones (naringenin, hesperetin), flavonolols, chalcones and aurones. In addition to the antitumor, antimitotic and antiviral properties of flavonoids, they are also able to inhibit kinases, and are active in radical scavenging and metal ion chelating. Most of these natural product modulators act by binding to the drug-binding pocket of ABC transporters. These compounds can thus change the absorption, distribution, metabolism and excretion (ADME) properties of chemotherapeutic drugs by modulating the activity of ABC transporters. A few of them might affect the ATP-binding or hydrolysis at the NBDs or alter the cell-surface expression of ABC transporters. Since most of these natural compounds are components of the human diet, it may be presumed that they would be non-toxic even at higher doses. A list of important natural product modulators for the three ABC drug transporters is given in Table 5. Although many innovative approaches are currently available, studies to design and find a modulator that is selective, relatively less toxic and highly potent appear to be the most likely way to resolve the problem of MDR [123, 124].

**Table 5** Plant natural products as modulators of ABC transporters

| Natural product | ABC transporter | | | References |
|---|---|---|---|---|
| | P-gp | ABCG2 | MRP1 | |
| 3′-4′-7′-trimethoxyflavone | – | + | – | [145] |
| 5-bromotetrandrine | + | – | – | [146] |
| 6-prenylchrysin | – | + | – | [124] |
| Abietane diterpene | + | – | – | [147] |
| Acacetin | + | + | – | [148] |
| Afrormosin | + | – | + | [149] |
| Agosterol A and derivatives | + | – | – | [150] |
| Alisol B 23-acetate | + | – | – | [151] |
| Amooranin | + | – | – | [152] |
| Apigenin | + | + | + | [153, 154] |
| Asarum heteropoides var. mandshuricum | + | – | – | [155] |
| Ayanin | – | + | – | [156] |
| Baicalein and derivatives | + | – | – | [124] |
| Biochanin A | + | + | – | [157] |
| Bitter melon extract | + | – | – | [158] |
| Cantharidin | + | – | – | [159] |
| Cannabinoids | + | + | + | [124] |
| Carotenoids | + | + | – | [160] |
| Catechins | + | – | – | [161] |
| Cepharanthine | + | – | + | [124] |
| Chrysin | + | + | – | [149, 153] |
| Citronellal | + | – | – | [162] |
| Coumarins | + | – | – | [163] |
| Curcumin | + | + | + | [124] |
| Cycleanine | + | – | – | [164] |
| Cycloartanes | + | – | – | [165] |
| Cyclopamine | + | – | – | [166] |
| Deoxyschizandrin | + | – | – | [167] |
| Dimethoxyaurone | + | + | ND | [168, 169] |
| Diadzein | – | + | – | [145] |
| Eudesmin | + | – | – | [170] |
| Eupatin | – | + | – | [171] |
| Euphocharacins A-L | + | – | – | [172] |
| Formononetin | + | – | + | [149] |
| Fumitremorgin C | – | + | – | [124] |
| Galangin | – | + | – | [122] |
| Genistein | + | + | + | [148] |
| Ginkgo biloba extract | + | – | + | [124] |
| Ginsenoside Rg | + | – | – | [173] |
| Ginsenosides | – | + | – | [174] |
| Grapefruit juice extracts | + | – | – | [175] |
| Hapalosin | + | – | – | [176] |
| Harmine | – | + | – | [177] |

(continued)

**Table 5** (continued)

| Natural product | ABC transporter | | | References |
|---|---|---|---|---|
| | P-gp | ABCG2 | MRP1 | |
| Hesperetine | + | + | – | [145] |
| Hypericin | + | – | – | [124] |
| Hyperforin | + | – | – | [124] |
| Ibogaine | – | + | – | [178] |
| Isoquinoline alkaloid | + | – | – | [124] |
| Isotetrandrine | + | – | – | [124] |
| Jatrophanes | + | – | – | [179] |
| Kaempferol | – | + | + | [148] |
| Kaempferia parviflora extracts | + | – | + | [124] |
| Kavalactones | + | – | – | [124] |
| Kendarimide A | + | – | – | [180] |
| Limonin | + | – | – | [181] |
| Luteolin | + | + | + | [153] |
| Morin | + | – | – | [157] |
| Myricetin | – | – | + | [182] |
| Naringin, Naringenin and derivatives | + | + | + | [122] |
| Ningalin B and derivatives | + | – | – | [124] |
| Opiates | + | – | – | [183] |
| Phloretin | + | + | – | [122] |
| Piperine | + | – | – | [184] |
| Plumbagin | – | + | – | [185] |
| Polyoxygenated steroids | + | – | – | [124] |
| Protopanaxatriol ginsenosides | + | – | – | [124] |
| Pyranocoumarins | + | – | – | [186] |
| Quercetin | + | + | + | [124] |
| Resveratrol | – | + | – | [145] |
| Retusin | – | + | – | [156] |
| Rotenoids | – | + | – | [187] |
| Rutaecarpine | + | – | – | [122] |
| Sanguinarine | + | – | – | [188] |
| Schisandrol A | + | – | – | [189] |
| Sesquiterpines | + | – | – | [190] |
| Silymarin | + | + | – | [124] |
| Sinensetin | + | – | – | [191] |
| Stemona curtissi root extract | + | – | + | [124] |
| Stilbenoids | – | + | – | [192] |
| Taxane derivatives | + | – | – | [124] |
| Tectochrysin | – | + | – | [187] |
| Terpenoids | + | + | – | [124] |
| Tetrahydrocurcumin | – | + | – | [122] |
| Tryprostatin A | + | – | – | [193] |
| Tryptanthrin | + | – | – | [194] |
| Vitamin E TPGS | + | – | – | [124] |

+ modulation; – no effect; ND Not determined

## Conclusion

With the advent of targeted therapies and improved drugs, the treatment of cancer has made remarkable progress in the past few decades. However, the MDR phenomenon limits the benefits of chemotherapy. Studies deciphering the structural and functional aspects of ABC transporters help us to better understand the mechanism of these transporters and assist in designing targeted drugs as modulators for these efflux proteins. Furthermore, understanding changes in the pharmacokinetics and pharmacodynamics of drugs owing to SNPs in these transporters in certain ethnic populations helps us to design treatment plans for specific populations or patients, which is a step towards individualized medicine. This area is being recognized due to variation in the genetic makeup of individuals and, hopefully, soon we will be able to predict the patient-specific efficacy of certain drugs. Clinical MDR is multifaceted and the broad and overlapping substrate specificity of ABC drug transporters adds another dimension to this problem. Recently characterized targeted-therapy drugs including TKIs and fourth generation natural product modulators will have to be tested in clinical trials along with conventional chemotherapeutic drugs to overcome MDR in the clinic and to increase the efficiency of chemotherapy for cancer patients.

**Acknowledgments** We thank Dr. Eduardo E. Chufan for critical comments on the manuscript and acknowledge George Leiman for editorial assistance. This work was supported by the Intramural Research Program of the NIH, National Cancer Institute, Center for Cancer Research.

**Conflicts of Interest** No potential conflicts of interest were disclosed.

## References

1. Ambudkar SV, Dey S, Hrycyna CA, Ramachandra M, Pastan I, Gottesman MM. Biochemical, cellular, and pharmacological aspects of the multidrug transporter. Annu Rev Pharmacol Toxicol. 1999;39:361–98.
2. Lage H. An overview of cancer multidrug resistance: a still unsolved problem. Cell Mol Life Sci. 2008;65:3145–67.
3. Mellor HR, Callaghan R. Resistance to chemotherapy in cancer: a complex and integrated cellular response. Pharmacology. 2008;81:275–300.
4. Gillet JP, Gottesman MM. Mechanisms of multidrug resistance in cancer. Methods Mol Biol. 2010;596:47–76.
5. Dean M, Fojo T, Bates S. Tumour stem cells and drug resistance. Nat Rev Cancer. 2005;5:275–84.
6. Baguley BC. Multiple drug resistance mechanisms in cancer. Mol Biotechnol. 2010;46:308–16.
7. Levchenko A, Mehta BM, Niu XL, Kang G, Villafania L, Way D, Polycarpe D, Sadeain M, Larson SM. Intercellular transfer of P-glycoprotein mediates acquired multidrug resistance in tumor cells. Proc Natl Acad Sci USA. 2005;102:1933–8.

8. Ambudkar SV, Sauna ZE, Gottesman MM, Szakacs G. A novel way to spread drug resistance in tumor cells: functional intercellular transfer of P-glycoprotein (ABCB1). Trends Pharmacol Sci. 2005;26:385–7.
9. Scagliotti GV, Novello S, Selvaggi G. Multidrug resistance in non-small-cell lung cancer. Ann Oncol. 1999;10:83–6.
10. Borst P. Cancer drug pan-resistance: pumps, cancer stem cells, quiescence, epithelial to mesenchymal transition, blocked cell death pathways, persisters or what? Open Biol. 2012;2:120066.
11. Hall MD, Handley MD, Gottesman MM. Is resistance useless? Multidrug resistance and collateral sensitivity. Trends Pharmacol Sci. 2009;30:546–56.
12. Pluchino KM, Hall MD, Goldsborough AS, Callaghan R, Gottesman MM. Collateral sensitivity as a strategy against cancer multidrug resistance. Drug Resist Updat. 2012;15:98–105. (Epub ahead of print).
13. Warr JR, Anderson M, Fergusson J. Properties of verapamil-hypersensitive multidrug-resistant Chinese-hamster ovary cells. Cancer Res. 1988;48:4477–83.
14. Nakagawa-Goto K, Bastow KF, Chen TH, Morris-Natschke SL, Lee KH. Antitumor agents 260. New desmosdumotin B analogues with improved in vitro anticancer activity. J Med Chem. 2008;51:3297–303.
15. Hall MD, Brimacombe KR, Varonka MS, Pluchino KM, Monda JK, Li JY, Walsh MJ, Boxer MB, Warren TH, Fales HM, Gottesman MM. Synthesis and structure-activity evaluation of Isatin-beta-thiosemicarbazones with improved selective activity toward multidrug-resistant cells expressing P-glycoprotein. J Med Chem. 2011;54:5878–89.
16. Goldsborough AS, Handley MD, Dulcey AE, Pluchino KM, Kannan P, Brimacombe KR, Hall MD, Griffiths G, Gottesman MM. Collateral sensitivity of multidrug-resistant cells to the orphan drug tiopronin. J Med Chem. 2011;54:4987–97.
17. Slot AJ, Molinski SV, Cole SPC. Mammalian multidrug-resistance proteins (MRPs). Essays Biochem: ABC Transport. 2011;50:179–207.
18. Wu CP, Hsieh CH, Wu YS. The emergence of drug transporter-mediated multidrug resistance to cancer chemotherapy. Mol Pharmaceut. 2011;8:1996–2011.
19. Vulevic B, Chen ZJ, Boyd JT, Davis W, Walsh ES, Belinsky MG, Tew KD. Cloning and characterization of human adenosine 5′-triphosphate-binding cassette, sub-family A, transporter 2 (ABCA2). Cancer Res. 2001;61:3339–47.
20. Chapuy B, Koch R, Radunski U, Corsham S, Cheong N, Inagaki N, Ban N, Wenzel D, Reinhardt D, Zapf A, Schweyer S, Kosari F, Klapper W, Truemper L, Wulf GG. Intracellular ABC transporter A3 confers multidrug resistance in leukemia cells by lysosomal drug sequestration. Leukemia. 2008;22:1576–86.
21. Smith AJ, van Helvoort A, van Meer G, Szabo K, Welker E, Szakacs G, Varadi A, Sarkadi B, Borst P. MDR3 P-glycoprotein, a phosphatidylcholine translocase, transports several cytotoxic drugs and directly interacts with drugs as judged by interference with nucleotide trapping. J Biol Chem. 2000;275:23530–9.
22. Kawanobe T, Kogure S, Nakamura S, Sato M, Katayama K, Mitsuhashi J, Noguchi K, Sugimoto Y. Expression of human ABCB5 confers resistance to taxanes and anthracyclines. Biochem Biophys Res Co. 2012;418:736–41.
23. Frank NY, Margaryan A, Huang Y, Schatton T, Waaga-Gasser AM, Gasser M, Sayegh MH, Sadee W, Frank MH. ABCB5-mediated doxorubicin transport and chemoresistance in human malignant melanoma. Cancer Res. 2005;65:4320–33.
24. Childs S, Yeh RL, Hui D, Ling V. Taxol resistance mediated by transfection of the liver-specific sister gene of P-glycoprotein. Cancer Res. 1998;58:4160–7.
25. Huisman MT, Chhatta AA, van Tellingen O, Beijnen JH, Schinkel AH. MRP2 (ABCC2) transports taxanes and confers paclitaxel resistance and both processes are stimulated by probenecid. Int J Cancer. 2005;116:824–9.
26. Zelcer N, Saeki T, Reid G, Beijnen JH, Borst P. Characterization of drug transport by the human multidrug resistance protein 3 (ABCC3). J Biol Chem. 2001;276:46400–7.

27. Ritter CA, Jedlitschky G, Schwabedissen HMZ, Grube M, Kock K, Kroemer HK. Cellular export of drugs and signaling molecules by the ATP-binding cassette transporters MRP4 (ABCC4) and MRP5 (ABCC5). Drug Metab Rev. 2005;37:253–78.
28. Hopper-Borge E, Chen ZS, Shchaveleva I, Belinsky MG, Kruh GD. Analysis of the drug resistance profile of multidrug resistance protein 7 (ABCC10): resistance to docetaxel. Cancer Res. 2004;64:4927–30.
29. Chen ZS, Guo YP, Belinsky MG, Kotova E, Kruh GD. Transport of bile acids, sulfated steroids, estradiol 17-beta-D-glucuronide, and leukotriene C4 by human multidrug resistance protein 8 (ABCC11). Mol Pharmacol. 2005;67:545–57.
30. Chen ZS, Tiwari AK. Multidrug resistance proteins (MRPs/ABCCs) in cancer chemotherapy and genetic diseases. FEBS J. 2011;278:3226–45.
31. Leslie EM, Deeley RG, Cole SPC. Multidrug resistance proteins: role of P-glycoprotein, MRP1, MRP2, and BCRP (ABCG2) in tissue defense. Toxicol Appl Pharmacol. 2005;204:216–37.
32. Koike K, Kawabe T, Tanaka T, Toh S, Uchiumi T, Wada M, Akiyama S, Ono M, Kuwano M. A canalicular multispecific organic anion transporter (cMOAT) antisense cDNA enhances drug sensitivity in human hepatic cancer cells. Cancer Res. 1997;57:5475–9.
33. Deeley RG, Westlake C, Cole SPC. Transmembrane transport of endo- and xenobiotics by mammalian ATP-binding cassette multidrug resistance proteins. Physiol Rev. 2006;86:849–99.
34. Nies AT, Keppler D. The apical conjugate efflux pump ABCC2 (MRP2). Pflugers Archiv-European J Physiol. 2007;453:643–59.
35. Frank NY, Pendse SS, Lapchak PH, Margaryan A, Shlain D, Doeing C, Sayegh MH, Frank MH. Regulation of progenitor cell fusion by ABCB5 P-glycoprotein, a novel human ATP-binding cassette transporter. J Biol Chem. 2003;278:47156–65.
36. Chen KG, Valencia JC, Gillet JP, Hearing VJ, Gottesman MM. Involvement of ABC transporters in melanogenesis and the development of multidrug resistance of melanoma. Pigm Cell Melanoma Res. 2009;22:740–9.
37. Dean M, Hamon Y, Chimini G. The human ATP-binding cassette (ABC) transporter superfamily. J Lipid Res. 2001;42:1007–17.
38. Borst P, Elferink RO. Mammalian ABC transporters in health and disease. Annu Rev Biochem. 2002;71:537–92.
39. Ambudkar SV, Kimchi-Sarfaty C, Sauna ZE, Gottesman MM. P-glycoprotein: from genomics to mechanism. Oncogene. 2003;22:7468–85.
40. Juliano RL, Ling V. Surface glycoprotein modulating drug permeability in Chinese-hamster ovary cell mutants. Biochim Biophys Acta. 1976;455:152–62.
41. Thiebaut F, Tsuruo T, Hamada H, Gottesman MM, Pastan I, Willingham MC. Cellular-localization of the multidrug-resistance gene-product p-glycoprotein in normal human-tissues. Proc Natl Acad Sci USA. 1987;84:7735–8.
42. Fojo AT, Ueda K, Slamon DJ, Poplack DG, Gottesman MM, Pastan I. Expression of a multidrug-resistance gene in human-tumors and tissues. Proc Natl Acad Sci USA. 1987;84:265–9.
43. Fromm MF. Importance of P-glycoprotein for drug disposition in humans. Eur J Clin Invest. 2003;33:6–9.
44. Thuerauf N, Fromm MF. The role of the transporter P-glycoprotein for disposition and effects of centrally acting drugs and for the pathogenesis of CNS diseases. Eur Arch Psy Clin Neurosci. 2006;256:281–6.
45. Callaghan R, Crowley E, Potter S, Kerr ID. P-glycloprotein: so many ways to turn it on. J Clin Pharmacol. 2008;48:365–78.
46. Lin TT, Islam O, Heese K. ABC transporters, neural stem cells and neurogenesis—a different perspective. Cell Res. 2006;16:857–71.
47. Sarkadi B, Homolya L, Szakacs G, Varadi A. Human multidrug resistance ABCB and ABCG transporters: participation in a chemoimmunity defense system. Physiol Rev. 2006;86:1179–236.

48. Doyle LA, Yang WD, Abruzzo LV, Krogmann T, Gao YM, Rishi AK, Ross DD. A multidrug resistance transporter from human MCF-7 breast cancer cells. Proc Natl Acad Sci USA. 1998;95:15665–70.
49. Chen YN, Mickley LA, Schwartz AM, Acton EM, Hwang J, Fojo AT. Characterization of adriamycin-resistant human breast-cancer cells which display overexpression of a novel resistance-related membrane-protein. J Biol Chem. 1990;265:10073–80.
50. Lee JS, Scala S, Matsumoto Y, Dickstein B, Robey R, Zhan ZR, Altenberg G, Bates SE. Reduced drug accumulation and multidrug resistance in human breast cancer cells without associated P-glycoprotein or MRP overexpression. J Cell Biochem. 1997;65:513–26.
51. Maliepaard M, Scheffer GL, Faneyte IF, van Gastelen MA, Pijnenborg ACLM, Schinkel AH, van de Vijver MJ, Scheper RJ, Schellens JHM. Subcellular localization and distribution of the breast cancer resistance protein transporter in normal human tissues. Cancer Res. 2001;61:3458–64.
52. Cooray HC, Blackmore CG, Maskell L, Barrand MA. Localisation of breast cancer resistance protein in microvessel endothelium of human brain. Neuroreport. 2002;13:2059–63.
53. Cole SPC, Bhardwaj G, Gerlach JH, Mackie JE, Grant CE, Almquist KC, Stewart AJ, Kurz EU, Duncan AMV, Deeley RG. Overexpression of a transporter gene in a multidrug-resistant human lung-cancer cell-line. Science. 1992;258:1650–4.
54. Evers R, Zaman GJR, van Deemter L, Jansen H, Calafat J, Oomen LCJM, Elferink RPJO, Borst P, Schinkel AH. Basolateral localization and export activity of the human multidrug resistance-associated protein in polarized pig kidney cells. J Clin Invest. 1996;97:1211–8.
55. Lorico A, Rappa G, Flavell RA, Sartorelli AC. Double knockout of the MRP gene leads to increased drug sensitivity in vitro. Cancer Res. 1996;56:5351–5.
56. Allen JD, Brinkhuis RF, van Deemter L, Wijnholds J, Schinkel AH. Extensive contribution of the multidrug transporters P-glycoprotein and Mrp1 to basal drug resistance. Cancer Res. 2000;60:5761–6.
57. Fletcher JI, Haber M, Henderson MJ, Norris MD. ABC transporters in cancer: more than just drug efflux pumps. Nat Rev Cancer. 2010;10:147–56.
58. Ho LL, Kench JG, Handelsman DJ, Scheffer GL, Stricker PD, Grygiel JG, Sutherland RL, Henshall SM, Allen JD, Horvath LG. Androgen regulation of multidrug resistance-associated protein 4 (MRP4/ABCC4) in prostate cancer. Prostate. 2008;68:1421–9.
59. Russel FGM, Koenderink JB, Masereeuw R. Multidrug resistance protein 4 (MRP4/ABCC4): a versatile efflux transporter for drugs and signalling molecules. Trends Pharmacol Sci. 2008;29:200–7.
60. Nies AT, Jedlitschky G, Konig J, Herold-Mende C, Steiner HH, Schmitt HP, Keppler D. Expression and immunolocalization of the multidrug resistance proteins, Mrp1-Mrp6 (ABCC1-ABCC6), in human brain. Neuroscience. 2004;129:349–60.
61. Nies AT, Spring H, Thon WF, Keppler D, Jedlitschky G. Immunolocalization of multidrug resistance protein 5 in the human genitourinary system. J Urology. 2002;167:2271–5.
62. Borst P, de Wolf C, de Wetering KV. Multidrug resistance-associated proteins 3, 4, and 5. Pflugers Archiv-Eur J Physiol. 2007;453:661–73.
63. Bortfeld M, Rius M, Konig J, Herold-Mende C, Nies AT, Keppler D. Human multidrug resistance protein 8 (MRP8/ABCC11), an apical efflux pump for steroid sulfates, is an axonal protein of the CNS and peripheral nervous system. Neuroscience. 2006;137:1247–57.
64. Bakos E, Evers R, Szakacs G, Tusnady GE, Welker E, Szabo K, de Haas M, van Deemter L, Borst P, Varadi A, Sarkadi B. Functional multidrug resistance protein (MRP1) lacking the N-terminal transmembrane domain. J Biol Chem. 1998;273:32167–75.
65. Robey RW, Polgar O, Deeken J, To KW, Bates SE. ABCG2: determining its relevance in clinical drug resistance. Cancer Metast Rev. 2007;26:39–57.
66. Loe DW, Almquist KC, Cole SPC, Deeley RG. ATP-dependent 17 beta-estradiol 17-(beta-D-glucuronide) transport by multidrug resistance protein (MRP)—inhibition by cholestatic steroids. J Biol Chem. 1996;271:9683–9.

67. Leier I, Jedlitschky G, Buchholz U, Cole SPC, Deeley RG, Keppler D. The Mrp gene encodes an atp-dependent export pump for leukotriene C-4, and structurally related conjugates. J Biol Chem. 1994;269:27807–10.
68. Muller M, Meijer C, Zaman GJR, Borst P, Scheper RJ, Mulder NH, Devries EGE, Jansen PLM. Overexpression of the gene encoding the multidrug resistance-associated protein results in increased Atp-dependent glutathione S-conjugate transport. Proc Natl Acad Sci USA. 1994;91:13033–7.
69. Kimchi-Sarfaty C, Oh JM, Kim IW, Sauna ZE, Calcagno AM, Ambudkar SV, Gottesman MM. A "silent" polymorphism in the MDR1 gene changes substrate specificity. Science. 2007;315:525–8.
70. Wolf SJ, Bachtiar M, Wang J, Sim T, Chong S, Lee CGL. An update on ABCB1 pharmacogenetics: insights from a 3D model into the location and evolutionary conservation of residues corresponding to SNPs associated with drug pharmacokinetics. Pharmacogenomics J. 2011;11:315–25.
71. Leschziner GD, Andrew T, Pirmohamed M, Johnson MR. ABCB1 genotype and PGP expression, function and therapeutic drug response: a critical review and recommendations for future research. Pharmacogenomics J. 2007;7:154–79.
72. Brant SR, Panhuysen CIM, Nicolae D, Reddy DM, Bonen DK, Karaliukas R, Zhang LL, Swanson E, Datta LW, Moran T, Ravenhill G, Duerr RH, Achkar JP, Karban AS, Cho JH. MDR1 Ala893 polymorphism is associated with inflammatory bowel disease. Am J Hum Genet. 2003;73:1282–92.
73. Potocnik U, Ferkolj I, Glavac D, Dean M. Polymorphisms in multidrug resistance 1 (MDR1) gene are associated with refractory Crohn disease and ulcerative colitis. Genes Immun. 2004;5:530–9.
74. Umbenhauer DR, Lankas GR, Pippert TR, Wise LD, Cartwright ME, Hall SJ, Beare CM. Identification of a P-glycoprotein-deficient subpopulation in the CF-1 mouse strain using a restriction fragment length polymorphism. Toxicol Appl Pharmacol. 1997;146:88–94.
75. Mealey KL, Bentjen SA, Gay JM, Cantor GH. Ivermectin sensitivity in collies is associated with a deletion mutation of the mdr1 gene. Pharmacogenetics. 2001;11:727–33.
76. Jeong H, Herskowitz I, Kroetz DL, Rine J. Function-altering SNPs in the human multidrug transporter gene ABCB1 identified using a Saccharomyces-based assay. PLoS Genet. 2007;3:367–76.
77. Nakagawa H, Tamura A, Wakabayashi K, Hoshijima K, Komada M, Yoshida T, Kometani S, Matsubara T, Mikuriya K, Ishikawa T. Ubiquitin-mediated proteasomal degradation of non-synonymous SNP variants of human ABC transporter ABCG2. Biochem J. 2008;411:623–31.
78. Furukawa T, Wakabayashi K, Tamura A, Nakagawa H, Morishima Y, Osawa Y, Ishikawa T. Major SNP (Q141K) variant of human ABC transporter ABCG2 undergoes lysosomal and proteasomal degradations. Pharmaceut Res. 2009;26:469–79.
79. Basseville A, Tamaki A, Ierano C, Trostel S, Ward Y, Robey RW, Hegde RS, Bates SE. Histone deacetylase inhibitors influence chemotherapy transport by modulating expression and trafficking of a common polymorphic variant of the ABCG2 efflux transporter. Cancer Res. 2012;72:3642–51.
80. Woodward OM, Kottgen A, Coresh J, Boerwinkle E, Guggino WB, Kottgen M. Identification of a urate transporter, ABCG2, with a common functional polymorphism causing gout. Proc Natl Acad Sci USA. 2009;106:10338–42.
81. Matsuo H, Takada T, Ichida K, Nakamura T, Nakayama A, Ikebuchi Y, Ito K, Kusanagi Y, Chiba T, Tadokoro S, Takada Y, Oikawa Y, Inoue H, Suzuki K, Okada R, Nishiyama J, Domoto H, Watanabe S, Fujita M, Morimoto Y, Naito M, Nishio K, Hishida A, Wakai K, Asai Y, Niwa K, Kamakura K, Nonoyama S, Sakurai Y, Hosoya T, Kanai Y, Suzuki H, Hamajima N, Shinomiya N. Common defects of ABCG2, a high-capacity urate exporter, cause gout: a function-based genetic analysis in a Japanese population. Sci Transla Med. 2009;1:5ra11.

82. Conseil G, Deeley RG, Cole SPC. Polymorphisms of MRP1 (ABCC1) and related ATP-dependent drug transporters. Pharmacogenet Genom. 2005;15:523–33.
83. Loo TW, Clarke DM. Identification of residues in the drug-binding site of human P-glycoprotein using a thiol-reactive substrate. J Biol Chem. 1997;272:31945–8.
84. Loo TW, Clarke DM. Identification of residues within the drug-binding domain of the human multidrug resistance P-glycoprotein by cysteine-scanning mutagenesis and reaction with dibromobimane. J Biol Chem. 2000;275:39272–8.
85. Loo TW, Bartlett MC, Clarke DM. Transmembrane segment 7 of human P-glycoprotein forms part of the drug-binding pocket. Biochem J. 2006;399:351–9.
86. Dey S, Ramachandra M, Pastan I, Gottesman MM, Ambudkar SV. Evidence for two nonidentical drug-interaction sites in the human P-glycoprotein. Proc Natl Acad Sci USA. 1997;94:10594–9.
87. Loo TW, Bartlett MC, Clarke DM. Simultaneous binding of two different drugs in the binding pocket of the human multidrug resistance P-glycoprotein. J Biol Chem. 2003;278:39706–10.
88. Sauna ZE, Andrus MB, Turner TM, Ambudkar SV. Biochemical basis of polyvalency as a strategy for enhancing the efficacy of P-glycoprotein (ABCB1) modulators: stipiamide homodimers separated with defined-length spacers reverse drug efflux with greater efficacy. Biochemistry. 2004;43:2262–71.
89. Lugo MR, Sharom FJ. Interaction of LDS-751 and rhodamine 123 with P-glycoprotein: evidence for simultaneous binding of both drugs. Biochemistry. 2005;44:14020–9.
90. Gutmann DAP, Ward A, Urbatsch IL, Chang G, van Veen HW. Understanding polyspecificity of multidrug ABC transporters: closing in on the gaps in ABCB1. Trends Biochem Sci. 2010;35:36–42.
91. Aller SG, Yu J, Ward A, Weng Y, Chittaboina S, Zhuo RP, Harrell PM, Trinh YT, Zhang QH, Urbatsch IL, Chang G. Structure of P-glycoprotein reveals a molecular basis for poly-specific drug binding. Science. 2009;323:1718–22.
92. Dawson RJP, Locher KP. Structure of a bacterial multidrug ABC transporter. Nature. 2006;443:180–5.
93. Jin MS, Oldham ML, Zhang QJ, Chen J. Crystal structure of the multidrug transporter P-glycoprotein from Caenorhabditis elegans. Nature. 2012;490:566–9.
94. Ward A, Reyes CL, Yu J, Roth CB, Chang G. Flexibility in the ABC transporter MsbA: alternating access with a twist. Proc Natl Acad Sci USA. 2007;104:19005–10.
95. Zolnerciks JK, Andress EJ, Nicolaou M, Linton KJ. Structure of ABC transporters. Essays Biochem ABC Transport. 2011;50:43–61.
96. Gottesman MM, Ambudkar SV, Xia D. Structure of a multidrug transporter. Nat Biotechnol. 2009;27:546–7.
97. Jones PM, George AM. Opening of the ADP-bound active site in the ABC transporter ATPase dimer: evidence for a constant contact, alternating sites model for the catalytic cycle. Proteins Struc Func Bioinformat. 2009;75:387–96.
98. Sauna ZE, Ambudkar SV. About a switch: how P-glycoprotein (ABCB1) harnesses the energy of ATP binding and hydrolysis to do mechanical work. Mol Cancer Therapeut. 2007;6:13–23.
99. Al-Shawi MK. Catalytic and transport cycles of ABC exporters. Essays Biochem ABC Transport. 2011;50:63–83.
100. Ambudkar SV, Kim IW, Xia D, Sauna ZE. The A-loop, a novel conserved aromatic acid subdomain upstream of the Walker A motif in ABC transporters, is critical for ATP binding. FEBS Lett. 2006;580:1049–55.
101. Urbatsch IL, Julien M, Carrier I, Rousseau ME, Cayrol R, Gros P. Mutational analysis of conserved carboxylate residues in the nucleotide binding sites of P-glycoprotein. Biochemistry. 2000;39:14138–49.
102. Pastan I, Gottesman MM. Molecular-biology of a multidrug transporter. FASEB J. 1992;6:A128.

103. Higgins CF, Gottesman MM. Is the multidrug transporter a flippase. Trends Biochem Sci. 1992;17:18–21.
104. Eckford PDW, Sharom FJ. P-glycoprotein (ABCB1) interacts directly with lipid-based anti-cancer drugs and platelet-activating factors. Biochem Cell Biol. 2006;84:1022–33.
105. Raggers RJ, Vogels I, van Meer G. Multidrug-resistance P-glycoprotein (MDR1) secretes platelet-activating factor. Biochem J. 2001;357:859–65.
106. Urbatsch IL, Sankaran B, Weber J, Senior AE. P-glycoprotein is stably inhibited by vanadate-induced trapping of nucleotide at a single catalytic site. J Biol Chem. 1995;270:19383–90.
107. Smith PC, Karpowich N, Millen L, Moody JE, Rosen J, Thomas PJ, Hunt JF. ATP binding to the motor domain from an ABC transporter drives formation of a nucleotide sandwich dimer. Mol Cell. 2002;10:139–49.
108. Callaghan R, Ford RC, Kerr ID. The translocation mechanism of P-glycoprotein. FEBS Lett. 2006;580:1056–63.
109. Sharom FJ. The P-glycoprotein multidrug transporter. Essays Biochem ABC Transport. 2011;50:161–78.
110. Jardetzk O. Simple allosteric model for membrane pumps. Nature. 1966;211:969–70.
111. Ambudkar SV, Kim IW, Sauna ZE. The power of the pump: mechanisms of action of P-glycoprotein (ABCB1). Eur J Pharmaceut Sci. 2006;27:392–400.
112. Sauna ZE, Kim IW, Nandigama K, Kopp S, Chiba P, Ambudkar SV. Catalytic cycle of ATP hydrolysis by P-glycoprotein: evidence for formation of the E-S reaction intermediate with ATP-gamma-S, a nonhydrolyzable analogue of ATP. Biochemistry. 2007;46:13787–99.
113. Senior AE, AlShawi MK, Urbatsch IL. The catalytic cycle of P-glycoprotein. FEBS Lett. 1995;377:285–9.
114. Sauna ZE, Ambudkar SV. Characterization of the catalytic cycle of ATP hydrolysis by human P-glycoprotein—the two ATP hydrolysis events in a single catalytic cycle are kinetically similar but affect different functional outcomes. J Biol Chem. 2001;276:11653–61.
115. Sauna ZE, Ambudkar SV. Evidence for a requirement for ATP hydrolysis at two distinct steps during a single turnover of the catalytic cycle of human P-glycoprotein. Proc Natl Acad Sci USA. 2000;97:2515–20.
116. Ford JM, Hait WN. Pharmacology of drugs that alter multidrug resistance in cancer. Pharmacol Rev. 1990;42:155–99.
117. Lampidis TJ, Krishan A, Planas L, Tapiero H. Reversal of intrinsic resistance to Adriamycin in normal-cells by verapamil. Cancer Drug Deliv. 1986;3:251–9.
118. Liscovitch M, Lavie Y. Cancer multidrug resistance: a review of recent drug discovery research. IDrugs. 2002;5:349–55.
119. Wu CP, Calcagno AM, Ambudkar SV. Reversal of ABC drug transporter-mediated multidrug resistance in cancer cells: evaluation of current strategies. Curr Mol Pharmacol. 2008;1:93–105.
120. Li S, Lei Y, Jia YJ, Li N, Wink M, Ma YG. Piperine, a piperidine alkaloid from Piper nigrum re-sensitizes P-gp, MRP1 and BCRP dependent multidrug resistant cancer cells. Phytomed. 2011;19:83–7.
121. Shukla S, Chen ZS, Ambudkar SV. Tyrosine kinase inhibitors as modulators of ABC transporter-mediated drug resistance. Drug Resist Updat. 2012;15:70–80.
122. Wink M, Ashour ML, El-Readi MZ. Secondary metabolites from plants inhibiting ABC transporters and reversing resistance of cancer cells and microbes to cytotoxic and antimicrobial agents. Front Microbiol. 2012;3:130.
123. Shukla S, Wu CP, Ambudkar SV. Development of inhibitors of ATP-binding cassette drug transporters—present status and challenges. Expert Opin Drug Metab Toxicol. 2008;4:205–23.
124. Wu CP, Ohnuma S, Ambudkar SV. Discovering natural product modulators to overcome multidrug resistance in cancer chemotherapy. Curr Pharmaceut Biotechnol. 2011;12:609–20.

125. Silverton L, Dean M, Moitra K. Variation and evolution of the ABC transporter genes ABCB1, ABCC1, ABCG2, ABCG5 and ABCG8: implication for pharmacogenetics and disease. Drug Metabol Drug Interact. 2011;26:169–79.
126. Shi Z, Tiwari AK, Shukla S, Robey RW, Kim IW, Parmar S, Bates SE, Si QS, Goldblatt CS, Abraham I, Fu LW, Ambudkar SV, Chen ZS. Inhibiting the function of ABCB1 and ABCG2 by the EGFR tyrosine kinase inhibitor AG1478. Biochem Pharmacol. 2009;77:781–93.
127. Mi YJ, Liang YJ, Huang HB, Zhao HY, Wu CP, Wang F, Tao LY, Zhang CZ, Dai CL, Tiwari AK, Ma XX, To KKW, Ambudkar SV, Chen ZS, Fu LW. Apatinib (YN968D1) reverses multidrug resistance by inhibiting the efflux function of multiple atp-binding cassette transporters. Cancer Res. 2010;70:7981–91.
128. Panischeva LA, Kakpakova ES, Rybalkina EY, Stavrovskaya AA. Influence of proteasome inhibitor bortezomib on the expression of multidrug resistance genes and Akt kinase activity. Biochem-Moscow. 2011;76:1009–16.
129. Erlichman C, Boerner SA, Hallgren CG, Spieker R, Wang XY, James CD, Scheffer GL, Maliepaard M, Ross DD, Bible KC, Kaufmann SH. The HER tyrosine kinase inhibitor CI1033 enhances cytotoxicity of 7-ethyl-10-hydroxycamptothecin and topotecan by inhibiting breast cancer resistance protein-mediated drug efflux. Cancer Res. 2001;61:739–48.
130. Tao LY, Liang YJ, Wang F, Chen LM, Yan YY, Dai CL, Fu LW. Cediranib (recentin, AZD2171) reverses ABCB1-and ABCC1-mediated multidrug resistance by inhibition of their transport function. Cancer Chemother Pharmacol. 2009;64:961–9.
131. Balabanov S, Gontarewicz A, Keller G, Raddrizzani L, Braig M, Bosotti R, Moll J, Jost E, Barett C, Rohe I, Bokemeyer C, Holyoake TL, Brummendorf TH. Abcg2 overexpression represents a novel mechanism for acquired resistance to the multi-kinase inhibitor Danusertib in BCR-ABL-positive cells in vitro. PLOS One. 2011;6:e19164.
132. Hegedus C, Ozvegy-Laczka C, Apati A, Magocsi M, Nemet K, Orfi L, Keri G, Katona M, Takats Z, Varadi A, Szakacs G, Sarkadi B. Interaction of nilotinib, dasatinib and bosutinib with ABCB1 and ABCG2: implications for altered anti-cancer effects and pharmacological properties. Br J Pharmacol. 2009;158:1153–64.
133. Shi Z, Peng XX, Kim IW, Shukla S, Si QS, Robey RW, Bates SE, Shen T, Ashby CR, Fu LW, Ambudkar SV, Chen ZS. Erlotinib (Tarceva, OSI-774) antagonizes ATP-bInding cassette subfamily B member 1 and ATP-binding cassette subfamily G member 2-mediated drug resistance. Cancer Res. 2007;67:11012–20.
134. Kuang YH, Shen T, Chen X, Sodani K, Hopper-Borge E, Tiwari AK, Lee JWKK, Fu LW, Chen ZS. Lapatinib and erlotinib are potent reversal agents for MRP7 (ABCC10)-mediated multidrug resistance. Biochem Pharmacol. 2010;79:154–61.
135. Yang CH, Huang CJ, Yang CS, Chu YC, Cheng AL, Whang-Peng J, Yang PC. Gefitinib reverses chemotherapy resistance in gefitinib-insensitive multidrug resistant cancer cells expressing ATP-binding cassette family protein. Cancer Res. 2005;65:6943–9.
136. Shen T, Kuang YH, Ashby CR, Lei Y, Chen A, Zhou Y, Chen X, Tiwari AK, Hopper-Borge E, Ouyang JY, Chen ZS. Imatinib and Nilotinib reverse multidrug resistance in cancer cells by inhibiting the efflux activity of the MRP7 (ABCC10). PLOS One. 2009;4:e7520.
137. Zhao XQ, Xie JD, Chen XG, Sim HM, Zhang X, Liang YJ, Singh S, Talele TT, Sun YL, Ambudkar SV, Chen ZS, Fu LW. Neratinib reverses ATP-binding cassette B1-mediated chemotherapeutic drug resistance in vitro, in vivo, and ex vivo. Mol Pharmacol. 2012;82:47–58.
138. Tiwari AK, Sodani K, Wang SR, Kuang YH, Ashby CR, Chen X, Chen ZS. Nilotinib (AMN107, Tasigna (R)) reverses multidrug resistance by inhibiting the activity of the ABCB1/Pgp and ABCG2/BCRP/MXR transporters. Biochem Pharmacol. 2009;78:153–61.
139. Sen R, Natarajan K, Bhullar J, Shukla S, Fang HB, Cai L, Chen ZS, Ambudkar SV, Baer MR. The novel BCR-ABL and FLT3 inhibitor ponatinib is a potent inhibitor of the MDR-associated ATP-binding cassette transporter ABCG2. Mol Cancer Ther. 2012;11:2033–44.

140. Liu KJ, He JH, Su XD, Sim HM, Xie JD, Chen XG, Wang F, Liang YJ, Singh S, Sodani K, Talele TT, Ambudkar SV, Chen ZS, Wu HY, Fu LW. Saracatinib (AZD0530) is a potent modulator of ABCB1-mediated multidrug resistance in vitro and in vivo. Int J Cancer. 2013;132:224–35.
141. Natarajan K, Bhullar J, Shukla S, Burcu M, Chen ZS, Ambudkar SV, Baer MR. The Pim kinase inhibitor SGI-1776 decreases cell surface expression of P-glycoprotein (ABCB1) and breast cancer resistance protein (ABCG2) and drug transport by Pim-1 dependent and -independent mechanisms. Biochem Pharmacol. 2013;85:514–24.
142. Shibayama Y, Nakano K, Maeda H, Taguchi M, Ikeda R, Sugawara M, Iseki K, Takeda Y, Yamada K. Multidrug resistance protein 2 implicates anticancer drug-resistance to sorafenib. Biol Pharmaceut Bull. 2011;34:433–5.
143. Shukla S, Robey RW, Bates SE, Ambudkar SV. Sunitinib (Sutent, SU11248), a small-molecule receptor tyrosine kinase inhibitor, blocks function of the ATP-binding cassette (ABC) transporters P-glycoprotein (ABCB1) and ABCG2. Drug Metab Dispos. 2009;37:359–65.
144. Wu CP, Sim HM, Huang YH, Liu YC, Hsiao SH, Cheng HW, Li YQ, Ambudkar SV, Hsu SC. Overexpression of ATP-binding cassette transporter ABCG2 as a potential mechanism of acquired resistance to vemurafenib in BRAF(V600E) mutant cancer cells. Biochem Pharmacol. 2013;85:325–34.
145. Cooray HC, Janvilisri T, van Veen HW, Hladky SB, Barrand MA. Interaction of the breast cancer resistance protein with plant polyphenols. Biochem Biophys Res Comm. 2004;317:269–75.
146. Fu LW, Liang YJ, Deng LW, Ding Y, Chen LM, Ye YL, Yang XP, Pan QC. Characterization of tetrandrine, a potent inhibitor of P-glycoprotein-mediated multidrug resistance. Cancer Chemother Pharmacol. 2004;53:349–56.
147. Madureira AM, Molnar A, Abreu PM, Molnar J, Ferreira MJU. A new sesquiterpene-coumarin ether and a new abietane diterpene and their effects as inhibitors of p-glycoprotein. Planta Med. 2004;70:828–33.
148. Imai Y, Tsukahara S, Asada S, Sugimoto Y. Phytoestrogens/flavonoids reverse breast cancer resistance protein/ABCG2-mediated multidrug resistance. Cancer Res. 2004;64:4346–52.
149. Gyemant N, Tanaka M, Antus S, Hohmann J, Csuka O, Mandoky L, Molnar J. In vitro search for synergy between flavonoids and epirubicin on multidrug-resistant cancer cells. In Vivo. 2005;19:367–74.
150. Mitsuo M, Noguchi T, Nakajima Y, Aoki S, Ren XQ, Sumizawa T, Haraguchi M, Kobayashi M, Baba M, Nagata Y, Akiyama S, Furukawa T. Binding site(s) on P-glycoprotein for a newly synthesized photoaffinity analog of agosterol A. Oncol Res. 2003;14:39–48.
151. Wang C, Zhang JX, Shen XL, Wan CK, Tse AKW, Fong WF. Reversal of P-glycoprotein-mediated multidrug resistance by Alisol B 23-acetate. Biochem Pharmacol. 2004;68:843–55.
152. Ramachandran C, Rabi TY, Fonseca HB, Melnick SJ, Escalon EA. Novel plant triterpenoid drug amooranin overcomes multidrug resistance in human leukemia and colon carcinoma cell lines. Int J Cancer. 2003;105:784–9.
153. Zhang SZ, Yang XN, Morris ME. Flavonoids are inhibitors of breast cancer resistance protein (ABCG2)-mediated transport. Mol Pharmacol. 2004;65:1208–16.
154. Leslie EM, Mao QC, Oleschuk CJ, Deeley RG, Cole SPC. Modulation of multidrug resistance protein 1 (MRP1/ABCC1) transport and ATPase activities by interaction with dietary flavonoids. Mol Pharmacol. 2001;59:1171–80.
155. Engi H, Hohmann J, Gang G, Pusztai R, Redei D, Kovacs O, Schelz Z, Molnar J. Chemoprevention and inhibition of P-glycoprotein in cancer cells by Chinese medicinal herbs. Phytother Res. 2008;22:1671–6.

156. Pick A, Muller H, Mayer R, Haenisch B, Pajeva IK, Weigt M, Bonisch H, Muller CE, Wiese M. Structure-activity relationships of flavonoids as inhibitors of breast cancer resistance protein (BCRP). Bioorganic Med Chem. 2011;19:2090–102.
157. Zhang SH, Morris ME. Effects of the flavonoids biochanin A, morin, phloretin, and silymarin on P-glycoprotein-mediated transport. J Pharmacol Exp Ther. 2003;304:1258–67.
158. Limtrakul P, Khantamat O, Pintha K. Inhibition of P-glycoprotein activity and reversal of cancer multidrug resistance by Momordica charantia extract. Cancer Chemother Pharmacol. 2004;54:525–30.
159. Zheng LH, Bao YL, Wu Y, Yu CL, Meng XY, Li YX. Cantharidin reverses multidrug resistance of human hepatoma HepG2/ADM cells via down-regulation of P-glycoprotein expression. Cancer Lett. 2008;272:102–9.
160. Gyemant N, Tanaka M, Molnar P, Deli J, Mandoky L, Molnar J. Reversal of multidrug resistance of cancer cells in vitro: modification of drug resistance by selected carotenoids. Anticancer Res. 2006;26:367–74.
161. Kitagawa S, Nabekura T, Kamiyama S. Inhibition of P-glycoprotein function by tea catechins in KB-C2 cells. J Pharm Pharmacol. 2004;56:1001–5.
162. Yoshida N, Takagi A, Kitazawa H, Kawakami J, Adachi I. Inhibition of P-glycoprotein-mediated transport by extracts of and monoterpenoids contained in Zanthoxyli fructus. Toxicol Appl Pharmacol. 2005;209:167–73.
163. Raad I, Terreux R, Richomme P, Matera EL, Dumontet C, Raynaud J, Guilet D. Structure-activity relationship of natural and synthetic coumarins inhibiting the multidrug transporter P-glycoprotein. Bioorganic Med Chem. 2006;14:6979–87.
164. Tian H, Pan QC. Modulation of multidrug resistance by three bisbenzyl-isoquinolines in comparison with verapamil. Acta Pharmacol Sinica. 1997;18:455–8.
165. Madureira AM, Spengler G, Molnar A, Varga A, Molnar J, Abreu PM, Ferreira MJU. Effect of cycloartanes on reversal of multidrug resistance and apoptosis induction on mouse lymphoma cells. Anticancer Res. 2004;24:859–64.
166. Lavie Y, Harel-Orbital T, Gaffield W, Liscovitch M. Inhibitory effect of steroidal alkaloids on drug transport and multidrug resistance in human cancer cells. Anticancer Res. 2001;21:1189–94.
167. Yoo HH, Lee M, Lee MW, Lim SY, Shin J, Kim DH. Effects of Schisandra lignans on P-glycoprotein-mediated drug efflux in human intestinal Caco-2 cells. Planta Med. 2007;73:444–50.
168. Sim HM, Lee CY, Ee PLR, Go ML. Dimethoxyaurones: potent inhibitors of ABCG2 (breast cancer resistance protein). Eur J Pharm Sci. 2008;35:293–306.
169. Sim HM, Loh KY, Yeo WK, Lee CY, Go ML. Aurones as modulators of ABCG2 and ABCB1: synthesis and structure-activity relationships. ChemMedChem. 2011;6:713–24.
170. Lim S, Grassi J, Akhmedjanova V, Debiton E, Balansard G, Beliveau R, Barthomeuf C. Reversal of p-glycoprotein-mediated drug efflux by eudesmin from Haplophyllum perforatum and cytotoxicity pattern versus diphyllin, podophyllotoxin and etoposide. Planta Med. 2007;73:1563–7.
171. Henrich CJ, Bokesch HR, Dean M, Bates SE, Robey RW, Goncharova EI, Wilson JA, McMahon JB. A high-throughput cell-based assay for inhibitors of ABCG2 activity. J Biomol Screen. 2006;11:176–83.
172. Corea G, Fattorusso E, Lanzotti V, Motti R, Simon PN, Dumontet C, Di Pietro A. Structure-activity relationships for euphocharacins A-L, a new series of jatrophane diterpenes, as inhibitors of cancer cell P-glycoprotein. Planta Med. 2004;70:657–65.
173. Kim SW, Kwon H, Chi DW, Shim JH, Park JD, Lee YH, Pyo S, Rhee DK. Reversal of P-glycoprotein-mediated multidrug resistance by ginsenoside Rg(3). Biochem Pharmacol. 2003;65:75–82.
174. Jin J, Shahi S, Kang HK, van Veen HW, Fan TP. Metabolites of ginsenosides as novel BCRP inhibitors. Biochem Biophys Res Comm. 2006;345:1308–14.

175. De Castro WV, Mertens-Talcott S, Derendorf H, Butterweck V. Grapefruit juice-drug interactions: grapefruit juice and its components inhibit p-glycoprotein (ABCB1) mediated transport of talinolol in caco-2 cells. J Pharmaceut Sci. 2007;96:2808–17.
176. Palomo C, Oiarbide M, Garcia JM, Gonzalez A, Pazos R, Odriozola JM, Banuelos P, Tello M, Linden A. A practical total synthesis of hapalosin, a 12-membered cyclic depsipeptide with multidrug resistance-reversing activity, by employing improved segment coupling and macrolactonization. J Organic Chem. 2004;69:4126–34.
177. Ma Y, Wink M. The beta-carboline alkaloid harmine inhibits BCRP and can reverse resistance to the anticancer drugs mitoxantrone and camptothecin in breast cancer cells. Phytother Res. 2010;24:146–9.
178. Tournier N, Chevillard L, Megarbane B, Pirnay S, Scherrmann JM, Decleves X. Interaction of drugs of abuse and maintenance treatments with human P-glycoprotein (ABCB1) and breast cancer resistance protein (ABCG2). Int J Neuropsychopharmacol. 2010;13:905–15.
179. Hohmann J, Redei D, Forgo P, Molnar J, Dombi G, Zorig T. Jatrophane diterpenoids from Euphorbia mongolica as modulators of the multidrug resistance of L5128 mouse lymphoma cells. J Nat Prod. 2003;66:976–9.
180. Aoki S, Cao LW, Matsui K, Rachmat R, Akiyama S, Kobayashi M. Kendarimide A, a novel peptide reversing P-glycoprotein-mediated multidrug resistance in tumor cells, from a marine sponge of *Haliclona* sp. Tetrahedron. 2004;60:7053–9.
181. El-Readi MZ, Hamdan D, Farrag N, El-Shazly A, Wink M. Inhibition of P-glycoprotein activity by limonin and other secondary metabolites from citrus species in human colon and leukaemia cell lines. Eur J Pharmacol. 2010;626:139–45.
182. van Zanden JJ, de Mul A, Wortelboer HM, Usta M, van Bladeren PJ, Rietjens IMCM, Cnubben NHP. Reversal of in vitro cellular MRP1 and MRP2 mediated vincristine resistance by the flavonoid myricetin. Biochem Pharmacol. 2005;69:1657–65.
183. Hemauer SJ, Patrikeeva SL, Nanovskaya TN, Hankins GDV, Ahmed MS. Opiates inhibit paclitaxel uptake by P-glycoprotein in preparations of human placental inside-out vesicles. Biochem Pharmacol. 2009;78:1272–8.
184. Han Y, Tan TMC, Lim LY. In vitro and in vivo evaluation of the effects of piperine on P-gp function and expression. Toxicol Appl Pharmacol. 2008;230:283–9.
185. Shukla S, Wu CP, Nandigama K, Ambudkar SV. The naphthoquinones, vitamin K3 and its structural analogue plumbagin, are substrates of the multidrug resistance-linked ATP binding cassette drug transporter ABCG2. Mol Cancer Ther. 2007;6:3279–86.
186. Choi CH, Kang G, Min YD. Reversal of P-glycoprotein-mediated multidrug resistance by protopanaxatriol ginsenosides from Korean red ginseng. Planta Med. 2003;69:235–40.
187. Ahmed-Belkacem A, Macalou S, Borrelli F, Capasso R, Fattorusso E, Taglialatela-Scafati O, Di Pietro A. Nonprenylated rotenoids, a new class of potent breast cancer resistance protein inhibitors. J Med Chem. 2007;50:1933–8.
188. Weerasinghe P, Hallock S, Tang SC, Trump B, Liepins A. Sanguinarine overcomes P-glycoprotein-mediated multidrug-resistance via induction of apoptosis and oncosis in CEM-VLB 1000 cells. Exp Toxicol Pathol. 2006;58:21–30.
189. Fong WF, Wan CK, Zhu GY, Chattopadhyay A, Dey S, Zhao ZZ, Shen XL. Schisandrol A from Schisandra chinensis reverses P-glycoprotein-mediated multidrug resistance by affecting Pgp-substrate complexes. Planta Med. 2007;73:212–20.
190. Munoz-Martinez F, Lu PH, Cortes-Selva F, Perez-Victoria JM, Jimenez IA, Ravelo AG, Sharom FJ, Gamarro F, Castanys S. Celastraceae sesquiterpenes as a new class of modulators that bind specifically to human P-glycoprotein and reverse cellular multidrug resistance. Cancer Res. 2004;64:7130–8.
191. Choi CH, Sun KH, An CS, Yoo JC, Hahm KS, Lee IH, Sohng JK, Kim YC. Reversal of P-glycoprotein-mediated multidrug resistance by 5,6,7,3′,4′-pentamethoxyflavone (sinensetin). Biochem Biophys Res Comm. 2002;295:832–40.
192. Morita H, Koyama K, Sugimoto Y, Kobayashi J. Antimitotic activity and reversal of breast cancer resistance protein-mediated drug resistance by stilbenoids from Bletilla striata. Bioorganic Med Chem Lett. 2005;15:1051–4.

193. Woehlecke H, Osada H, Herrmann A, Lage H. Reversal of breast cancer resistance protein-mediated drug resistance by tryprostatin A. Int J Cancer. 2003;107:721–8.
194. Yu ST, Chen TM, Tseng SY, Chen YH. Tryptanthrin inhibits MDR1 and reverses doxorubicin resistance in breast cancer cells. Biochem Biophys Res Comm. 2007;358:79–84.
195. Pedersen PL, Carafoli E. Ion motive Atpases. 2. Energy coupling and work output. Trends Biochemi Sci. 1987;12:186–9.
196. Gottesman MM, Pastan I. Biochemistry of multidrug-resistance mediated by the multidrug transporter. Ann Rev Biochem. 1993;62:385–427.
197. Sauna ZE, Kim IW, Ambudkar SV. Genomics and the mechanism of P-glycoprotein (ABCB1). J Bioenerg Biomembr. 2007;39:481–7.
198. Kerr KM, Sauna ZE, Ambudkar SV. Correlation between steady-state ATP hydrolysis and vanadate-induced ADP trapping in human P-glycoprotein—evidence for ADP release as the rate-limiting step in the catalytic cycle and its modulation by substrates. J Biol Chem. 2001;276:8657–64.

# Multidrug Resistance: A Role for Membrane Physics, pH and Drug Transporters

**Chloe Daniel and Cyril Rauch**

**Abstract** Cancer is the second cause of mortality worldwide ($\sim$8 million death a year) with a cost amounting to $\sim$\$900b a year. Early treatment is of paramount importance, and tissue susceptible to becoming cancerous is a target for preventative treatment. Multi-drug resistance (MDR) is a common cause of chemotherapeutic failure in the case of 90 % of metastatic cancer. Different valid theories have been suggested about MDR (drugs transporters, pH, membrane and so on) but there is now a need to provide a unified model concerning all these observations. Herein we show how the alteration in the membrane physical properties mediated by pH changes and the expression of drug transporters are paramount in MDR.

**Keywords** MDR · NHE1 · pH · Warburg's hypothesis

**Abbreviations**

| | |
|---|---|
| ABC | ATP-binding cassette |
| MDR | Multi-drug resistance |
| NHE1 | The sodium-hydrogen exchanger 1 |
| Pgp | P-glycoprotein |

## Introduction

The number of cases of cancer in the year 2030 has been estimated, via extrapolation of 2007 statistics [1], and is expected to surpass 430,000 new clinical cases. With a projected rise in all cancers of 45 % by the year 2030 [1], despite taking into the account the predicted increase in screening for breast and prostate cancer,

C. Daniel · C. Rauch (✉)
School of Veterinary Medicine and Science, University of Nottingham, College Road, Sutton Bonington LE12 5RD, UK
e-mail: Cyril.rauch@nottingham.ac.uk

B. Bonavida (ed.), *Molecular Mechanisms of Tumor Cell Resistance to Chemotherapy*, Resistance to Targeted Anti-Cancer Therapeutics 1, DOI: 10.1007/978-1-4614-7070-0_2,

research into potential therapeutic techniques is becoming more important than ever. A major obstacle in the treatment of cancer is the development of Multi-drug resistance (MDR) in both metastatic and benign tumours. This transformation, from a cell susceptible to treatment by drugs to a state of resistance has long been the focus of cancer research and continues to be to this day.

MDR involves necessarily the plasma membrane as drugs need to cross the plasma membrane leading to interactions between their physico-chemical properties and those of the bilayer structure. Furthermore, pH [2–4], proton pumps/transporters [5], Pgp [6, 7] and sodium channels [8] are just a few examples of other cancer cell features which have been discovered to have a role in tumour cell survival and/or MDR. All things considered, an interaction must exist between all these biological and physico-chemical components and the ability of a cancer cell to escape drugs treatment.

Understanding how the usual mechanisms of apoptosis are avoided, and how MDR is achieved, should facilitate our research into reversing these characteristics of cancer cells. Whilst the ultimate aim would be a cure for cancer, it is crucial that all researchers have the patient in mind; therefore, it is more realistic that research will enable us to provide palliative treatment options rather than a cure.

## Multi-Drug Resistance and the Fluid Mosaic Model of the Plasma Membrane

The plasma membrane of a cell is composed of a phospholipid bilayer, as well as many other lipids and proteins. Its structural integrity is maintained by the electrostatic forces between amphipathic phospholipids; outward facing hydrophilic heads protect the hydrophobic fatty acid tails. The membrane creates a selectively impermeable barrier to most water-soluble solutes. However, it is by no means a rigid, static structure; the many membrane proteins and lipids can laterally traverse the surface, and the numbers of these cell membrane features can alter dramatically, even between the inner and outer leaflet. This constant movement of the constituents, and the varying composition, of the plasma membrane has led to the term 'Fluid Mosaic Model' [9] being used to describe the cell membrane structure. In addition to the make-up of the membrane, the presence, or absence, of certain membrane proteins can radically modify the permeability of a cell to a range of molecules.

In the treatment of cancer, as in many other diseases, a drug molecule must first negotiate the cell membrane before acting on its target inside the cell. For that reason, the phospholipid bilayer can be seen as the first barrier a drug must encounter at the level of the cell. It therefore follows, that alterations in the composition, fluidity and, as a result, the permeability of the membrane can have a significant effect on the efficacy of a drug.

Multi-drug resistance can be characterised by cells ability to avoid apoptosis when treated with a substance which aims to bring about its' demise. MDR is known to occur at the level of the cell membrane, it has been shown to manifest itself through increased efflux of a drug as well as through complex interactions with the components of the membrane itself, leading to decreased permeability [10–12]. This suggests that part of the MDR cause may reside in a change of the plasma membrane physical properties.

## MDR and Lipid Packing Theory

The density of phospholipids in the membrane has been shown to have a noticeable effect on membrane biomechanics [13, 14] such as fluid-phase endocytosis [15, 16] and the overall permeability of the membrane.

A densely populated cell membrane leads to a more compact packing of phospholipids and this, in turn, augments the rigidity of the membrane [17]. The stiffer the membrane, the less permeable it is to drugs, for example [18].

A higher density of phospholipids in the inner membrane of a cell, compared to the outer leaflet, has been shown to increase the rate of endocytosis [15, 16]. In an intracellular vesicle, the number of phospholipids in the outer membrane is greater than that in the inner membrane, causing a spherical structure to form. This means, therefore, that more lipid was present in the inner leaflet of the plasma membrane at the onset of vesiculation. Increasing the number of phospholipids in the inner leaflet of a cell membrane causes compression, and an out-pouching must occur to alleviate the pressure and rectify the imbalance between the two layers.

Endocytosis of anticancer drugs envelops them in intracellular compartments where they are not free to interact with other cytosolic components of the cell. This can be easily observed in culture conditions [19]. It therefore follows that an increased rate of endocytosis will render more of the drug futile. It has been shown that in cancer cells that are resistant to drugs, the inner leaflet of the membrane is more densely compacted [11, 20], thus a level of MDR can be achieved through the packing of lipids within the cell membrane impeding the transverse movement of drugs. Let us think 1 min about the changes that an increase in membrane tension (i.e. lipid packing) can generate. To interact with its intracellular target a drug needs to traverse the membrane but this event can take some time usually defined by classical rules of thermodynamics and energy barriers involved. If the membrane is "stiffer", the drug will remain trapped in the lipid bilayer for longer period of time. During its residency time, the drug is not static but does diffuse. This residency time will likely increase the likelihood of a drug interacting with any partner on the membrane. Here comes an essential aspect as MDR is also related to the expression of drug transporters among those the archetypal Pgp. Interaction between drugs and Pgp may thus fundamentally rely on the membrane physical state.

## Pgp

P-glycoprotein (Pgp) is a member of the ATP-binding cassette (ABC) transporter family, and is also commonly known as the Multi-drug resistance protein 1 (MDR1). As the name suggests, the protein is involved in controlling what substances cross the membrane. A low level of Pgp is present in many tissues of the body [21], however, higher levels of expression occur in tissues such as the blood–brain barrier [22] and excretory luminal epithelium [21, 23]. Distribution of this protein suggests a role in protection of tissues from xenobiotics and toxins of the circulatory system. In cancer, Pgp is responsible for the efflux of xenobiotic drugs [6], and hence has an important role to play in the development of multi-drug resistance in cancer cells. Pgp actively transports the drug back out of the cell and acts as a "flippase" to eject the drug from the membrane [24, 25].

Pgp expression is known to be upregulated in the MDR phenotype [7, 26, 27] in cancer cells and as such is a key target for improving the efficacy of chemotherapy. Pgp inhibitors, such as guggulsterone and CJX1, are capable of re-sensitising cells to drugs [28, 29]. They are known to achieve this through inhibiting Pgp drug expulsion mechanisms in addition to preventing expression of the channel protein in the lipid bilayer.

The Pgp theory looks adequate on its own but there is one point that needs further investigation. Why would drugs interact with a transporter like Pgp and be expelled efficiently? Remind that by definition MDR violates the law of enzyme specificity. The question is to envision a mechanism whereby chemical efficiency without the need for chemical affinity occurs.

## The Role of Space Dimensions in MDR

There is one beautiful mathematical theorem known as Polya's theorem that shows that when a particle diffuses in a space with a dimension smaller or equal to two, all the space will be visited and this possibly more than one time [30]. This means that a drug diffusing in the membrane over a long period of time will repeatedly collide with drug transporters. According to Polya's theorem the amount of collision should be far greater in 2D than 3D. In a sense the dimensionality of the membrane (as a 2D object) is paramount for MDR. What we see here is that the low dimensionality of a membrane plays a significant role similar (but certainly not identical) to the one that the temperature plays. This brings in an interesting point that is that if we were living in a different universe with a different number of space dimensions, say 4D for example, MDR would be "physically" impossible.

In any case, Polya's theorem reinforces the role of membrane stiffness in MDR with the presence of drug transporters. The next question is to find out the main biological process that by changing the membrane physical properties increases its stiffness to allow repeated collisions between drugs and transporters (as dictated by Polya's theorem). An important contributor to membrane stiffness is the pH.

## pH

Intracellular pH is intricately regulated by all animal cells, with the exception of erythrocytes, and a small change in $[H^+]$ can dramatically change the ionisation state of all weak bases and acids, affecting most internal cellular processes. Although the actual mechanisms for $pH_i$ regulation are complex and diverse, the concept of pH management is relatively simple. As reviewed by Boron et al. [31] transporters known as "Acid extruders" and "Acid loaders" are responsible for the alkalinisation and acidification, respectively, of the cytosol. For a competent $pH_i$ regulation mechanism to exist, there must be a pH sensor, a signal-transduction system and effector proteins—the 'extruders' and 'loaders'. In this review the membrane-bound protein NHE1 is extensively reviewed as an example.

A reversal in the pH gradient of the cell membrane is now recognised as the foundation for the neoplastic phenotype. Pioneering the research into the involvement of pH in MDR, Otto Warburg's 'Theory of Cancer' of 1927 [32], more commonly known as 'Warburg's Hypothesis', is to this day, the most widely accepted explanation for the aforementioned reversal of the pH gradient. Intracellular pH is known to increase from 6.2–6.9 to 7.3–7.4, alongside a decrease in extracellular pH from 7.12–7.7 to 6.99–7.05 [33]. pH is known to be regulated via many membrane proteins [34], including NHE1 [35], Carbonic anhydrases [36], Monocarboxylic transporters [37], V-ATPase [38, 39] and $Cl^-/HCO_3^-$ [40]. However, the role of NHE1 in particular, has been thoroughly examined [5, 33, 35, 41] and its importance validated in the absence of other membrane transporters. The role of NHE1 will be considered in more detail later in this review. Both intracellular and extracellular pH have a role to play in cancer cell survival, metastasis and invasion.

## Warburg's Hypothesis

In 1924, Otto Heinrich Warburg was the first to propose that tumorigenesis is driven by the upregulation of glycolysis [32], an anaerobic mechanism for respiration which promotes cell survival in the often hypoxic environment in the centre of a solid tumour. However, Warburg discovered that this switch to glycolysis is observed even under aerobic conditions. The level of proliferation in cancer cells is much higher than in other cells [42], and therefore they have an increased metabolic rate and subsequent demand for ATP. Given the relative inefficiency of glycolysis, a yield of just 2 ATP molecules in contrast to the 36–38 ATP molecules generated via oxidative phosphorylation, upregulation would appear to be counter-intuitive. This is indeed the case when the cells are under aerobic conditions, however, as previously mentioned, when cells are subjected to the hypoxic conditions typical of solid tumours. Aerobic respiration via oxidative phosphorylation can therefore not take place and glycolysis must provide the ATP necessary. Warburg demonstrates that the upregulation of glycolysis as seen in cancer cells is initiated by damage to the mitochondria [32], this alleviates mitochondrial inhibition of glycolysis [43]. Bonnet et al. [44] reversed the damage to mitochondrial respiration in cancer cells and thus observed a decline in growth and proliferation, validating the role of the mitochondria in the switch to glycolysis.

The Warburg Effect, not to be mistaken for the Warburg hypothesis, is a measurement of this switch to glycolysis, and is calculated as the ratio of glycolysis to respiration. The impact of this switch to glycolysis is the build-up of lactate and its dissociation products: lactic acid and $H^+$. Accumulation of $H^+$ ions intracellularly is detrimental to cell survival and so the cell must extrude excess $H^+$ ions via proton pumps and transporters. As is described earlier, the mechanism for this externalisation of $H^+$ ions is via membrane proteins, especially NHE1. It therefore follows that it is the reversal of the pH gradient which must occur first, or indeed in parallel, in order to permit tumour cells to undergo a switch to glycolysis.

## NHE1: The Archetypal Example in Cancer

The Sodium-Hydrogen Exchanger 1 (NHE1) is an antiporter which is involved in the regulation of cell pH and volume; it achieves this through mediating the influx and efflux of the cations $Na^+$ and $H^+$ respectively, across the cell membrane [34, 35]. NHE1 is incredibly pH sensitive [45]. Its activity is triggered by a drop in intracellular pH, whereby the membrane-bound protein causes the efflux of $H^+$ ions, to bring the intracellular pH back towards normal. An allometric binding-site for protons is able to detect the smallest in changes of $pH_{i.}$ In normal cells, NHE1 is calibrated to the physiological pH of the cytosol and remains dormant until such a time as a fall in $pH_i$ is measured [35]. This is then quickly rectified via extrusion of positively-charged hydrogen ions. Present in virtually all tissues [46], the $Na^+$/$H^+$ exchanger also plays a particularly important role in the cardiomyocyte action potential.

The role of NHE1 is crucial to the development of the reversed pH gradient and subsequent cancer cell survival. Tumour cells undergo a transformation propelled by the oncogenes ras and v-mas [47, 48]. Activity of the transmembrane protein, NHE1, is amplified through increasing the affinity for its proton substrate [33]. In addition, increased transcription of the gene *SLC9A1* upregulates the expression of NHE1 in tumour cells, one factor implicated in the increased transcription of this gene is epidermal growth factor (EGF) [49]. It has recently been hypothesised that, for NHE1 to be functioning at full capacity and efficiency, the sodium-ion channel $Na_V1.5$ must also be present [50, 51]. The theory behind this describes the role of

$Na_V1.5$ in replenishing the intracellular $Na^+$ stock which NHE1 uses in exchange for $H^+$ ions to bring about intracellular alkalinisation [35].

The combination of increased expression and activity of NHE1 augments the efflux of hydrogen ions, acidifying the external microenvironment with the excess of positively charged ions. The resulting depletion of hydrogen ions from the cytosol creates an intracellular alkalinisation. This reversal of the pH gradient across the membrane is necessary for the further transformations a tumour cell must undergo, including, and arguably most importantly, the metabolic switch to glycolysis.

## pH and Tumour Cell Survival

Once a reversed pH gradient has been established, tumour cells are free to invade surrounding tissues and commence metastasis. The efflux of $H^+$ ions, mediated by transporters such as NHE1, not only permits glycolysis to take place at an elevated rate, but also enhances tumour cell aggressiveness [52, 53]. Extruded $H^+$ ions contribute to sufficiently lowering the $pH_e$ such that the immediate tumour microenvironment is toxic to surrounding tissues [54, 55]. Indeed, proteases secreted by the tumour cell itself have been suggested to become activated by acidic $pH_e$ and therefore bring about the destruction of the extracellular matrix [56]. Furthermore, specific proteases such as Cathepsin B and metalloproteinases have been implicated in cancer cell invasion [57]. Death of adjoining cells creates space for tumour cell advancement, and therefore permits metastasis.

The role of pH in cancer cell survival has been thoroughly researched. Reviewed recently by Harguindey et al. [41] it can be summarised as follows. An increase in $pH_i$ alkalinity drives cells into the S-phase of the cell cycle, promoting a state of perpetual disorganised cell division, and therefore tumour growth. Inhibition of NHE1, an important pH regulator, can prevent tumour cell growth by reducing cancer cell proliferation. Neoangiogenesis is an important feature of tumours, as it allows oxygen and other important nutrients to perfuse solid tumours. Peptides which promote angiogenesis are known to be mediated via NHE1, therefore upregulation and intracellular alkalinisation promote the growth of new blood vessels. $pH_i$ has also been implicated in the functioning of p53, tumour suppressor gene; alkaline pH inhibits p53 therefore promoting cell survival. Suppression of p53 is widely recognised as being an important factor implicated in MDR; loss of function is key to resistance to anticancer drugs.

The effect of intracellular and extracellular pH in cancer cell invasion and metastasis is summarised in Fig. 1.

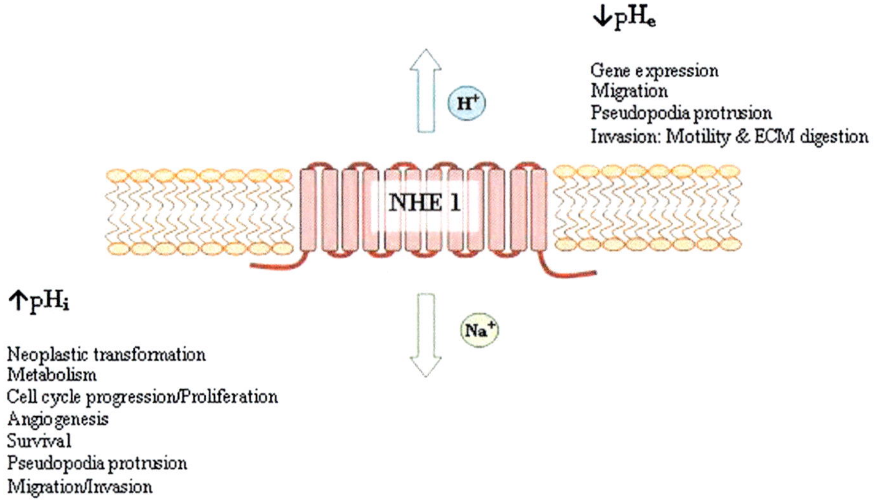


**Fig. 1** This figure highlights the role of NHE1 in the acidification of the extracellular tumoural microenvironment and alkalinisation of the cellular cytosol. It also summarises the effects of the aforementioned reverse in membrane potential [35]

## pH and MDR

To consider the role of pH in MDR, both intracellular and extracellular pH must be reviewed in turn. Whilst the role of extracellular pH has been examined in detail, intracellular pH has only recently been considered a more significant participant in MDR [58].

The pH partition theory is a widely accepted model which explains how $pH_e$ affects weakly acidic and weakly basic drug entry into a cell. From this theory, it is understood that acidic $pH_e$ has the effect of reducing uptake of weakly basic drugs, such as Doxorubicin, due to protonation making the compound impermeable to the cell membrane barrier [59].

The significance of $pH_i$ has an impact on current potential anticancer therapies. Re-establishing a normal pH gradient across cells would decrease $pH_{i,}$ with the potential to increase weak base drugs binding with their intracellular targets.

In addition to its effect on the drug itself, it has been proposed that pH can have an effect on the lipid-packing of the cell membrane, which will contribute towards MDR [20]. In a drug-sensitive cell, the fluidity of the inner leaflet of the membrane is maintained via accumulation of protons in the cytosol in close proximity to the membrane. It is hypothesised that these positively-charged ions could have the effect of diminishing the electrostatic repulsion between neighbouring, negatively-charged, phospholipids. However, in a drug-resistant cell, as previously discussed, a proton-efflux has been established which leads to depletion of intracellular protons.

Therefore, neutralisation of the repulsive charges between the phospholipids could not be achieved to the same extent, resulting in a more rigid and less permeable membrane.

## Conclusion

MDR is of paramount importance when considering treating a patient with any disease. This is especially so in cancer where often palliative care, and not a cure, is the prevailing objective. MDR is responsible for anticancer drug therapy failure, and the focus of research today is to reverse this phenomenon with the aim of re-sensitising cells to current therapeutic treatments.

Research into the influence of the cell membrane in MDR is extensive and related now to the pioneering research of Otto Warburg and his discovery of the switch in cancer cell metabolism. It is now widely accepted that MDR can be characterised at the level of the cell membrane, and that the key determining factor in the neoplastic phenotype relies predominantly on the pH gradient across the membrane. The $Na^+/H^+$ antiporter NHE1 is perhaps one of the most important membrane-bound transporters involved in the maintenance of the reversed pH gradient, and therefore it has been the focus of many studies.

Research into MDR will continue indefinitely, with the aim being: to identify other parameters involved in MDR, and ultimately to discover therapeutic methods that disrupt the mechanisms cancer cells employ to achieve resistance to multiple drugs.

**Conflicts of Interest** No potential conflicts of interest were disclosed.

## References

1. Mistry M, Parkin DM, Ahmad AS, Sasieni P. Cancer incidence in the United Kingdom: projections to the year 2030. Br J Cancer. 2011;105:1795–803.
2. Gillies RJ, Rgahunand N, Karczmar GS, Bhujwalla Z. MRI of the tumor microenvironment. J Magn Reson Imaging. 2002;16:430–50.
3. Keizer HG, Joenje H. Increased cytosolic ph in multidrug-resistant human lung tumor cells: effect of verapamil. J Natl Cancer Inst. 1989;81:706–9.
4. Boscoboinik D, Gupta RS, Epand RM. Investigation of the relationship between altered intracellular pH and multidrug resistance in mammalian cells. Br J Cancer. 1990;61:568–72.
5. Cardone RA, Casavola V, Reshkin SJ. The role of distrubed pH dynamics and the Na+/H+ exchanger in metastasis. Nat Rev Cancer. 2005;5:786–95.
6. Shen F, Chu S, Bence AK, Bailey B, Xue X, Erickson PA, Montrose MH, Beck WT, Erickson LC. Quantitation of doxorubicin uptake, efflux, and modulation of multidrug resistance (MDR) in MDR human cancer cells. J Pharmacol Exp Ther. 2008;324:95–102.
7. Endicott JA, Ling V. The biochemicstry of P-glycoprotein-mediated multidrug resistance. Annu Rev Biochem. 1989;58:137–71.

8. Fiske JL, Fomin VP, Brown ML, Duncan RL, Sikes RA. Voltage-senstive ion channels and cancer. Cancer Metastasis Rev. 2006;25:493–500.
9. Singer SJ, Nicolson GL. The fluid mosaic model of the structure of cell membranes. Science. 1972;175:103–31.
10. Rauch C. The "multi" of drug resistance explained by oscillating drug transporters, drug-membrane physical interactions and spatial dimensionality. Cell Biochem Biophys. 2011;61:103–13.
11. Rauch C. On the relationship between drug's size, cell membrane mechanical properties and high levels of multi drug resistance: a comparison to published data. Eur Biophys J. 2009;38:537–46.
12. Ferte J. Analysis of the tangled relationships between P-glycoprotein-mediated multidrug resistance and the lipid phase of the cell membrane. Eur J Biochem. 2000;267:277–94.
13. Chapman D. Phase transistions and fluidity characteristics of lipids and cell membranes. Q Rev Biophys. 1975;8:185–235.
14. Bretscher MS. Membrane structure: some general principles. Science. 1973;181:622–9.
15. Farge E, Ojcius DM, Subtil A, Dautry-Varsat A. Enhancement of endocytosis due to aminophospholipid transport across the plasma membrane of living cells. Am J Physiol. 1999;276:C725–33.
16. Rauch C, Farge E. Endocytosis switch controlled by transmembrane osmotic pressure and phospholipid number asymmetry. Biophys J. 2000;78:3036–47.
17. Edidin M. Lipids on the frontier: a century of cell-membrane bilayers. Nat Rev Mol Cell Biol. 2003;4:414–8.
18. Rauch C, Pluen A. Multi drug resistance-dependent "vacuum cleaner" functionality potentially driven by the interactions between endocytosis, drug size and Pgp-like transporters surface density. Eur Biophys J. 2007;36:121–31.
19. Altan N, Chen Y, Schindler M, Simon SM. Defective acidification in human breast tumor cells and implications for chemotherapy. J Exp Med. 1998;187:1583–98.
20. Rauch C. Toward a mechanical control of drug delivery. On the relationship between Lipinski's 2nd rule and cytosolic pH changes in doxorubicin resistance levels in cancer cells: a comparison to published data. Eur Biophys J. 2009;38:829–46.
21. Thiebaut F, Tsuruo T, Hamada H, Gottesman MM, Pastan I, Willingham MC. Cellular localization of the multidrug-resistance gene product P-glycoprotein in normal human tissues. Proc Natl Acad Sci USA. 1987;84:7735–8.
22. Beaulieu E, Demeule M, Ghitescu L, Beliveau R. P-glycoprotein is strongly expressed in the luminal membranes of the endothelium of blood vessels in the brain. Biochem J. 1997;326:539–44.
23. Arceci RJ, Croop JM, Horwitz SB, Housman D. The gene encoding multidrug resistance is induced and expressed at high levels during pregnancy in the secretory epithelium of the uterus. Proc Natl Acad Sci USA. 1988;85:4350–4.
24. Higgins CF, Gottesman MM. Is the multidrug transporter a flippase? Trends Biochem Sci. 1992;17:18–21.
25. Homolya L, Hollo Z, Germann UA, Pastan I, Gottesman MM, Sarkadi B. Fluorescent cellular indicators are extruded by the multidrug resistance protein. J Biol Chem. 1993;268:21493–6.
26. Broxterman HJ, Giaccone G, Lankelma J. Multidrug resistance proteins and other drug transport-related resistance to natural product agents. Curr Opin Oncol. 1995;7:532–40.
27. Ferreira MJ, Gyemant N, Madureira AM, Tanaka M, Koos K, Didziapetris R, Molnar J. The effects of jatrophane derivatives on the reversion of MDR1- and MRP-mediated multidrug resistance in the MDA-MB-231 (HTB-26) cell line. Anticancer Res. 2005;25:4173–8.
28. Xu HB, Li L, Liu GQ. Reversal of p-glycoprotein-mediated multidrug resistance by guggulsterone in doxorubicin-resistance human myelogenous leukemia (K562/DOX) cells. Pharmazie. 2009;64:660–5.
29. Ji BS, He L, Liu GQ. Reversal of P-glycoprotein-mediated multidrug resistance by CJX1, an amlodipine derivative, in doxorubicin-resistant human myelogenous leukemia (K562/DOX) cells. Life Sci. 2005;77:2221–32.

30. Panagiotopoulou V, Richardson G, Jensen OE, Rauch C. ON a biophysical and mathematical model of Pgp-mediated multidrug resistance: understanding the "space-time" dimension of MDR. Eur Biophys J. 2010;39:201–11.
31. Boron WF. Regulation of intracellular pH. Adv Physiol Educ. 2004;28:160–79.
32. Warburg O, Posener K, Negelein E. On the metabolism of tumours. London: Constable; 1930.
33. Reshkin SJ, Bellizzi A, Caldeira S. Albarani V, Malanchi I, Poignee M, Alunni-Fabbroni M, Casavola V, Tommasino M. Na +/H + exchanger-dependent intracellular alkalinization is an early event in malignant transformation and plays an essential role in the development of subsequent tranformation-associated phenotypes. FASEB J. 2000;14:2185–2197.
34. Casey JR, Grinstein S, Orlowski J. Sensors and regulators of intracellular pH. Nat Rev Mol Cell Biol. 2010;11:50–61.
35. Reshkin SJ, Cardone RA, Harguindey S. $Na^+$ -$H^+$ exchanger, pH regulation and cancer. Recent Pat Anticancer Drug Discov. 2013;8:85–99.
36. Swietach P, Vaughan-Jones RD, Harris AL. Regulation of tumor pH and the role of carbonic anhydrase 9. Cancer Metastasis Rev. 2007;26:299–310.
37. Pinheiro C, Longatto-Filho A, Azevedo-Silva J, Casal M, Schmitt FC, Baltazar F. Role of monocarboxylate transporters in human cancers: state of the art. J Bioenerg Biomembr. 2012;44:127–39.
38. Hinton A, Sennoune SR, Bond S, Fang M, Reuveni M, Sahagian GG, Jay D, Martinez-Zaguilan R, Forgac M. Functions of a subunit isoforms of the V-ATPase in pH homeostasis and in vitro invasion of MDA-MB231 human breast cancer cells. J Biol Chem. 2009;284:16400–8.
39. Harvey WR. Physiology of V-ATPase. J Exp Biol. 1992;172:1–17.
40. Boyer MJ, Tannock IF. Regulation of intracellular pH in tumor cell lines: influence of microenvironmental conditions. Cancer Res. 1992;52:4441–7.
41. Harguindey S, Orive G, Luis Pedraz J, Paradiso A, Reshkin SJ. The role of pH dynamics and the Na +/H + antiporter in the etiopathogenesis and treatment of cancer. Two faces of the same coin–one single nature. Biochim Biophys Acta. 2005;1756:1–24.
42. Andreeff M, Goodrich D, Pardee AB, editors. Holland-Frei cancer medicine: cell proliferation, differentiation, and apoptosis. 5th ed. Hamilton: BC Decker Inc; 2000.
43. Tennant DA, Duran RV, Boulahbel H, Gottlieb E. Metabolic transformation in cancer. Carcinogenesis. 2009;30:1269–80.
44. Bonnet S, Archer SL, Allalunis-Turner J, Haromy A, Beaulieu C, Thompson R, Lee CT, Lopaschuk GD, Puttagunta L, Harry G, Hashimoto K, Porter CJ, Andrade MA, Thebaud B, Michelakis ED. A mitochondra-$K^+$ channel axis is suppressed in cancer and its normalization promotes apoptosis and inhibits cancer growth. Cancer Cell. 2007;11:37–51.
45. Boedtkjer E, Bunch L, Pedersen SF. Physiology, pharmacology and pathophysiology of the pH regulatory transport proteins NHE1 and NBCn1: similarities, differences, and implications for cancer therapy. Curr Pharm Des. 2012;18:1345–71.
46. Slepkov E, Fliegel L. Structure and function of the NHE1 isoform of the $Na^+/H^+$ exchanger. Biochem Cell Biol. 2002;80:499–508.
47. Doppler W, Jaggi R, Groner B. Induction of v-mos and activated Ha-ras oncogene expression in quiescent NIH 3T3 cells causes intracellular alkalinisation and cell-cycle progression. Gene. 1987;54:147–53.
48. Hagag N, Lacal JC, Graber M, Aaronson S, Viola MV. Microinjection of ras p21 induces a rapid rise in intracellular pH. Mol Cell Biol. 1987;7:1984–8.
49. Chiang Y, Chou CY, Hsu KF, Huang YF, Shen MR. EGF upregulates $Na^+/H^+$ exchanger NHE1 by post-translational regulation that is important for cervical cancer cell invasiveness. J Cell Physiol. 2008;214:810–9.
50. Brisson L, Gillet L, Calaghan S, Besson P, Le Guennec JY, Roger S, Gore J. Na(V)1.5 enhances breast cancer cell invasiveness by increasing NHE1-dependent H(+) efflux in caveolae. Oncogene. 2011;30:2070–2076.

51. Gillet L, Roger S, Besson P, Lecaille F, Gore J, Bougnoux P, Lalmanach G, LE Guennec JY. Voltage-gated sodium channel activity promotes cysteine cathespin-dependent invasiveness and colony growth of human cancer cells. J Biol Chem. 2009;284:8680–91.
52. Arias-Pulido H, Royce M, Gong Y, Joste N, Lomo L, Lee SJ, Chaher N, Verschraegen C, Lara J, Prossnitz ER, Cristofanilli M. GPR30 and estrogen receptor expression: new insights into hormone dependence of inflammatory breast cancer. Breast Cancer Res Treat. 2010;123:51–8.
53. Heerdt AS, Borgen PI. Current status of tamoxifen use: an update for the surgical oncologist. J Surg Oncol. 1999;72:42–9.
54. Gatenby RA, Gawlinski ET, Gmitro AF, Kaylor B, Gillies RJ. Acid-mediated tumor invasion: a multidisciplinary study. Cancer Res. 2006;66:5216–23.
55. Kraus M, Wolf B. Implications of acidic tumor microenvironment for neoplastic growth and cancer treatment: a computer analysis. Tumour Biol. 1996;17:133–54.
56. Martin NK, Gaffney EA, Gatenby RA, Maini PK. Tumour-stromal interactions in acid-mediated invasion: a mathematical model. J Theor Biol. 2010;267:461–70.
57. Mason SD, Joyce JA. Proteolytic networks in cancer. Trends Cell Biol. 2011;21:228–37.
58. Swietach P, Hulikova A, Patiar S, Vaughan-Jones RD, Harris AL. Importance of intracellular pH in determining the uptake and efficacy of the weakly basic chemotherapeutic drug, doxorubicin. PLoS ONE. 2012;7:e35949.
59. Gerweck LE, Kozin SV, Stocks SJ. The pH partition theory predicts the accumulation and toxicity of doxorubicin in normal and low-pH adapted cells. Br J Cancer. 1999;79:838–42.

# Mechanisms and Potential Therapies for Acquired Resistance to Inhibitors Targeting the Raf or MEK Kinases in Cancer

**Alyson K. Freeman and Deborah K. Morrison**

**Abstract** Aberrant activation of the Ras/Raf/MEK/ERK signaling pathway often occurs in human cancer through the acquisition of oncogenic mutations in key pathway components. In particular, Ras mutations are found in about 20 % of human cancers, and B-Raf mutations occur in more than half of all melanomas. Thus, this pathway has become an attractive target for cancer therapies. Inhibitors targeting either the Raf or MEK kinases have shown initial success in treating cancers that depend on the Ras/Raf/MEK/ERK signaling pathway; unfortunately, resistance to the inhibitors eventually develops. For both sets of inhibitors, resistance most commonly occurs via reactivation of the Ras/Raf/MEK/ERK pathway or upregulated signaling from an alternate pathway, such as the PI3K/AKT pathway. Here, we discuss the mechanisms for acquired resistance to inhibitors targeting the Raf or MEK kinases and possible combination therapies to overcome or delay drug resistance.

**Keywords** Raf inhibitor · MEK inhibitor · Inhibitor resistance · Combination therapy · Signaling pathways

**Abbreviations**

| | |
|---|---|
| ATP | Adenosine-5′-triphosphate |
| AML | Acute myeloid leukemia |
| CML | Chronic myelogenous leukemia |
| ERK | Extracellular regulated kinase |
| EGFR | Epidermal growth factor receptor |
| FGFR3 | Fibroblast growth factor receptor 3 |
| HGF | Hepatocyte growth factor |
| IGF-1R | Insulin-like growth factor 1 receptor |
| MEK | MAP or ERK kinase |

A. K. Freeman · D. K. Morrison (✉)
Laboratory of Cell and Developmental Signaling, National Cancer Institute at Frederick, Frederick, MD 21702, USA
e-mail: morrisod@mail.nih.gov

B. Bonavida (ed.), *Molecular Mechanisms of Tumor Cell Resistance to Chemotherapy*, Resistance to Targeted Anti-Cancer Therapeutics 1, DOI: 10.1007/978-1-4614-7070-0_3,

| | |
|---|---|
| NSCLC | Non-small-cell lung carcinoma |
| PDGFR-$\beta$ | Platelet-derived growth factor receptor-$\beta$ |
| PI3K | Phosphatidylinositide 3-kinase |
| PTEN | Phosphatase and tensin homolog |
| RTK | Receptor tyrosine kinase |
| SAHA | Suberoylanilide hydroxamic acid |
| STAT3 | Signal transducer and activator of transcription 3 |
| VEGFR | Vascular endothelial growth factor receptor |

## Introduction

### *The Ras/Raf/MEK/ERK Pathway*

The Ras/Raf/MEK/ERK pathway is a critical regulator of cellular processes that include proliferation, differentiation, and senescence [1]. Upon receptor tyrosine kinase (RTK) activation, Ras (K-Ras, N-Ras, or H-Ras) cycles to its active GTP-bound form and recruits the Raf kinases (A-Raf, B-Raf, and C-Raf) to the membrane. At the cell surface, the Rafs dimerize and become active, thus initiating a phosphorylation cascade in which Raf activates MEK, and MEK activates ERK [2, 3]. Once activated, ERK phosphorylates numerous substrates that are then needed for the desired cellular response. Active ERK also instigates a negative feedback loop, whereby ERK phosphorylates the Rafs at several sites, thus attenuating Raf signaling by disrupting the Ras–Raf interaction and dissociating the Raf dimers [4, 5]. ERK activation also results in the upregulation of protein phosphatase activity, which in combination with ERK's inhibitory phosphorylation on Raf as well as the Ras guanine exchange factor SOS1, provides tight control of pathway signaling [6, 7].

The biological importance of proper Ras/Raf/MEK/ERK pathway regulation is highlighted by the frequent mutation of its components in human cancer. Ras is mutated in 20 % of all human cancers, and *B-Raf* is mutated in ~50 % of melanomas, with a single valine-to-glutamic acid substitution at position 600 (V600E) accounting for 90 % of the B-Raf mutations in melanoma [8–10]. Two recent reviews have examined in detail how activating mutations in Ras and Raf can affect various hallmarks of cancer [11, 12]. Given the role of the Ras/MEK/ERK pathway in cancer, the kinase components of this signaling cascade have become attractive targets for pharmaceutical inhibition [13]. A number of Raf inhibitors as well as MEK inhibitors have been developed and are in clinical use. Although these inhibitors are showing promise in the treatment of certain cancers, as has been observed with other kinase-targeted therapeutics, drug-resistance to the inhibitor eventually develops [14, 15].

## *Raf Inhibitors*

Given that B-Raf is a prominent cancer driver and that signaling initiated by activated RTK or Ras proteins is dependent on the Rafs, considerable effort has been focused on the development of Raf kinase inhibitors (Table 1). These include drugs that target all three Raf family members (B-Raf, C-Raf, and A-Raf) as well as those with specificity towards the V600E-B-Raf mutant. Most of the reported Raf inhibitors are ATP-competitive kinase inhibitors, including sorafenib (BAY 43-9006), vemurafenib (PLX4032 and PLX472, a tool compound for in vitro studies), dabrafenib (GSK2118436), SB590885, AZ628, and RAF265. Of these, sorafenib, vemurafenib, and dabrafenib have been studied most extensively in clinical trials.

Sorafenib is a multi-kinase inhibitor that exhibits activity towards wild-type Raf kinases and V600E-B-Raf, as well as the RTKs VEGFR and PDGFR-$\beta$, among others [16]. Because B-Raf is so highly mutated in melanoma, several clinical trials have investigated sorafenib's effects in treating this cancer. Unfortunately, this drug appears ineffective at treating melanoma, with a beneficial effect only observed in a subset of patients that have yet to be characterized [17]. Nonetheless, sorafenib has been approved as a standard therapy for hepatocellular carcinoma and shows potential for the treatment of renal cancer [18, 19].

**Table 1** Kinase inhibitors that target either the Raf or MEK kinases

| | Generic name | Target(s) | Mechanism | References |
|---|---|---|---|---|
| *Raf inhibitors* | | | | |
| BAY 43-9006 | Sorafenib | C-Raf, VEGFR-2, VEGFR-3, B-Raf, V600E-B-Raf, PDGFRb, Flt-3, c-KIT | ATP competitive | [16] |
| PLX4032/PLX4720 | Vemurafenib | V600E-B-Raf, BRK, B-Raf | ATP competitive | [20] |
| GSK2118436 | Dabrafenib | V600E-B-Raf, B-Raf, C-Raf | ATP competitive | [21] |
| SB590885 | | V600E-B-Raf, C-Raf | ATP competitive | [88] |
| AZ628 | | V600E-B-Raf, B-Raf | ATP competitive | [89] |
| RAF265 | | C-Raf, B-Raf, V600E-B-Raf, VEGFR-2 | ATP competitive | [90] |
| *MEK inhibitors* | | | | |
| AZD6244/ARRY142886 | Selumetinib | MEK1/2 | Allosteric | [91] |
| GSK1120212/JTP-74057 | Trametinib | MEK1/2 | Allosteric | [92] |
| RO4927350/RG7167 | | MEK1/2 | Allosteric | [93] |
| PD0325901 | | MEK1/2 | Allosteric | [94] |
| CI-1040/PD184352 | | MEK1/2 | Allosteric | [94] |

Unlike sorafenib, vemurafenib and dabrafenib primarily target the highly active V600E-B-Raf [20, 21]. Both of these inhibitors have exhibited clinical efficacy for the treatment of metastatic melanomas that express the B-Raf V600E mutant [14, 15]. However, a frequently reported adverse effect has been the appearance of cutaneous squamous-cell carcinomas and keratoacanthomas that, in many cases, develop because the patient's skin cells harbor an oncogenic mutation in H-Ras [22–25]. In addition, vemurafenib treatment of one melanoma patient was found to promote the proliferation of a previously undetected Ras-mutant leukemia [26]. Surprisingly, all ATP-competitive Raf inhibitors generated to date promote and stabilize Raf dimer formation in the presence of activated Ras, and in these secondary cancers, inhibitor treatment causes a paradoxical activation of the pathway, which results in tumor progression [27–29].

## *MEK Inhibitors*

Although MEK mutations are rare in human cancer, MEK kinases, like the Rafs, are requisite mediators of Ras/Raf/MEK/ERK signaling. Thus, numerous MEK inhibitors have also been developed (Table 1), all of which are allosteric inhibitors and bind to a site other than the ATP-binding pocket. These drugs include selumetinib (AZD6244/ARRY-142886), trametinib (GSK1120212/JTP-74057), RO4927350 (RG7167), PD0325901, CI-1040 (PD184352), and U0126 (a tool compound). Of the MEK inhibitors, trametinib and selumetinib have progressed the furthest in clinical trials, as discussed below.

During inhibitor development, several groups set out to determine predictive markers for cellular responsiveness to MEK inhibition in a variety of Ras pathway-dependent tumor types. Overall, the presence of *Ras* or *B-Raf* mutations was found to correlate with sensitivity to various MEK inhibitors when tested in cancer cell lines of diverse lineages [30–35]. Although one study examining colorectal cancer lines did not find a significant correlation between the presence of *K-Ras* or *B-Raf* mutations and sensitivity to the MEK inhibitor selumetinib, the seven lines they deemed sensitive did have mutations in *K-Ras* or *B-Raf* [36]. Interestingly, sensitivity to the MEK inhibitors has not been found to correlate with levels of phosphorylated ERK or AKT in several studies [32, 33, 36]. In addition, some cell types may be inherently more sensitive to MEK inhibition than others. For example, basal-type and triple-negative breast cancer lines are quite sensitive to MEK inhibition as are the hematologic malignancies AML and CML, whereas breast cancer lines of luminal-origin and cancer cells that have undergone an epithelial-to-mesenchymal transition are less sensitive [35, 37].

MEK inhibitors have shown mixed responses in clinical trials. Similar to Raf inhibitor therapy, one study found that trametinib treatment led to improved overall and progression-free survival in melanoma patients containing a V600E or V600K *B-Raf* mutation [38]. One important difference, however, was that unlike the Raf inhibitors, treatment with trametinib did not induce secondary squamous cell carcinomas [38]. In regard to selumetinib, although several clinical trials indicate that it has no significant benefit as a general cancer therapy, other studies find that it may be useful for the treatment of melanoma or other cancers that contain oncogenic *Ras* or *B-Raf* mutations [39–45].

## Raf Inhibitor Resistance

Despite the reported success of the Raf inhibitors vemurafenib and dabrafenib in the treatment of metastatic melanoma, drug resistance typically develops within months [14, 15]. As a consequence, much effort has been focused on determining the mechanism(s) of acquired resistance. Vemurafenib is the best characterized of these drugs and has been approved by the Food and Drug Administration for use in unresectable or metastatic melanoma, many of which contain activating *B-Raf* mutations [46]. Therefore, much of the work investigating Raf inhibitor resistance has been conducted with vemurafenib and melanoma cell lines or patient tumors samples expressing V600E-B-Raf. To date, the majority of resistance mechanisms identified involve either reactivation of the Ras/Raf/MEK/ERK pathway or sustained upregulation of an alternative pathway, such as the PI3K/AKT pathway (Fig. 1).

### *Reactivation of the Ras/Raf/MEK/ERK Pathway in Raf Inhibitor Resistance*

A common mechanism conferring resistance to numerous ATP-competitive kinase inhibitors is mutation of the gatekeeper residue in the protein kinase domain, which disrupts drug binding [47]. Although experimentally-induced mutation of the B-Raf gatekeeper site can confer the V600E-B-Raf or V600K-B-Raf proteins resistance to many Raf inhibitors [48], mutation of the gatekeeper site does not appear to be a naturally occurring mechanism of Raf inhibitor resistance, in that DNA sequencing has failed to identify any B-Raf gatekeeper mutations in patient samples or melanoma lines exhibiting drug resistance [49–55]. However, a B-Raf alteration that has been identified in patient samples and melanoma lines resistant to vemurafenib involves a splice variant of V600E-B-Raf [56]. This mutant protein lacks exons coding for the Ras binding domain, and it dimerizes in a Ras-independent manner [56]. Thus, due to its enhanced dimerization potential, this mutant increases ERK signaling in the presence of Raf inhibitor [56].

Another mechanism of Raf inhibitor resistance involves the upregulation of Raf protein levels. In a report by Shi et al. [57], whole exon sequencing of tumor samples from four patients with disease progression following vemurafenib treatment revealed copy number gains in the mutant *B-Raf* gene. Increased

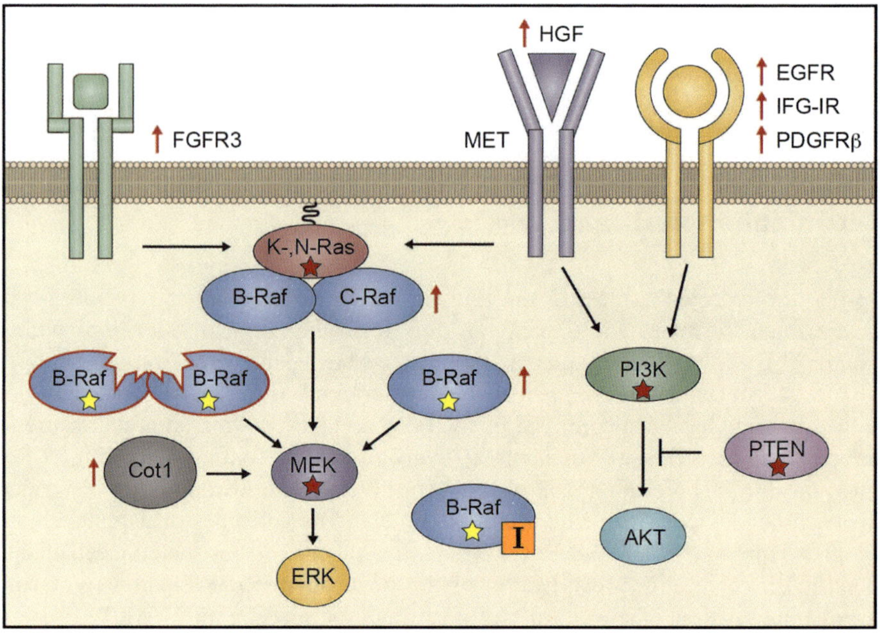


**Fig. 1** *Mechanisms of Raf inhibitor resistance.* Several mechanisms of Raf inhibitor resistance are highlighted in *red*, all of which can increase signaling through the Ras/Raf/MEK/ERK or PI3K/AKT pathways. These include increased signaling through FGFR3; higher levels of HGF causing increased signaling through MET; an increase in the number of EGF, IGF-1, or PDGF receptors; mutations (indicated by *stars*) in K-Ras, N-Ras, MEK, PI3K, or PTEN; an increase in C-Raf, V600E-B-Raf, or COT1 protein levels; and a splice variant of V600E-B-Raf

expression of the V600E-B-Raf mutant has also been observed in a vemurafenib-resistant melanoma cell line [51]. Interestingly, further investigation by Shi et al. [57] revealed that the vemurafenib-resistance caused by upregulation of V600E-B-Raf could be overcome by increasing the dose of the Raf inhibitor. An increase in C-Raf protein levels has also been reported to promote drug resistance in melanoma cell lines treated with the Raf inhibitors PLX4720, AZ628 or SB590885 [50, 51, 58]. Moreover, another study using shRNAs to deplete each Raf family member independently and then simultaneously showed that, although the sustained ERK activation in drug-resistant cells required the Rafs, the signal could be generated by any of the Raf family members [53]. Taken together, these studies indicate that, in the presence of a Raf inhibitor, increased signaling from mutant B-Raf or through A-Raf or C-Raf can increase the activation of ERK, thus conferring drug resistance.

Reactivation of the Ras pathway despite inhibitor treatment can also occur via increased signaling from components upstream of Raf, such as RTKs and Ras. A recent study found that FGFR3 showed heightened activity in vemurafenib-resistant melanoma cell lines, potentially due to an autocrine loop generated by an increase in cellular bFGF protein and secreted FGF2 ligand [55]. Upregulated signaling from the

PDGFR and the IGF-1R has also been observed in resistant melanomas [49, 53]. In addition, a case study of a patient whose cancer progressed after vemurafenib treatment reported two different activating mutations in N-Ras at metastatic sites: Q61K and Q61R [49]. Several other studies have linked mutations at N-Ras Q61 to vemurafenib and dabrafenib resistance [49, 52, 57, 59], which may be dependent, in part, on C-Raf and the SHOC2/Sur8 scaffold protein, which contributes to C-Raf activation [60]. Two additional Ras mutations, N-Ras A146T and K-Ras K117N, have also been implicated as mediators of vemurafenib resistance in B-Raf mutant melanomas [51].

Alterations in Ras pathway components that function downstream or parallel to Raf can also confer Raf inhibitor resistance. For example, activating mutations in the downstream MEK1 kinase, such as C121S, Q65P, and K59del, have been reported to mediate PLX4720/vemurafenib resistance in melanoma cells [61–63]. In a study conducted by Shi and colleagues [61], pre-existing MEK1 mutations, P124S or I111S, were identified in baseline tumors from five melanoma patients. However, when a doxycycline-repressible promoter was used to regulate the expression of these MEK1 mutants in cells, neither mutation was found to alter ERK signaling or confer vemurafenib-resistance [61]. In contrast, a different group found that MEK1 P124S or P124L could confer modest resistance to vemurafenib [62]. The cause for this difference is currently unknown, but may be related to different experimental conditions or the influence of other cellular factors.

Raf inhibitor resistance has also been observed in cells with increased expression of Cot1/Tpl2, a kinase that functions parallel to Raf to phosphorylate and activate MEK. By stably expressing cDNA clones encoding various protein kinases in a melanoma cell line, Johannessen et al. [58] found that increased Cot1 protein levels could confer resistance to PLX4720 by increasing MEK and ERK phosphorylation in a Raf-independent manner. High Cot1 expression levels and increased ERK signaling have also been observed in a dabrafenib-resistant melanoma cell line [59]. Moreover, in the study by Johannessen et al. [58], two out of three samples from patients treated with vemurafenib showed increased expression of the mRNA for Cot1, and in a specimen from a patient whose cancer had recurred, the expression was further increased.

## *Raf Inhibitor Resistance Through Alternative Pathways*

Another major mechanism for resistance to Raf inhibitor therapy is the upregulation of signaling through the PI3K/AKT pathway. In particular, an increase in phosphorylated AKT and AKT activity appears to play a role in promoting vemurafenib resistance in melanoma lines [51, 64]. The mechanism is likely to be due, in part, to the pro-survival/anti-apoptotic activity of activated AKT, given that the expression of a constitutively active AKT can protect melanoma cells from PLX4720-induced apoptosis in collagen growth assays [65]. In further support of this mechanism, melanoma cells lacking PTEN, a negative regulator of the PI3K/

AKT pathway, showed reduced apoptosis in response to PLX4720 treatment [66]. Moreover, a failure to upregulate the pro-apoptotic factors Bim and Bmf was found to accompany the reduction in apoptosis, indicating that increased AKT signaling may block the initiation of the apoptotic pathway [65–67].

One manner by which the PI3K/AKT pathway can be engaged is through enhanced RTK signaling. In one study, upregulation of IGF-1R led to increased PI3K/AKT signaling in SB590885-resistant melanoma cells, with the upregulation of IGF-1R due, at least in part, to the downregulation of its negative regulator, IGFBP-3 [53]. In addition, increases in IGF-1R expression and phosphorylated AKT levels were observed in a tumor biopsy from a post-relapse melanoma patient treated with vemurafenib [53]. The upregulation of PDGFR-$\beta$ can also promote increased levels of phosphorylated AKT and reduced apoptosis in vemurafenib-resistant melanoma lines, and in one study, four of eleven resistant tumors from patients showed higher levels of PDGFR-$\beta$ expression as compared to the baseline tumor [49, 68].

Although activation of the PI3K/AKT pathway via EGFR upregulation has not been reported in drug resistant melanoma cells, activation of EGFR appears to contribute to the efficacy of the Raf inhibitors in certain cancer types. More specifically, while vemurafenib is beneficial in the treatment of metastatic melanoma harboring the B-Raf V600E mutation, it has little effect on colon cancers with the same mutation [69]. The inhibition of V600E-B-Raf in colon cancer cells reportedly causes feedback activation of EGFR, thus increasing AKT activation and reducing the tumor's dependence on mutant B-Raf to ERK signaling [70]. Interestingly, melanoma cells express relatively low levels of EGFR, and thus cannot deploy this mechanism of inhibitor resistance, at least at the time of initial treatment [70].

Finally, signals from the tumor microenvironment can also impact Raf inhibitor therapy. For example, melanoma cells exposed to the HGF ligand in a cell culture setting had increased resistance to the PLX4720 tool compound and showed elevated ERK and AKT signaling that was initiated by the MET RTK [71, 72]. Similarly, melanoma patients with high plasma levels of HGF or whose stromal cells secreted HGF, showed poorer responses to vemurafenib than patients lacking these characteristics [71, 72].

## MEK Inhibitor Resistance

MEK inhibitor resistance is mediated by many of the same mechanisms as Raf inhibitor resistance. These include reactivation of Ras/Raf/MEK/ERK pathway signaling, via MEK mutations or increases in Ras or Raf protein levels, as well as increased signaling through alternative pathways, including the PI3K/AKT and STAT3 pathways (Fig. 2).

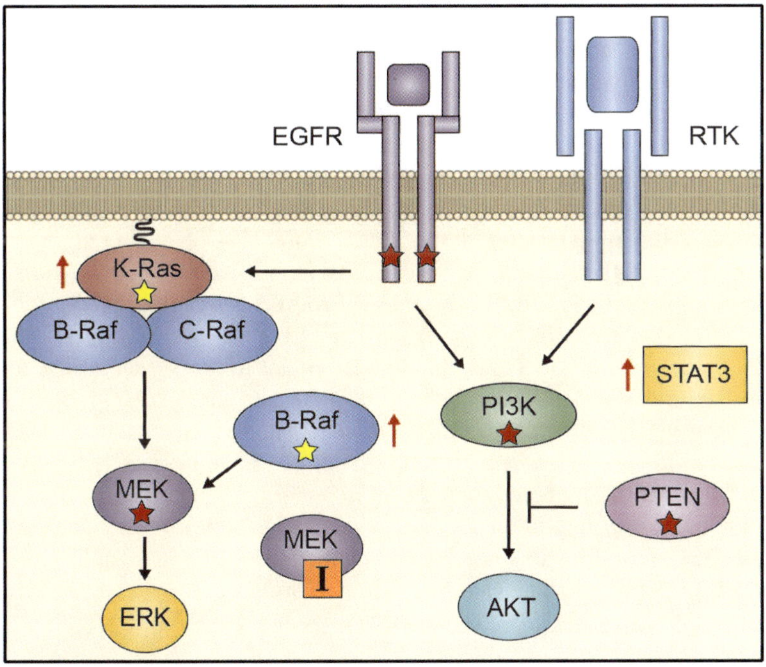


**Fig. 2** *Mechanisms of MEK inhibitor resistance*. Several mechanisms of MEK inhibitor resistance are highlighted in *red*. These include mutations (indicated by *stars*) in EGFR, MEK, PI3K, or PTEN, upregulation of STAT3 expression, and an increase in mutant K-Ras or mutant B-Raf protein levels. These alterations can all cause increased signaling through either the Ras/Raf/MEK/ERK, PI3K/AKT, or STAT3 pathways

## *MEK Inhibitor Resistance Via Reactivation of the Ras/Raf/MEK/ERK Pathway*

Although secondary mutations that disrupt drug binding to oncogenic B-Raf have not been identified as a naturally occurring mechanism of Raf inhibitor resistance, this type of mutation does promote acquired resistance to the MEK inhibitors in tumor cell lines and patients. By expressing a saturating cDNA library of MEK1 mutations in melanoma cells containing V600E-B-Raf, a study by Emery et al. [62] identified two classes of mutations that conferred resistance to the MEK inhibitors selumetinib or CI-1040: one set that may enhance MEK1 activity (Q56P and P124S) and another set that alters binding of the drug to MEK1 either through direct mutation of the binding site or by altering the conformation of the αC helix. Mutations in the MEK allosteric binding pocket, including L115P, F129L, and V215E, have been found to promote resistance against the MEK inhibitors RO4927350 or PD0325901 in breast cancer cells harboring an activated K-Ras mutation and in colorectal cells with a B-Raf mutation [73, 74]. Of note, the MEK

F129L mutant also showed increased kinase activity over wild-type MEK1 in vitro, as well as stronger binding to C-Raf [74].

MEK inhibitor resistance can also arise from gene amplification that upregulates the expression of mutant Ras or Raf proteins. In colorectal cancer lines expressing V600E-B-Raf, an increased copy number of mutant *B-Raf* has been found to upregulate mutant B-Raf protein levels, and thereby confer selumetinib resistance [72, 75]. An increase in *K-Ras* gene copy number has also been observed in selumetinib or CI-1040 resistant colorectal lines that contain a mutant K-Ras [73]. Strikingly, when the driver oncogene in these cases was *K-Ras*, *K-Ras* amplification increased both ERK and PI3K/AKT signaling; however, when V600E-B-Raf was the amplified driver, only the ERK signaling was activated [72].

Finally, upregulation of RTK signaling can affect tumor susceptibility to MEK inhibition. More specifically, intrinsic/*de novo* resistance to MEK inhibition (via PD0325901 or CI-1040) has been observed in non-small-cell lung carcinoma (NSCLC) cell lines with mutations in the EGFR kinase domain [31], and in several breast cancer cell lines where increased EGFR and AKT activation were detected [37]. As with the Raf inhibitors, further analysis of patient samples and tumor cell lines will be required to determine whether increased EGFR signaling will be a mechanism of acquired resistance to MEK inhibition.

## *Upregulation of Alternative Pathways Can Cause MEK Inhibitor Resistance*

Several studies have focused on the role of increased signaling through the PI3K/AKT pathway in MEK inhibitor resistance. In one report, four selumetinib-resistant lung cancer cell lines exhibited high levels of phosphorylated AKT and high AKT activity [76]. In another study, increased signaling through the PI3K/AKT pathway in a panel of colorectal cancer cell lines was found to correlate with a decrease in sensitivity to the MEK inhibitor selumetinib [77]. Moreover, in diverse cancer cell lines, K-Ras mutant cells that lacked PTEN or contained activating mutations in PI3K exhibited resistance to MEK inhibition by PD0325901 or selumetinib treatment [78, 79]. Taken together, these findings suggest that increased signaling through the PI3K/AKT pathway can mediate both *de novo* and acquired resistance to MEK inhibitors.

Additional factors may also contribute to MEK inhibitor resistance. In particular, an increase in STAT3 activation has been reported to promote selumetinib resistance in NSCLC cell lines by upregulating miR-17, a microRNA cluster that modulates the expression of certain genes including the pro-apoptotic gene *Bim* [80]. As a result, STAT3 activation prevented apoptosis by blocking the upregulation of Bim and, in turn, PARP cleavage [79, 80]. Treatment of colorectal cancer and melanoma cells with TNF$\alpha$ has also been shown to block cell death in response to the MEK inhibitor CI-1040 [81]. In one study, a 13-gene signature that predicts selumetinib resistance mediated by Ras effectors other than PI3K was identified [30]. In addition, gene array analysis has revealed that upregulation of the Wnt pathway can contribute to selumetinib resistance in K-Ras mutant colorectal cancer cell lines [36]. Finally, intrinsic resistance to selumetinib in a panel of NSCLC or colorectal cancer cell lines has also been linked to cAMP-dependent PKA activation [82].

## Treatment Strategies to Overcome RAF or MEK Inhibitor Resistance

Current studies with cancer cell lines and patient samples indicate that the administration of two or more targeted inhibitors may be more effective than single-drug therapies in treating human malignancies. In particular, combination therapies are proving beneficial for the treatment of tumors that have developed resistance to a particular inhibitor. Moreover, using combination therapy from the start of treatment may prevent or delay the development of drug resistance. Selecting combination therapies to treat Raf or MEK inhibitor resistance has largely focused on the two most common mechanisms of resistance, namely, reactivation of the Ras/Raf/MEK/ERK pathway and upregulation of the PI3K/AKT pathway (for details, see above sections).

### *Dual Therapies Co-Targeting Raf and MEK*

Dual therapies that co-target the Raf and MEK kinases are proving successful for suppressing growth in cancer lines that have become resistant to either Raf or MEK inhibition via reactivated ERK signaling. In melanoma cells, the combination of vemurafenib and selumetinib has been reported to act synergistically and to inhibit the proliferation of vemurafenib-resistant cells that have increased FGFR3 activation or the amplification of mutant *N-Ras* or *B-Raf* genes [55, 57]. Vemurafenib and the MEK inhibitor RO5068760 were also found to promote apoptosis in B-Raf mutant melanoma cells with acquired resistance to vemurafenib and to have significant antitumor activity in mouse xenograft assays [51]. In dabrafenib-resistant melanoma cells containing *N-Ras* or *B-Raf* mutations, the combination of dabrafenib and trametinib was found to inhibit growth and induce apoptosis, although it may only partially restore sensitivity to dabrafenib [52]. Combined treatment with PLX4720 and a MEK inhibitor (CI-1040 or selumetinib) has also been reported to be more effective at reducing cell growth than either single treatment in melanoma cells with elevated Cot1 expression [58, 59].

It is likely that dual targeting of Raf and MEK will also be useful for the treatment of some colorectal cancers. Studies indicate that colorectal cell lines resistant to a MEK inhibitor (RO4927350 or selumetinib) due to a MEK1 mutation or amplification of the mutant *B-Raf* gene are sensitive to combination therapy with the MEK inhibitor and a Raf inhibitor (RG7204 or AZ628) [72, 74, 75]. In some cases, both the resistant and parental cells were shown to display increased apoptosis with dual treatment [75]. These findings indicate that combination therapy can be effective even in MEK-inhibitor-naïve cells, and that dual treatment may prevent inhibitor resistance from developing. Indeed, the emergence of resistant colonies was inhibited when B-Raf mutant melanoma cells with or without a MEK1 mutation were treated with both Raf and MEK inhibitors (PLX4720 and selumetinib or U0126) [62, 83].

Dual therapies with Raf and MEK inhibitors are currently ongoing in the clinic. Recently, the results of a phase 1/2 combined trial revealed that patients receiving dual dabrafenib and trametinib therapy had significantly improved progression-free survival when compared with dabrafenib monotherapy (an average of 9.4 vs. 5.8 months) [84]. Moreover, the percentage of progression-free survival at one year was also greater with the combined therapy (41 % as opposed to 9 %) [84]. Of note, the secondary squamous cell carcinomas seen in single-therapy trials with dabrafenib, but not with trametinib, also occurred with the dual therapy [84]. Should resistance to the dual inhibition of Raf and MEK develop, it will be interesting to determine whether the mechanisms are similar to those seen in single inhibitor therapies.

## *Combination Therapies Co-Targeting Raf and RTKs*

In certain cases, the co-inhibition of Raf and an upstream RTK may also be an effective cancer therapy. For example, treatment of NSCLC and colorectal cancer cell lines containing *K-Ras* or *B-Raf* mutations with a Raf inhibitor (sorafenib or vemurafenib) plus an EGFR inhibitor (erlotinib, cetuximab, or gefitinib) caused synergistic anti-growth effects and enhanced apoptosis [70, 85]. This combined approach was also effective in reducing tumor growth in xenograft mouse models [70, 85]. The combined inhibition of Raf and MET (using PLX4720 and crizotinib or gefitinib) also exhibited a synergistic effect in the treatment of B-Raf mutant non-melanoma cell lines that secret HGF [71]. Finally, melanoma cells with vemurafenib resistance through increased FGFR3 activity were re-sensitized to vemurafenib via treatment with the FGFR inhibitor LY2874455 [55].

## *Combination Therapies Co-Targeting Raf and the PI3K/AKT Pathway*

Simultaneously targeting the Ras/Raf/MEK/ERK and PI3K/AKT pathways may be an additional method for combating Raf-inhibitor resistance. Studies have found that B-Raf mutant melanoma cells resistant to the Raf inhibitor vemurafenib/

PLX4720 can be re-sensitized to drug by co-inhibition of Raf and PI3K (using LY294002 or GDC-0941), Raf and AKT (using MK-2206), or Raf and mTOR (a kinase component of both the mTORC1 and mTORC2 complexes in the PI3K/AKT pathway) [51, 64, 66, 68]. In addition, the dual PI3K and mTOR inhibitor BEZ235 has been reported to have a strongly synergistic anti-growth effect when combined with vemurafenib to treat resistant melanoma cells expressing V600E-B-Raf [68]. Villanueva et al. [53] also observed increased cell cycle arrest and apoptosis when Raf inhibitor (SB590885)-resistant melanoma cells were treated with a MEK inhibitor (trametinib or selumetinib) and the PI3K inhibitor GSK2126458. Two more studies have provided further support that treatment with inhibitors for PI3K/AKT and either Raf or MEK may be effective against Raf inhibitor-resistant cancers. In one study, combination treatment with trametinib or dabrafenib and the PI3K/mTOR inhibitor GSK2126458 synergistically inhibited cell growth in dabrafenib-resistant melanoma cells expressing mutant N-Ras or B-Raf [52]. In a second study, the addition of an AKT inhibitor to treatment with vemurafenib or selumetinib synergistically blocked the growth of patient-derived melanoma cell lines previously exposed to vemurafenib [54]. Finally, Staussman et al. [71] found that HGF-induced PLX4720 resistance in melanomas could be suppressed by the inhibition of MEK and AKT (using PD184352 and MK-2206).

As well, cells resistant to MEK inhibitors may be sensitive to the co-targeting of MEK and PI3K or AKT. U0126 and the PI3K inhibitor LY294002 were found to have an additive effect when used on a cell line derived from a metastatic melanoma that was partially resistant to these inhibitors alone [86]. Some breast cancer cell lines resistant to the MEK inhibitor CI-1040 have also shown sensitivity to dual MEK/PI3K inhibition using CI-1040 and PIK90 or PI103 [37]. However, targeting MEK and the PI3K/AKT pathway may not work in all MEK inhibitor-resistant cells, given that selumetinib-resistant colorectal cancer cells with upregulation of the mutant *K-Ras* gene were also resistant to the combination of selumetinib and a PI3K/mTOR inhibitor, AZ12321046 [72].

## *Other Potential Therapeutic Combination Strategies*

The use of other targeted inhibitors is also being explored as therapies for cancers resistant to Raf or MEK inhibition. For example, HSP90 inhibitors may be of value in dual therapy strategies, given that AZ628-resistant melanoma cells expressing elevated C-Raf protein levels were reported to be significantly more sensitive to the HSP90 inhibitor geldanamycin than were parental cells [50]. In addition, the HSP90 inhibitor XL888 could effectively induce apoptosis in a variety of vemurafenib-resistant melanoma cell lines, regardless of their mechanism of resistance [87]. In a study by Shao et al. [67], treating PLX4720-resistant melanoma cells with the histone deacetylase inhibitor SAHA (suberoylanilide hydroxamic acid) and either PLX4720 or selumetinib re-sensitized cells and inhibited the re-emergence of PLX4720-resistant colonies in a long-term survival assay. The combined inhibition of MEK using selumetinib and STAT3 with JSI-124 has also been found to re-sensitize selumetinib-resistant NSCLC cell lines to selumetinib and increase the level of selumetinib-induced apoptosis [79, 80]. Finally, other combination therapies that may be beneficial, depending on the mechanism of resistance, include treatment with a Raf or MEK inhibitor and a Cot1, IGF-1R, or ERK inhibitor [53, 58, 73].

## Conclusions

There are currently hundreds of active clinical trials evaluating the use of drugs targeting the MEK or Raf kinase in cancer treatment (for details, see www.clinicaltrials.gov). New inhibitors are also in development, and may prove even more effective. With the knowledge coming from these trials, we are moving ever forward into the era of personalized medicine. It is now clear that knowing the genotype of a tumor prior to treatment is critical for selecting the proper kinase inhibitor therapy and for predicting the possible mechanisms of inhibitor resistance. With time, it may be possible to prevent kinase inhibitor resistance from developing, which may end the danger of recurrence for at least some tumor types.

**Acknowledgments** This project was supported by Federal funds from the National Cancer Institute.

**Conflicts of Interest** No potential conflicts of interest were disclosed.

## References

1. Shaul YD, Seger R. The MEK/ERK cascade: from signaling specificity to diverse functions. Biochim Biophys Acta. 2007;1773:1213–26.
2. Weber CK, Slupsky JR, Kalmes HA, Rapp UR. Active Ras induces heterodimerization of cRaf and BRaf. Cancer Res. 2001;61:3595–8.
3. Rushworth LK, Hindley AD, O'Neill E, Kolch W. Regulation and role of Raf-1/B-Raf heterodimerization. Mol Cell Biol. 2006;26:2262–72.
4. Ritt DA, Monson DM, Specht SI, Morrison DK. Impact of feedback phosphorylation and Raf heterodimerization on normal and mutant B-Raf signaling. Mol Cell Biol. 2010;30:806–19.
5. Dougherty MK, Muller J, Ritt DA, Zhou M, Zhou XZ, Copeland TD, Conrads TP, Veenstra TD, Lu KP, Morrison DK. Regulation of Raf-1 by direct feedback phosphorylation. Mol Cell. 2005;17:215–24.
6. Dong C, Waters SB, Holt KH, Pessin JE. SOS phosphorylation and disassociation of the Grb2-SOS complex by the ERK and JNK signaling pathways. J Biol Chem. 1996;271:6328–32.
7. Roskoski R Jr. ERK1/2 MAP kinases: structure, function, and regulation. Pharmacol Res. 2012;66:105–43.

8. Davies H, Bignell GR, Cox C, Stephens P, Edkins S, Clegg S, Teague J, Woffendin H, Garnett MJ, Bottomley W, Davis N, Dicks E, Ewing R, Floyd Y, Gray K, Hall S, Hawes R, Hughes J, Kosmidou V, Menzies A, Mould C, Parker A, Stevens C, Watt S, Hooper S, Wilson R, Jayatilake H, Gusterson BA, Cooper C, Shipley J, Hargrave D, Pritchard-Jones K, Maitland N, Chenevix-Trench G, Riggins GJ, Bigner DD, Palmieri G, Cossu A, Flanagan A, Nicholson A, Ho JW, Leung SY, Yuen ST, Weber BL, Seigler HF, Darrow TL, Paterson H, Marais R, Marshall CJ, Wooster R, Stratton MR, Futreal PA. Mutations of the BRAF gene in human cancer. Nature. 2002;417:949–54.
9. Downward J, Targeting RAS. Signalling pathways in cancer therapy. Nat Rev Cancer. 2003;3:11–22.
10. Roring M, Brummer T. Aberrant B-Raf signaling in human cancer: 10 years from bench to bedside. Crit Rev Oncog. 2012;17:97–121.
11. Maurer G, Tarkowski B, Baccarini M. Raf kinases in cancer-roles and therapeutic opportunities. Oncogene. 2011;30:3477–88.
12. Pylayeva-Gupta Y, Grabocka E, Bar-Sagi D. RAS oncogenes: weaving a tumorigenic web. Nat Rev Cancer. 2011;11:761–74.
13. Hingorani SR, Jacobetz MA, Robertson GP, Herlyn M, Tuveson DA. Suppression of BRAF (V599E) in human melanoma abrogates transformation. Cancer Res. 2003;63:5198–202.
14. Flaherty KT, Puzanov I, Kim KB, Ribas A, McArthur GA, Sosman JA, O'Dwyer PJ, Lee RJ, Grippo JF, Nolop K, Chapman PB. Inhibition of mutated, activated BRAF in metastatic melanoma. N Engl J Med. 2010;363:809–19.
15. Hauschild A, Grob JJ, Demidov LV, Jouary T, Gutzmer R, Millward M, Rutkowski P, Blank CU, Miller WH Jr, Kaempgen E, Martin-Algarra S, Karaszewska B, Mauch C, Chiarion-Sileni V, Martin AM, Swann S, Haney P, Mirakhur B, Guckert ME, Goodman V, Chapman PB. Dabrafenib in BRAF-mutated metastatic melanoma: a multicentre, open-label, phase 3 randomised controlled trial. Lancet. 2012;380:358–65.
16. Wilhelm SM, Carter C, Tang L, Wilkie D, McNabola A, Rong H, Chen C, Zhang X, Vincent P, McHugh M, Cao Y, Shujath J, Gawlak S, Eveleigh D, Rowley B, Liu L, Adnane L, Lynch M, Auclair D, Taylor I, Gedrich R, Voznesensky A, Riedl B, Post LE, Bollag G, Trail PA. BAY 43–9006 exhibits broad spectrum oral antitumor activity and targets the RAF/MEK/ERK pathway and receptor tyrosine kinases involved in tumor progression and angiogenesis. Cancer Res. 2004;64:7099–109.
17. Mangana J, Levesque MP, Karpova MB, Dummer R. Sorafenib in melanoma. Expert Opin Investig Drugs. 2012;21:557–68.
18. Strumberg D. Sorafenib for the treatment of renal cancer. Expert Opin Pharmacother. 2012;13:407–19.
19. Woo HY, Heo J. Sorafenib in liver cancer. Expert Opin Pharmacother. 2012;13:1059–67.
20. Tsai J, Lee JT, Wang W, Zhang J, Cho H, Mamo S, Bremer R, Gillette S, Kong J, Haass NK, Sproesser K, Li L, Smalley KS, Fong D, Zhu YL, Marimuthu A, Nguyen H, Lam B, Liu J, Cheung I, Rice J, Suzuki Y, Luu C, Settachatgul C, Shellooe R, Cantwell J, Kim SH, Schlessinger J, Zhang KY, West BL, Powell B, Habets G, Zhang C, Ibrahim PN, Hirth P, Artis DR, Herlyn M, Bollag G. Discovery of a selective inhibitor of oncogenic B-Raf kinase with potent antimelanoma activity. Proc Natl Acad Sci USA. 2008;105:3041–6.
21. Laquerre S, Arnone M, Moss K, Yang J, Fisher K, Kane-Carson LS, Smitheman K, Ward J, Heidrich B, Rheault T, Adjabeng G, Hornberger K, Stellwagen J, Waterson A, Han C, Mook RA, Uehling D, King AJ. A selective Raf kinase inhibitor induces cell death and tumor regression of human cancer cell lines encoding B-Raf V600E mutation. Mol Cancer Ther. 2009;8:B88.
22. Sosman JA, Kim KB, Schuchter L, Gonzalez R, Pavlick AC, Weber JS, McArthur GA, Hutson TE, Moschos SJ, Flaherty KT, Hersey P, Kefford R, Lawrence D, Puzanov I, Lewis KD, Amaravadi RK, Chmielowski B, Lawrence HJ, Shyr Y, Ye F, Li J, Nolop KB, Lee RJ, Joe AK, Ribas A. Survival in BRAF V600-mutant advanced melanoma treated with vemurafenib. N Engl J Med. 2012;366:707–14.

23. Chapman PB, Hauschild A, Robert C, Haanen JB, Ascierto P, Larkin J, Dummer R, Garbe C, Testori A, Maio M, Hogg D, Lorigan P, Lebbe C, Jouary T, Schadendorf D, Ribas A, O'Day SJ, Sosman JA, Kirkwood JM, Eggermont AM, Dreno B, Nolop K, Li J, Nelson B, Hou J, Lee RJ, Flaherty KT, McArthur GA. Improved survival with vemurafenib in melanoma with BRAF V600E mutation. N Engl J Med. 2011;364:2507–16.
24. Falchook GS, Long GV, Kurzrock R, Kim KB, Arkenau TH, Brown MP, Hamid O, Infante JR, Millward M, Pavlick AC, O'Day SJ, Blackman SC, Curtis CM, Lebowitz P, Ma B, Ouellet D, Kefford RF. Dabrafenib in patients with melanoma, untreated brain metastases, and other solid tumours: a phase 1 dose-escalation trial. Lancet. 2012;379:1893–901.
25. Su F, Viros A, Milagre C, Trunzer K, Bollag G, Spleiss O, Reis-Filho JS, Kong X, Koya RC, Flaherty KT, Chapman PB, Kim MJ, Hayward R, Martin M, Yang H, Wang Q, Hilton H, Hang JS, Noe J, Lambros M, Geyer F, Dhomen N, Niculescu-Duvaz I, Zambon A, Niculescu-Duvaz D, Preece N, Robert L, Otte NJ, Mok S, Kee D, Ma Y, Zhang C, Habets G, Burton EA, Wong B, Nguyen H, Kockx M, Andries L, Lestini B, Nolop KB, Lee RJ, Joe AK, Troy JL, Gonzalez R, Hutson TE, Puzanov I, Chmielowski B, Springer CJ, McArthur GA, Sosman JA, Lo RS, Ribas A, Marais R. RAS mutations in cutaneous squamous-cell carcinomas in patients treated with BRAF inhibitors. N Engl J Med. 2012;366:207–15.
26. Callahan MK, Rampal R, Harding JJ, Klimek VM, Chung YR, Merghoub T, Wolchok JD, Solit DB, Rosen N, Abdel-Wahab O, Levine RL, Chapman PB. Progression of RAS-mutant leukemia during RAF inhibitor treatment. N Engl J Med. 2012;367:2316–21.
27. Heidorn SJ, Milagre C, Whittaker S, Nourry A, Niculescu-Duvas I, Dhomen N, Hussain J, Reis-Filho JS, Springer CJ, Pritchard C, Marais R. Kinase-dead BRAF and oncogenic RAS cooperate to drive tumor progression through CRAF. Cell. 2010;140:209–21.
28. Hatzivassiliou G, Song K, Yen I, Brandhuber BJ, Anderson DJ, Alvarado R, Ludlam MJ, Stokoe D, Gloor SL, Vigers G, Morales T, Aliagas I, Liu B, Sideris S, Hoeflich KP, Jaiswal BS, Seshagiri S, Koeppen H, Belvin M, Friedman LS, Malek S. RAF inhibitors prime wild-type RAF to activate the MAPK pathway and enhance growth. Nature. 2010;464:431–5.
29. Poulikakos PI, Zhang C, Bollag G, Shokat KM, Rosen N. RAF inhibitors transactivate RAF dimers and ERK signalling in cells with wild-type BRAF. Nature. 2010;464:427–30.
30. Dry JR, Pavey S, Pratilas CA, Harbron C, Runswick S, Hodgson D, Chresta C, McCormack R, Byrne N, Cockerill M, Graham A, Beran G, Cassidy A, Haggerty C, Brown H, Ellison G, Dering J, Taylor BS, Stark M, Bonazzi V, Ravishankar S, Packer L, Xing F, Solit DB, Finn RS, Rosen N, Hayward NK, French T, Smith PD. Transcriptional pathway signatures predict MEK addiction and response to selumetinib (AZD6244). Cancer Res. 2010;70:2264–73.
31. Pratilas CA, Hanrahan AJ, Halilovic E, Persaud Y, Soh J, Chitale D, Shigematsu H, Yamamoto H, Sawai A, Janakiraman M, Taylor BS, Pao W, Toyooka S, Ladanyi M, Gazdar A, Rosen N, Solit DB. Genetic predictors of MEK dependence in non-small cell lung cancer. Cancer Res. 2008;68:9375–83.
32. Yeh JJ, Routh ED, Rubinas T, Peacock J, Martin TD, Shen XJ, Sandler RS, Kim HJ, Keku TO, Der CJ. KRAS/BRAF mutation status and ERK1/2 activation as biomarkers for MEK1/2 inhibitor therapy in colorectal cancer. Mol Cancer Ther. 2009;8:834–43.
33. Garon EB, Finn RS, Hosmer W, Dering J, Ginther C, Adhami S, Kamranpour N, Pitts S, Desai A, Elashoff D, French T, Smith P, Slamon DJ. Identification of common predictive markers of in vitro response to the Mek inhibitor selumetinib (AZD6244; ARRY-142886) in human breast cancer and non-small cell lung cancer cell lines. Mol Cancer Ther. 2010;9:1985–94.
34. Solit DB, Garraway LA, Pratilas CA, Sawai A, Getz G, Basso A, Ye Q, Lobo JM, She Y, Osman I, Golub TR, Sebolt-Leopold J, Sellers WR, Rosen N. BRAF mutation predicts sensitivity to MEK inhibition. Nature. 2006;439:358–62.
35. Jing J, Greshock J, Holbrook JD, Gilmartin A, Zhang X, McNeil E, Conway T, Moy C, Laquerre S, Bachman K, Wooster R, Degenhardt Y. Comprehensive predictive biomarker analysis for MEK inhibitor GSK1120212. Mol Cancer Ther. 2012;11:720–9.
36. Tentler JJ, Nallapareddy S, Tan AC, Spreafico A, Pitts TM, Morelli MP, Selby HM, Kachaeva MI, Flanigan SA, Kulikowski GN, Leong S, Arcaroli JJ, Messersmith WA,

Eckhardt SG. Identification of predictive markers of response to the MEK1/2 inhibitor selumetinib (AZD6244) in K-ras-mutated colorectal cancer. Mol Cancer Ther. 2010;9:3351–62.
37. Mirzoeva OK, Das D, Heiser LM, Bhattacharya S, Siwak D, Gendelman R, Bayani N, Wang NJ, Neve RM, Guan Y, Hu Z, Knight Z, Feiler HS, Gascard P, Parvin B, Spellman PT, Shokat KM, Wyrobek AJ, Bissell MJ, McCormick F, Kuo WL, Mills GB, Gray JW, Korn WM. Basal subtype and MAPK/ERK kinase (MEK)-phosphoinositide 3-kinase feedback signaling determine susceptibility of breast cancer cells to MEK inhibition. Cancer Res. 2009;69:565–72.
38. Flaherty KT, Robert C, Hersey P, Nathan P, Garbe C, Milhem M, Demidov LV, Hassel JC, Rutkowski P, Mohr P, Dummer R, Trefzer U, Larkin JM, Utikal J, Dreno B, Nyakas M, Middleton MR, Becker JC, Casey M, Sherman LJ, Wu FS, Ouellet D, Martin AM, Patel K, Schadendorf D. Improved survival with MEK inhibition in BRAF-mutated melanoma. N Engl J Med. 2012;367:107–14.
39. Bekaii-Saab T, Phelps MA, Li X, Saji M, Goff L, Kauh JS, O'Neil BH, Balsom S, Balint C, Liersemann R, Vasko VV, Bloomston M, Marsh W, Doyle LA, Ellison G, Grever M, Ringel MD, Villalona-Calero MA. Multi-institutional phase II study of selumetinib in patients with metastatic biliary cancers. J Clin Oncol. 2011;29:2357–63.
40. Hayes DN, Lucas AS, Tanvetyanon T, Krzyzanowska MK, Chung CH, Murphy BA, Gilbert J, Mehra R, Moore DT, Sheikh A, Hoskins J, Hayward MC, Zhao N, O'Connor W, Weck KE, Cohen RB, Cohen EE. Phase II efficacy and pharmacogenomic study of Selumetinib (AZD6244; ARRY-142886) in iodine-131 refractory papillary thyroid carcinoma with or without follicular elements. Clin Cancer Res Official J Am Assoc Cancer Res. 2012;18:2056–65.
41. Hainsworth JD, Cebotaru CL, Kanarev V, Ciuleanu TE, Damyanov D, Stella P, Ganchev H, Pover G, Morris C, Tzekova V. A phase II, open-label, randomized study to assess the efficacy and safety of AZD6244 (ARRY-142886) versus pemetrexed in patients with non-small cell lung cancer who have failed one or two prior chemotherapeutic regimens. J Thorac Oncol. 2010;5:1630–6.
42. Bodoky G, Timcheva C, Spigel DR, La Stella PJ, Ciuleanu TE, Pover G, Tebbutt NC. A phase II open-label randomized study to assess the efficacy and safety of selumetinib (AZD6244 [ARRY-142886]) versus capecitabine in patients with advanced or metastatic pancreatic cancer who have failed first-line gemcitabine therapy. Invest New Drugs. 2011;30:1216–23.
43. Banerji U, Camidge DR, Verheul HM, Agarwal R, Sarker D, Kaye SB, Desar IM, Timmer-Bonte JN, Eckhardt SG, Lewis KD, Brown KH, Cantarini MV, Morris C, George SM, Smith PD, van Herpen CM. The first-in-human study of the hydrogen sulfate (Hyd-sulfate) capsule of the MEK1/2 inhibitor AZD6244 (ARRY-142886): a phase I open-label multicenter trial in patients with advanced cancer. Clin Cancer Res Official J Am Assoc Cancer Res. 2010;16:1613–23.
44. Kirkwood JM, Bastholt L, Robert C, Sosman J, Larkin J, Hersey P, Middleton M, Cantarini M, Zazulina V, Kemsley K, Dummer R. Phase II open-label, randomized trial of the MEK1/2 inhibitor selumetinib as monotherapy versus temozolomide in patients with advanced melanoma. Clin Cancer Res Official J Am Assoc Cancer Res. 2011;18:555–67.
45. Patel SP, Lazar AJ, Papadopoulos NE, Liu P, Infante JR, Glass MR, Vaughn CS, Lorusso PM, Cohen RB, Davies MA, Kim KB. Clinical responses to selumetinib (AZD6244; ARRY-142886)-based combination therapy stratified by gene mutations in patients with metastatic melanoma. Cancer. 2012. doi:10.1002/cncr.27790.
46. Flaherty KT, Yasothan U, Kirkpatrick P. Vemurafenib. Nat Rev Drug Discov. 2011;10:811–2.
47. Gibbons DL, Pricl S, Kantarjian H, Cortes J, Quintas-Cardama A. The rise and fall of gatekeeper mutations? The BCR-ABL1 T315I paradigm. Cancer. 2012;118:293–9.

48. Whittaker S, Kirk R, Hayward R, Zambon A, Viros A, Cantarino N, Affolter A, Nourry A, Niculescu-Duvaz D, Springer C, Marais R. Gatekeeper mutations mediate resistance to BRAF-targeted therapies. Sci Transl Med. 2010;2:35ra41.
49. Nazarian R, Shi H, Wang Q, Kong X, Koya RC, Lee H, Chen Z, Lee MK, Attar N, Sazegar H, Chodon T, Nelson SF, McArthur G, Sosman JA, Ribas A, Lo RS. Melanomas acquire resistance to B-RAF (V600E) inhibition by RTK or N-RAS upregulation. Nature. 2010;468:973–7.
50. Montagut C, Sharma SV, Shioda T, McDermott U, Ulman M, Ulkus LE, Dias-Santagata D, Stubbs H, Lee DY, Singh A, Drew L, Haber DA, Settleman J. Elevated CRAF as a potential mechanism of acquired resistance to BRAF inhibition in melanoma. Cancer Res. 2008;68:4853–61.
51. Su F, Bradley WD, Wang Q, Yang H, Xu L, Higgins B, Kolinsky K, Packman K, Kim MJ, Trunzer K, Lee RJ, Schostack K, Carter J, Albert T, Germer S, Rosinski J, Martin M, Simcox ME, Lestini B, Heimbrook D, Bollag G. Resistance to selective BRAF inhibition can be mediated by modest upstream pathway activation. Cancer Res. 2012;72:969–78.
52. Greger JG, Eastman SD, Zhang V, Bleam MR, Hughes AM, Smitheman KN, Dickerson SH, Laquerre SG, Liu L, Gilmer TM. Combinations of BRAF, MEK, and PI3K/mTOR inhibitors overcome acquired resistance to the BRAF inhibitor GSK2118436 dabrafenib, mediated by NRAS or MEK mutations. Mol Cancer Ther. 2012;11:909–20.
53. Villanueva J, Vultur A, Lee JT, Somasundaram R, Fukunaga-Kalabis M, Cipolla AK, Wubbenhorst B, Xu X, Gimotty PA, Kee D, Santiago-Walker AE, Letrero R, D'Andrea K, Pushparajan A, Hayden JE, Brown KD, Laquerre S, McArthur GA, Sosman JA, Nathanson KL, Herlyn M. Acquired resistance to BRAF inhibitors mediated by a RAF kinase switch in melanoma can be overcome by cotargeting MEK and IGF-1R/PI3K. Cancer Cell. 2010;18:683–95.
54. Atefi M, von Euw E, Attar N, Ng C, Chu C, Guo D, Nazarian R, Chmielowski B, Glaspy JA, Comin-Anduix B, Mischel PS, Lo RS, Ribas A. Reversing melanoma cross-resistance to BRAF and MEK inhibitors by co-targeting the AKT/mTOR pathway. PLoS ONE. 2011;6:e28973.
55. Yadav V, Zhang X, Liu J, Estrem S, Li S, Gong XQ, Buchanan S, Henry JR, Starling JJ, Peng SB. Reactivation of mitogen-activated protein kinase (MAPK) pathway by FGF receptor 3 (FGFR3)/Ras mediates resistance to vemurafenib in human B-RAF V600E mutant melanoma. J Biol Chem. 2012;287:28087–98.
56. Poulikakos PI, Persaud Y, Janakiraman M, Kong X, Ng C, Moriceau G, Shi H, Atefi M, Titz B, Gabay MT, Salton M, Dahlman KB, Tadi M, Wargo JA, Flaherty KT, Kelley MC, Misteli T, Chapman PB, Sosman JA, Graeber TG, Ribas A, Lo RS, Rosen N, Solit DB. RAF inhibitor resistance is mediated by dimerization of aberrantly spliced BRAF (V600E). Nature. 2011;480:387–90.
57. Shi H, Moriceau G, Kong X, Lee MK, Lee H, Koya RC, Ng C, Chodon T, Scolyer RA, Dahlman KB, Sosman JA, Kefford RF, Long GV, Nelson SF, Ribas A, Lo RS. Melanoma whole-exome sequencing identifies (V600E) B-RAF amplification-mediated acquired B-RAF inhibitor resistance. Nature Commun. 2012;3:724.
58. Johannessen CM, Boehm JS, Kim SY, Thomas SR, Wardwell L, Johnson LA, Emery CM, Stransky N, Cogdill AP, Barretina J, Caponigro G, Hieronymus H, Murray RR, Salehi-Ashtiani K, Hill DE, Vidal M, Zhao JJ, Yang X, Alkan O, Kim S, Harris JL, Wilson CJ, Myer VE, Finan PM, Root DE, Roberts TM, Golub T, Flaherty KT, Dummer R, Weber BL, Sellers WR, Schlegel R, Wargo JA, Hahn WC, Garraway LA. COT drives resistance to RAF inhibition through MAP kinase pathway reactivation. Nature. 2010;468:968–72.
59. Gowrishankar K, Snoyman S, Pupo GM, Becker TM, Kefford RF, Rizos H. Acquired resistance to BRAF inhibition can confer cross-resistance to combined BRAF/MEK inhibition. J Invest Dermatol. 2012;132:1850–9.
60. Kaplan FM, Kugel CH, Dadpey N, Shao Y, Abel EV, Aplin AE. SHOC2 and CRAF mediate ERK1/2 reactivation in mutant NRAS-mediated resistance to RAF inhibitor. J Biol Chem. 2012;287:41797–807.

61. Shi H, Moriceau G, Kong X, Koya RC, Nazarian R, Pupo GM, Bacchiocchi A, Dahlman KB, Chmielowski B, Sosman JA, Halaban R, Kefford RF, Long GV, Ribas A, Lo RS. Preexisting MEK1 exon 3 mutations in V600E/KBRAF melanomas do not confer resistance to BRAF inhibitors. Cancer Discovery. 2012;2:414–24.
62. Emery CM, Vijayendran KG, Zipser MC, Sawyer AM, Niu L, Kim JJ, Hatton C, Chopra R, Oberholzer PA, Karpova MB, MacConaill LE, Zhang J, Gray NS, Sellers WR, Dummer R, Garraway LA. MEK1 mutations confer resistance to MEK and B-RAF inhibition. Proc Natl Acad Sci USA. 2009;106:20411–6.
63. Wagle N, Emery C, Berger MF, Davis MJ, Sawyer A, Pochanard P, Kehoe SM, Johannessen CM, Macconaill LE, Hahn WC, Meyerson M, Garraway LA. Dissecting therapeutic resistance to RAF inhibition in melanoma by tumor genomic profiling. J Clin Oncol Official J Am Soc Clin Oncol. 2011;29:3085–96.
64. Jiang CC, Lai F, Thorne RF, Yang F, Liu H, Hersey P, Zhang XD. MEK-independent survival of B-RAFV600E melanoma cells selected for resistance to apoptosis induced by the RAF inhibitor PLX4720. Clin Cancer Res Official J Am Assoc Cancer Res. 2011;17:721–30.
65. Shao Y, Aplin AE. Akt3-mediated resistance to apoptosis in B-RAF-targeted melanoma cells. Cancer Res. 2010;70:6670–81.
66. Paraiso KH, Xiang Y, Rebecca VW, Abel EV, Chen YA, Munko AC, Wood E, Fedorenko IV, Sondak VK, Anderson AR, Ribas A, Palma MD, Nathanson KL, Koomen JM, Messina JL, Smalley KS. PTEN loss confers BRAF inhibitor resistance to melanoma cells through the suppression of BIM expression. Cancer Res. 2011;71:2750–60.
67. Shao Y, Aplin AE. BH3-only protein silencing contributes to acquired resistance to PLX4720 in human melanoma. Cell Death Differ. 2012;19:2029–39.
68. Shi H, Kong X, Ribas A, Lo RS. Combinatorial treatments that overcome PDGFRbeta-driven resistance of melanoma cells to V600EB-RAF inhibition. Cancer Res. 2011;71:5067–74.
69. Kopetz S, Desai J, Chan E, Hecht JR, O'Dwyer PJ, Lee RJ, Nolop NB, Saltz L. PLX4032 in metastatic colorectal cancer patients with mutant BRAF tumors. J Clin Oncol. 2010;28 Suppl; abstr 3534.
70. Prahallad A, Sun C, Huang S, Di Nicolantonio F, Salazar R, Zecchin D, Beijersbergen RL, Bardelli A, Bernards R. Unresponsiveness of colon cancer to BRAF (V600E) inhibition through feedback activation of EGFR. Nature. 2012;483:100–3.
71. Straussman R, Morikawa T, Shee K, Barzily-Rokni M, Qian ZR, Du J, Davis A, Mongare MM, Gould J, Frederick DT, Cooper ZA, Chapman PB, Solit DB, Ribas A, Lo RS, Flaherty KT, Ogino S, Wargo JA, Golub TR. Tumour micro-environment elicits innate resistance to RAF inhibitors through HGF secretion. Nature. 2012;487:500–4.
72. Little AS, Balmanno K, Sale MJ, Newman S, Dry JR, Hampson M, Edwards PA, Smith PD, Cook SJ. Amplification of the driving oncogene, KRAS or BRAF, underpins acquired resistance to MEK1/2 inhibitors in colorectal cancer cells. Sci Signal. 2011;4:ra17.
73. Hatzivassiliou G, Liu B, O'Brien C, Spoerke JM, Hoeflich KP, Haverty PM, Soriano R, Forrest WF, Heldens S, Chen H, Toy K, Ha C, Zhou W, Song K, Friedman LS, Amler LC, Hampton GM, Moffat J, Belvin M, Lackner MR. ERK inhibition overcomes acquired resistance to MEK inhibitors. Mol Cancer Ther. 2012;11:1143–54.
74. Wang H, Daouti S, Li WH, Wen Y, Rizzo C, Higgins B, Packman K, Rosen N, Boylan JF, Heimbrook D, Niu H. Identification of the MEK1 (F129L) activating mutation as a potential mechanism of acquired resistance to MEK inhibition in human cancers carrying the B-RafV600E mutation. Cancer Res. 2011;71:5535–45.
75. Corcoran RB, Dias-Santagata D, Bergethon K, Iafrate AJ, Settleman J, Engelman JA. BRAF gene amplification can promote acquired resistance to MEK inhibitors in cancer cells harboring the BRAF V600E mutation. Sci Signal. 2010;3:ra84.
76. Meng J, Peng H, Dai B, Guo W, Wang L, Ji L, Minna JD, Chresta CM, Smith PD, Fang B, Roth JA. High level of AKT activity is associated with resistance to MEK inhibitor AZD6244 (ARRY-142886). Cancer Biol Ther. 2009;8:2073–80.
77. Balmanno K, Chell SD, Gillings AS, Hayat S, Cook SJ. Intrinsic resistance to the MEK1/2 inhibitor AZD6244 (ARRY-142886) is associated with weak ERK1/2 signalling and/or

strong PI3 K signalling in colorectal cancer cell lines. Int J Cancer J Int du Cancer. 2009;125:2332–41.
78. Wee S, Jagani Z, Xiang KX, Loo A, Dorsch M, Yao YM, Sellers WR, Lengauer C, Stegmeier F. PI3K pathway activation mediates resistance to MEK inhibitors in KRAS mutant cancers. Cancer Res. 2009;69:4286–93.
79. Yoon YK, Kim HP, Han SW, Oh do Y, Im SA, Bang YJ, Kim TY. KRAS mutant lung cancer cells are differentially responsive to MEK inhibitor due to AKT or STAT3 activation: implication for combinatorial approach. Mol Carcinog. 2010;49:353–362.
80. Dai B, Meng J, Peyton M, Girard L, Bornmann WG, Ji L, Minna JD, Fang B, Roth JA. STAT3 mediates resistance to MEK inhibitor through microRNA miR-17. Cancer Res. 2011;71:3658–68.
81. Gray-Schopfer VC, Karasarides M, Hayward R, Marais R. Tumor necrosis factor-alpha blocks apoptosis in melanoma cells when BRAF signaling is inhibited. Cancer Res. 2007;67:122–9.
82. Troiani T, Vecchione L, Martinelli E, Capasso A, Costantino S, Ciuffreda LP, Morgillo F, Vitagliano D, D'Aiuto E, De Palma R, Tejpar S, Van Cutsem E, De Lorenzi M, Caraglia M, Berrino L, Ciardiello F. Intrinsic resistance to selumetinib, a selective inhibitor of MEK1/2, by cAMP-dependent protein kinase A activation in human lung and colorectal cancer cells. Br J Cancer. 2012;106:1648–59.
83. Paraiso KH, Fedorenko IV, Cantini LP, Munko AC, Hall M, Sondak VK, Messina JL, Flaherty KT, Smalley KS. Recovery of phospho-ERK activity allows melanoma cells to escape from BRAF inhibitor therapy. Br J Cancer. 2010;102:1724–30.
84. Flaherty KT, Infante JR, Daud A, Gonzalez R, Kefford RF, Sosman J, Hamid O, Schuchter L, Cebon J, Ibrahim N, Kudchadkar R, Burris HA 3rd, Falchook G, Algazi A, Lewis K, Long GV, Puzanov I, Lebowitz P, Singh A, Little S, Sun P, Allred A, Ouellet D, Kim KB, Patel K, Weber J. Combined BRAF and MEK inhibition in melanoma with BRAF V600 mutations. N Engl J Med. 2012;367:1694–703.
85. Martinelli E, Troiani T, Morgillo F, Rodolico G, Vitagliano D, Morelli MP, Tuccillo C, Vecchione L, Capasso A, Orditura M, De Vita F, Eckhardt SG, Santoro M, Berrino L, Ciardiello F. Synergistic antitumor activity of sorafenib in combination with epidermal growth factor receptor inhibitors in colorectal and lung cancer cells. Clinical Cancer Res Official J Am Assoc Cancer Res. 2010;16:4990–5001.
86. Smalley KS, Haass NK, Brafford PA, Lioni M, Flaherty KT, Herlyn M. Multiple signaling pathways must be targeted to overcome drug resistance in cell lines derived from melanoma metastases. Mol Cancer Ther. 2006;5:1136–44.
87. Paraiso KH, Haarberg HE, Wood E, Rebecca VW, Chen YA, Xiang Y, Ribas A, Lo RS, Weber JS, Sondak VK, John JK, Sarnaik AA, Koomen JM, Smalley KS. The HSP90 inhibitor XL888 overcomes BRAF inhibitor resistance mediated through diverse mechanisms. Clinical Cancer Res Official J Am Assoc Cancer Res. 2012;18:2502–14.
88. King AJ, Patrick DR, Batorsky RS, Ho ML, Do HT, Zhang SY, Kumar R, Rusnak DW, Takle AK, Wilson DM, Hugger E, Wang L, Karreth F, Lougheed JC, Lee J, Chau D, Stout TJ, May EW, Rominger CM, Schaber MD, Luo L, Lakdawala AS, Adams JL, Contractor RG, Smalley KS, Herlyn M, Morrissey MM, Tuveson DA, Huang PS. Demonstration of a genetic therapeutic index for tumors expressing oncogenic BRAF by the kinase inhibitor SB-590885. Cancer Res. 2006;66:11100–5.
89. Shen M, Lyne P, Aquila B, Drew L. Linking molecular characteristics to the pharmacological response of a panel of cancer cell lines to the BRAF inhibitor, AZ628. 98th AACR Annual Meeting 2007; Los Angeles.
90. Amiri P, Aikawa M, Dove J, Stuart D, Poon D, Pick T, Ramurthy S, Subramanian S, Levine B, Costales A, Harris A, Paul R. CHIR-265 is a potent selective inhibitor of c-Raf/B-Raf/mutB-Raf that effectively inhibits proliferation and survival of cancer cell lines with Ras/Raf pathway mutations. 97th AACR Annual Meeting 2006; Washington, DC.
91. Yeh TC, Marsh V, Bernat BA, Ballard J, Colwell H, Evans RJ, Parry J, Smith D, Brandhuber BJ, Gross S, Marlow A, Hurley B, Lyssikatos J, Lee PA, Winkler JD, Koch K, Wallace E.

Biological characterization of ARRY-142886 (AZD6244), a potent, highly selective mitogen-activated protein kinase kinase 1/2 inhibitor. Clinical Cancer Res Official J Am Assoc Cancer Res. 2007;13:1576–83.

92. Gilmartin AG, Bleam MR, Groy A, Moss KG, Minthorn EA, Kulkarni SG, Rominger CM, Erskine S, Fisher KE, Yang J, Zappacosta F, Annan R, Sutton D, Laquerre SG. GSK1120212 (JTP-74057) is an inhibitor of MEK activity and activation with favorable pharmacokinetic properties for sustained in vivo pathway inhibition. Clinical Cancer Res Official J Am Assoc Cancer Res. 2011;17:989–1000.
93. Daouti S, Wang H, Li WH, Higgins B, Kolinsky K, Packman K, Specian A Jr, Kong N, Huby N, Wen Y, Xiang Q, Podlaski FJ, He Y, Fotouhi N, Heimbrook D, Niu H. Characterization of a novel mitogen-activated protein kinase kinase 1/2 inhibitor with a unique mechanism of action for cancer therapy. Cancer Res. 2009;69:1924–32.
94. Barrett SD, Bridges AJ, Dudley DT, Saltiel AR, Fergus JH, Flamme CM, Delaney AM, Kaufman M, LePage S, Leopold WR, Przybranowski SA, Sebolt-Leopold J, Van Becelaere K, Doherty AM, Kennedy RM, Marston D, Howard WA Jr, Smith Y, Warmus JS, Tecle H. The discovery of the benzhydroxamate MEK inhibitors CI-1040 and PD 0325901. Bioorg Med Chem Lett. 2008;18:6501–4.

# Mechanisms of Resistance to Targeted B-Raf Therapies

**Ramana V. Tantravahi, Benjamin Hoffman and E. Premkumar Reddy**

**Abstract** *Targeted Cancer Therapies and Resistance* Our modern search for effective cancer therapies began after the First World War and the discovery of the myelosuppressive properties of nitrogen mustards. Since that time, adjuvant and chemo-radiotherapy have become the standard of care for many cancers. These non-targeted therapies have produced remissions in many patients. As our technical capacity to target radiation to tumors and our ability to perform surgical interventions improves, it is likely that these approaches will gain in effectiveness. Nevertheless, the dose limiting toxicities of chemotherapy drugs and ionizing radiation have served to expose vulnerabilities in this clinical approach. Modern translational research has sought to develop novel, targeted approaches to cancer therapy. These new approaches are based upon our understanding of cellular growth control, and bring with them the promise of greater potency and safety. The concept of targeted cancer therapy began with the identification and characterization of growth regulatory proteins in normal cells. These proteins were initially identified in acute transforming retroviruses. During the 1970s and 1980s, investigators working in a number of laboratories learned that the transforming genes of these viruses were actually derived from the host genome. In subsequent years, investigators came to understand that biochemical or genetic inactivation of these deleterious proteins in cultured cell lines leads to tumor cell death. This new understanding led to the search for specific proteins, whose activities drive the malignant transformation process. These proteins are now referred to as *druggable targets*. Chemical and molecular biologists are translating their understanding of the cell proliferation control into molecular therapeutics directed at these targets. Targeted therapy development for cancer is still in its infancy. Despite this,

E. Premkumar Reddy (✉)
Department of Oncological Sciences, Mount Sinai School of Medicine, Room 15-20 Icahn Medical Institute, 1425 Madison Avenue, New York, NY 10029, USA
e-mail: ep.reddy@mssm.edu

R. V. Tantravahi · B. Hoffman
Onconova Therapeutics Inc., 375 Pheasant Run, Newtown, PA 18940, USA

B. Bonavida (ed.), *Molecular Mechanisms of Tumor Cell Resistance to Chemotherapy*, Resistance to Targeted Anti-Cancer Therapeutics 1, DOI: 10.1007/978-1-4614-7070-0_4,

clinical trials for a number of small molecules have provided positive outcomes in patients who may have been given dire prognoses only a few years ago. Two FDA-approved drugs, Gleevec (Novartis) and Vemurafinib (Plexxikon) have provided some of the most dramatic results. Unfortunately, these drugs have also revealed a new challenge in cancer therapy—the development of resistance and relapse. Here, we describe the B-raf growth signaling pathway associated with the development of melanoma, summarize the mechanisms associated with resistance to targeted B-raf therapies, and discuss the future of this field of inquiry.

**Keywords** B-Raf mutations • Melanoma • Vemurafenib • Resistance

**Abbreviations**

| | |
|---|---|
| BCR | Breakpoint cluster region |
| CML | Chronic myelogenous leukemia |
| Src | The transforming gene of the Rous Sarcoma Virus (RSV), a non-receptor tyrosine kinase that represents a family of such kinases that regulate cell proliferation and play a prominent role in malignant transformation |
| Abl | The transforming gene of Abelson Murine Leukemia Virus (A-MuLV). Abl is a src family kinase associated with transformation of the lymphoid compartment of the hematopoietic system |
| MAP | Originally microtubule-associated protein, now also mitogen-activated protein. MAP kinases refer to a group of cytoplasmic protein kinases that act downstream of growth factor receptors to transmit proliferative signals to cells |
| FTI | Farnesyl transferase inhibitor |
| JNK | c-Jun N terminal kinase |
| PI-3K | Phosphatidyl inositol 3′ kinase |
| MEK | Mitogen activated/extracellular regulatory kinase |
| ERK | Extracellular regulatory kinase |

## Introduction: Growth Associated Signaling Pathways

The development of cancer has long been understood to have a genetic basis. Investigators have come to understand the clonal nature of malignancy from the earliest observations of tumors arising in avian and murine systems. Studies in human populations led Nordling [1] and later Knudson [2] to assert that a minimum of two mutational events within a single cell is necessary for cancer development [1, 2]. Our modern understanding of cancer is now considerably more sophisticated than it was at the turn of the 20th century. Since then, we have assembled and tested elegantly constructed experimental model systems, identified numerous growth control genes and elucidated complex regulatory pathways that control cell growth, differentiation and death.

The characterization of complex signal transduction networks lies at the heart of our current understanding of cancer; it shapes many of the strategies with which therapeutics are conceived and developed. One of the most transformative developments in modern cancer therapy was the development of Gleevec® (Imatinib mesylate, STI57, Novartis), a targeted treatment for chronic myelogenous leukemia (CML) [3]. The synthesis of Gleevec followed on from the accumulated understanding of decades of basic science research into the identity and biological function of a subset of growth control proteins known as the *src* family tyrosine kinases. The oncogenic activation of the *src* family kinase c-Abl via chromosomal translocation was determined to be the critical step in CML development [4]. Gleevec is the result of structure-based design of a molecule that could effectively inhibit the kinase activity of the deleterious BCR-Abl protein [5]. The structure-based design paradigm led to a drug that acts specifically and with great efficacy in the clinic.

The development of Gleevec is a useful case study in the current development of targeted cancer therapies. CML is treated with Gleevec routinely across the world where once less effective and more highly toxic interventions served as the only therapeutic avenues [6]. Despite its widespread success, Gleevec does not cure CML. Many patients, after years in remission, develop resistance to the drug and relapse [7]. So while our greater understanding of the molecular events underlying the malignant transformation has led to significant advances in therapeutics development, the enthusiasm over these advances must be tempered with the understanding that much work remains to be done.

The present chapter addresses the identification and development of novel therapies directed at a different growth control protein, B-raf, as well as the observed mechanisms of therapeutic resistance to B-raf inhibitors. B-raf sits at the center of a highly complex series of regulatory nodes that control proliferation, differentiation, and programmed death (apoptosis). As was the case with the Abl protein in CML, the B-raf protein, and one of its mutant isoforms, B-raf$^{V600E}$, is associated with a particular form of cancer, melanoma [8]. Here, we discuss the identification and function of B-raf and its role in cellular homeostasis. We will also discuss the mechanistic effects of B-raf mutation on cell proliferation control and the current state of therapeutics development.

## *MAP Kinase Pathway Signaling*

For many decades, investigators in a variety of disciplines have tried to come to an understanding as to how extracellular mediators can guide changes in patterns of gene expression. Molecular biologists addressed this issue within the past 30 years through identification, and structure–function characterization of gene products associated with cell growth and proliferation. These gene products were found to form interdependent regulatory networks. The key proteins involved in these networks were found to have discrete biochemical functions, and their dysregulation in a

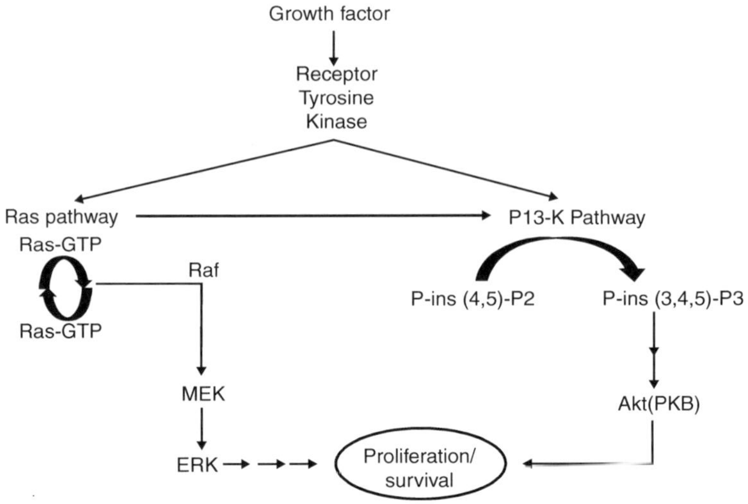


**Fig. 1** Signaling Pathways that control Proliferation and Survival–Extracellular mediators such as soluble growth factors bind with high affinity to their cognate receptors, leading to the activation of downstream kinase cascades. Certain receptor tyrosine kinases (*RTK*)

variety of cellular contexts was often found to contribute to malignant transformation.

Figure 1 is a representation of the Mitogen-Activated Protein (MAP) kinase network, which serves to couple extracellular signals in the form of soluble growth factors with changes in gene expression associated with proliferation and survival. Growth factor receptors—cell surface proteins with extracellular ligand-binding and intracellular tyrosine kinase domains initiate this process. Ligand binding results in receptor dimerization coupled with auto- and trans-phosphorylation of receptor dimers. Phosphorylated dimers are then primed to bind intracellular mediators, including single subunit GTP-binding proteins of the *ras* family, and cytoplasmic protein kinases. The MAP kinase cascade plays a central role in cellular growth dynamics because of the association of its terminal kinases, $p38^{MAPK}$, c-Jun N-Terminal Kinase (JNK), Extracellular Regulatory Kinase (ERK), and Phosphatidyl Inositol 3′ kinase (PI-3 kinase) with DNA-binding transcription factors. These transcription factors regulate genes involved in the control of proliferation and survival. The ERK1/2 pathway is the most commonly involved in the proliferative response. ERK1/2 activation is also associated with malignant transformation, and the components of the MAP kinase pathway most frequently involved in dysregulated activation occur upstream of ERK1/2. Constitutively upregulated activity of RTK, or overexpression of growth factors can lead to upregulation of ERK1/2. Also, mutations in Ras and Raf isoforms can similarly lead to constitutive proliferation signals mediated by ERK1/2.

## Ras

The Ras proteins comprise a subset of small membrane-associated single polypeptides that bind to and hydrolyze GTP. Four isoforms of Ras have been identified in humans, Ha-ras, N-ras, Ki-ras 4A and Ki-Ras 4B [9]. Each Ras isoform has a distinct pattern of expression and a characteristic ability to affect specific downstream signaling pathways, such as the MAP kinase cascade, or the PI-3 kinase/protein kinase B (Akt) pathway via differential interaction with Raf isoforms [10]. Each Ras isoform is encoded on a single gene, except for Ki-ras 4A and 4B, which arise from differential splicing of a single parental transcript [9].

### *History*

The Ras genes were identified in isolates from three acute transforming retroviruses. The Harvey and Kirstein sarcoma viruses (named for their discoverers) were able to induce ***RA***t ***S***arcomas, and were thus named *ras* genes (Ha-ras, and Ki-ras) [11, 12]. A third virus that transduced the mouse ortholog of *ras* was isolated from a BALB/c mouse tumor and was named as BALB-MSV [13]. Sequence analysis of these three viral oncogenes revealed that BALB-MSV encodes the mouse homolog of the *Ha-ras* oncogene [14–16]. Human cellular Ras orthologs were discovered later by several groups [17]. As is the case with many of the oncogenes first identified in the genomes of sarcoma viruses, the *ras* gene has been the subject of intense scrutiny for the past 30 years. *Ras* is the most commonly mutated gene in all human cancers, and mutation of the coding sequence has been observed in a variety of human malignancies. The identification of specific cellular *Ras* mutations at codons 12 and 61 remains a fundamental contribution to our understanding of the neoplastic process [18–24].

### *Structure and Function*

The human genome encodes three *Ras* genes, termed H-, K- and N-*Ras*. These three *Ras* genes encode proteins of approximately 200 amino acids characterized by a GTP-binding domain. This domain is encoded on 5 so-called G box elements spread throughout the primary amino acid sequence [25]. The C-terminus of Ras encodes a membrane-association sequence, CAAX. Either farnesyl or geranyl-geranyl transferase enzymes covalently modify the cysteine residue within the CAAX sequence [26]. These two isoprenoid lipid moieties enable Ras association with the plasma membrane. Ras isoforms that cannot associate with membranes fail to transmit mitogenic signals as efficiently as does wild type Ras [27]. Consequently, inhibition of lipid modification has been considered a viable strategy for therapy development. Many attempts have been made to develop small molecules that target farnesyl transferases [28, 29]. Some of these molecules have been assessed in clinical trials, although their efficacy has not had the dramatic effect that had been originally hoped. Since certain Ras proteins can be modified by both farnesylation and geranylgeranylation, inhibition of one process does not necessarily prevent Ras association with the plasma membrane. Farnesyl transferase inhibitors (FTI) are largely unable to inhibit geranylgeranyl transfer. Consequently, Ras can associate with the plasma membrane even in the presence of these compounds. Also, FTI affect H-Ras disproportionately. K-*Ras* and N-*Ras* are, however, more commonly mutated in human cancer [27]. While activated Ras remains a high profile target for drug design and development, it has become clear in recent years that the Ras downstream effector, Raf, can also contribute to the neoplastic process.

## Raf

The Raf proteins function as intracellular conduits through which growth, differentiation and apoptosis signals are transmitted. Raf proteins are serine/threonine protein kinases that reside proximal to the membrane and participate in complex multi-protein kinase relays that ultimately activate powerful programs of gene expression. Raf proteins are now known to participate in membrane-associated signaling events, such as transduction of growth factor receptor signals via Ras. Recent studies have tied mutations in *raf* genes, particularly *B-raf*, with melanoma as well as carcinomas of the thyroid, ovary and colon [30–32].

### *History*

The v-*raf* oncogene was first identified in the murine acute transforming retrovirus, 3611-MSV by Ulf Rapp in 1983 [33, 34]. The name *raf* derives itself from the ability of 3611-MSV to induce ***RA***pid murine ***F***ibrosarcomas. Jansen et al. [35] working in avian systems, isolated the MH2 acute transforming virus in the same year. The MH2 transforming gene was named *v-mil*. In the following year, two groups would report that *v-mil* and *v-raf* are sequence orthologs derived from murine and avian genomes and that both genes encode serine/threonine protein kinase activities [36, 37]. Cellular *raf* genes were later cloned, and the first isoform cloned was named *c-raf* [38]. C-Raf is highly conserved in evolution, with orthologs reported in *D. melanogaster*, and *C. elegans* [39, 40]. There are 2 mammalian *c-raf* genes, *c-raf-1* and *c-raf-2* (c-*raf*-1 is now referred to as *Raf*-1). Raf-1 is a protein coding gene, while *c-raf-2* is a pseudogene [32, 37]. In 1986 and 1987, two different groups reported the sequence of A-*raf*, a highly similar, but not identical isoform of *Raf*-1 [35, 40] Humans possess two A-*raf* gene sequences,

A-*Raf-1* and A-*Raf-2*. As was the case with *Raf-1*, A-*Raf-2* is a pseudogene [41]. The B-*raf* ortholog was identified by Marx et al. [42] in 1988. The original study was designed to identify acute transforming viruses following infection with the Rous-associated virus-1 (RAV-1) in chicken neuroretinal cells. The isolation of proliferating cells revealed the presence of a novel virus isolate, Ic10, that had transduced the coding sequence from the host genome between the RAV-1 *gag* and *env* genes. When sequenced, the host-derived gene bore significant homology to *Raf-1*, but was considered to be a related gene due to sequence divergence at the N and C termini of the predicted protein and the presence of the genomic sequence in an unrelated area of the genome to the avian *Raf-1* gene.

## Structure and Function

Raf proteins share significant structural homology befitting their similar roles in cell homeostasis and their similar biochemical functions. Raf isoforms differ in size and patterns of expression. Human A-Raf is a 68 kD protein. Raf-1, with a slightly longer N-terminus, is approximately 74 kD. B-Raf protein sizes are heterogeneous, appearing anywhere from 75 to 100 kD. Multiple B-Raf species arise due to alternative splicing of the B-*raf* message [43]. *Raf-1* is expressed ubiquitously both during development and in the adult organism; B-*Raf* expression predominates in neuronal tissues. A-*Raf* expression predominates in the epididymis and ovary [44].

Each Raf protein isoform contains three principle regions of homology denoted as *conserved regions* (CR) 1, 2 and 3 [45]. Each of these regions contains structural and/or regulatory properties of interest. CR1 contains two sub-regions, the RAS-binding domain, and the cysteine-rich region, both of which are responsible directly or indirectly for the movement of Raf toward the plasma membrane. Upstream of the CR1 region, Raf-1 contains a regulatory serine (S43), which acts as an inhibitor of Ras/Raf association [45]. CR2 contains docking sites for the 14-3-3 class of adapter proteins. This further reinforces the notion that Raf signaling occurs against the backdrop of formation of large multifactorial protein complexes that can regulate activity both spatially and temporally. CR2 also contains the protein's N region, a site of overall negative charge. Phosphorylation of the N region is considered to be a mechanism by which Raf kinase activity is primed to overcome its intramolecular negative regulation by its C-terminus [45]. The Raf CR3 contains the ATP-binding site and kinase domain. Serine and threonine residues within CR3, especially those near the ATP binding activation segment, are of particular interest from a regulatory perspective. In the case of B-raf, mutations within this region are particularly informative with regard to cancer development [45].

## BRAF Mutations in Cancer

In 1988, *BRAF* was identified as an oncogene through the NIH/3T3 assay system [46]. Following this discovery, attempts were made to characterize *BRAF* mutations in human tumor samples. Yet, it was not until 2002 when a team of researchers were able to identify specific abnormalities in the gene encoding B-*Raf* [47]. In this seminal study, investigators used high-throughput sequencing technology on 923 cancer samples and ultimately identified 43 presumptive mutations. Strikingly, they also found that 70 % of the melanoma samples tested harbored mutations in the *BRAF* gene.

### *BRAF Mutations in Melanoma*

A preponderance of the lesions, first identified by Davies and colleagues [47], were found to occur in exons 11 and 15 of the *BRAF* gene. At the structural level, nearly all of the mutations were immediately adjacent to the B-Raf kinase domain, and one particular mutation, a V600E substitution (initially called V599E), was highly represented in many of the melanoma samples tested. In the same study, it was found that the V600E lesion caused a mutated version of B-Raf that had increased enzymatic activity [47].

The V600E lesion occurs in a critical coding region and is the result of a single point mutation involving a thymine to adenine switch (T1796A) [47]. This point mutation alters the coding sequence and leads to a glutamate –> valine substitution (at the peptide level) causing a structural change in the activation segment of the B-raf catalytic domain [48]. Ultimately, this structural change introduces a negative charge that mimics critical activation events such as the phosphorylation of threonine 598 and serine 601 [48, 49]. In this way, the V600E mutation allows for a B-Raf mutant, that is independent of Ras and, thus, can signal through its downstream effectors in the absence of any upstream stimuli (see Fig. 2).

The added kinase activity of V600E B-raf has been well characterized and its functionality in establishing and maintaining the melanoma phenotype has been studied by a number of groups including Pollock et al. Their work, in particular, was the first to provide evidence that V600E B-Raf was critical for all of the steps in melanoma pathogenesis [50]. By examining benign pre-melanocytic nevi, primary malignant melanoma, and metastatic disease, Pollock and colleagues were able to demonstrate that the V600E mutation was pervasive in each step in the natural history of melanoma. While this study validated the importance of the B-Raf/MEK pathway in disease maintenance, it also suggested a role for additional oncogenic signals in the transformation process.

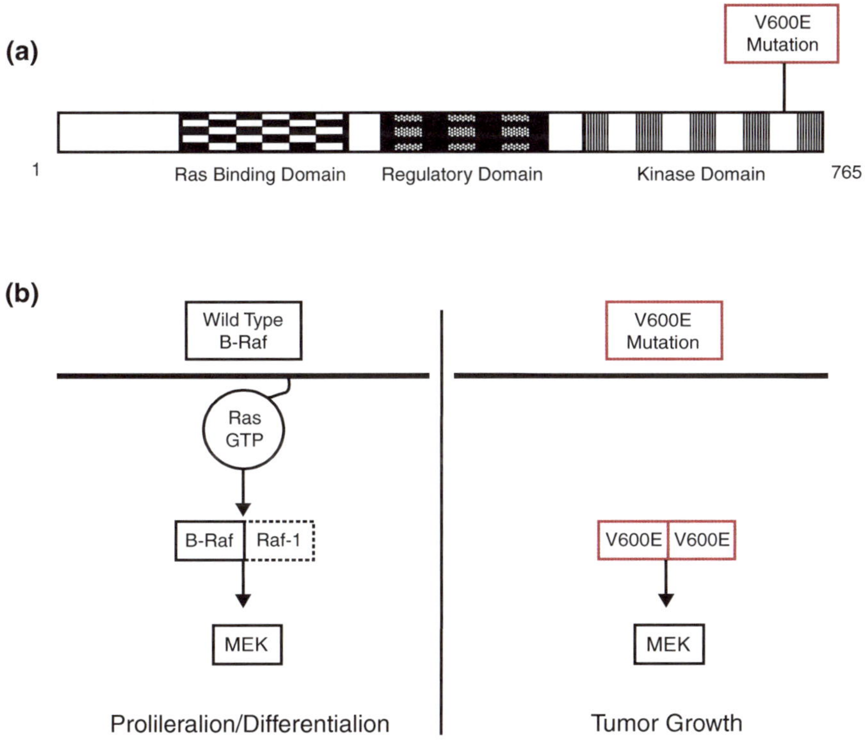


**Fig. 2** B-Raf$^{V600E}$ stimulates tumor cell proliferation independently of Ras

## *Prevalence of BRAF Mutations in Melanoma*

According to the SEER database (SEER.cancer.gov), 76,250 people are expected to be diagnosed with melanoma in the United States in 2012. Nearly 4 % of newly diagnosed melanoma patients have distal metastases and 9 % have disease that has spread regionally to their lymph nodes (SEER.cancer.gov). Therefore, almost 10,000 new cases of un-resectable melanoma are expected to be diagnosed this year in the US alone. In a recent clinical study, 56 % of un-resectable melanoma cases tested positive for the V600E mutation [51].

## *BRAF Mutations in Other Malignancies*

Mutations in *BRAF* have been identified in other human malignancies. In addition to melanoma, *BRAF* is mutated in other solid tumors such as: papillary thyroid carcinoma, non-small cell lung cancer, and colorectal cancer [52–55]. Also, a recent a clinico-pathological report suggested a novel role for *BRAF* in hematological malignancies [56]. In this study, a team of investigators used whole exome sequencing technology to test paired malignant/normal tissues from 47 patients that had been diagnosed with Hairy Cell Leukemia (HCL). Using the advanced sequencing technology, they were able to positively identify V600E mutations in 100 % of the patients that were tested. Interestingly, this mutation was not present in any of the other leukemias or B cell disorders that were studied. In light of this report, more work will be required to fully characterize the role of *BRAF* mutations in the many different types of human hematological malignancies.

## B-Raf as a Target for Anti-Cancer Agents

The discovery that *BRAF* was highly mutated in human cancer spurred an intense effort by both industry and academia to develop targeted B-Raf agents. To date, a number of different B-Raf agents with varying molecular mechanisms of action have been tested in the clinic for the treatment of both solid and hematological malignancies.

### *Sorafenib*

Bay 43-9006 (sorafenib or Nexavar) was the first Raf-targeting agent to achieve clinical success in treating human cancers. Sorafenib is a biaryl urea ATP-competitive protein kinase inhibitor [57]. It was first identified as a potent small molecule inhibitor of the Raf family of kinases and in particular as a C-Raf inhibitor [58]. However, it was later shown that sorafenib also had a high affinity for wild type B-Raf, B-Raf$^{V600E}$, and a number of other tyrosine kinases including some of the VEGFR and the PDGFR family members [59]. The elucidation of this dual mechanism has helped to explain both the anti-angiogenic and pro-apoptotic effects of sorafenib [58, 60]. Moreover, the *in vitro* activities and molecular mechanism of sorafenib may also help explain its strong clinical activity. To date, sorafenib has been approved for use in the treatment of advanced renal cell carcinoma (RCC) as well as primary hepatocellular carcinoma (HCC). While sorafenib has shown clinical efficacy in the treatment of HCC and RCC, it lacked substantial activity when treating malignant melanoma [61]. While this ended the development of sorafenib for this indication, it did provide critical information regarding the class of B-Raf inhibitors.

## *PLX-4032*

The failure of sorafenib in melanoma provided an important data point for both scientists and physicians to build upon. Many in the field hypothesized that sorafenib's promiscuous nature or lack of selectivity for B-Raf and B-Raf$^{V600E}$ explained the nominal clinical efficacy in malignant melanoma [62]. The first test of this principle occurred when researchers, using crystallographic techniques, identified a 7-azaindole derivative (PLX4270) which was highly selective for the active B-Raf kinase or the V600E mutant [63]. Not only was this molecule shown to be active in the test tube, but it was also active against melanoma cell lines in the laboratory and in melanoma-based xenograft models. Moreover, the high degree of selectivity for the active B-Raf kinase allowed for a minimal amount of preclinical toxicity. Ultimately, an analogue of PLX4270, called PLX4032 (vemurafenib), was carried into the clinic for testing in human subjects.

The translation into the clinic followed a unique development strategy. To provide proof of principle, developers designed clinical trials to enrich for malignant melanoma patients that harbored the V600E B-Raf mutation. This strategy allowed clinicians to observe efficacy at a very early point in clinical development, and the proof of principle was provided in the first of these trials [64]. Vemurafenib is now approved for the treatment of mutant B-Raf$^{V600E}$ metastatic melanoma.

## *Other B-Raf Agents in Development*

Sorafenib and vemurafenib provided direct evidence that inhibition of the Raf family of proteins, and in particular B-Raf, was a highly effective strategy for the development of targeted anti-tumor agents. A number of investigational agents has added to this class of molecules. Follow-on-drugs such as dabrafenib are being advanced in the clinic, and while these agents should provide differentiation with respect to toxicity profile and general tolerability, they do not alleviate any of the clinical resistance that has been seen with the first-in-class agents [65]. Therefore, strategies need to be developed that deal with the mechanisms which are associated with resistance to the B-Raf class of inhibitors.

# Clinical Observations of Resistance to B-Raf Inhibitors

Although clinical trials have shown that select tumors are extremely sensitive to mutant B-Raf inhibition, resistance has also been observed [66]. In the first clinical publication of PLX4032, Flaherty et al. [64] demonstrated complete or partial regressions in 81 % of V600E melanoma patients. While this response was observed to last 2–18 months, ultimately patients were desensitized to B-Raf inhibition and their disease progressed. Moreover, 19 % of the patients tested in these trials had no significant clinical response (indicated as >30 % tumor regression).

## *Mechanisms of Resistance*

Understanding the mechanisms that underlie desensitization and resistance have become an area of intense interest. A number of molecular mechanisms have been proposed and subsequently tested in the laboratory setting. Many of the proposed mechanisms have been studied using cultured melanoma cell lines and melanoma patient samples, and therefore these mechanisms may well be cell-type specific. Ultimately, other tissue-specific resistance mechanisms may be identified, but for the scope of this review, we will focus on the resistance mechanisms that have hitherto been elucidated in melanoma.

The canonical Ras-Raf-MEK-ERK1/2 pathway is well characterized. The result of Raf activation is phosphorylation of ERK1/2, and the translocation of activated ERKs into the nucleus to activate substrates that are involved in cellular proliferation. Melanoma cells, in which the predominance of signaling is mediated through V600E B-Raf, are highly dependent on the MEK/ERK signaling, and mutant B-Raf inhibitors can affect this signaling in a mutation-specific manner [67]. Given the critical nature of the MEK/ERK signaling, it is not surprising that all of the resistance mechanisms, hitherto identified, involve the reactivation of the tumorigenic MEK/ERK signaling.

## Reactivation of the MEK-ERK Signaling Pathway by Overexpression of RAF Isomers

In 2008, Monagut et al. [68] were the first to demonstrate a specific resistance mechanism. Their research efforts predated any observations of clinical resistance and involved the establishment of drug-insensitive clones using human melanoma cells that harbored the V600E mutation. While the investigators did not test for resistance to PLX4032, they did use a similar compound (AZ268 a V600E inhibitor) for selection. Monagut and colleagues followed clonal selection with DNA sequencing and discovered that resistance and desensitization resulted from an increase in CRAF expression and reactivation of the MEK-ERK signaling axis.

## Reactivation of the MEK-ERK Signaling Pathway by Upstream Activation

Resistance and desensitization can also occur in response to activation of signals that are upstream of the Raf proteins [69]. In 2010, Nazarian and colleagues used PLX-4032 drug-resistant cell lines, as well as tumor-matched short term cultures from melanoma patients, to identify two mechanisms for acquired resistance. The first was an activating mutation in NRAS (Q61 K), and in a series of elegant experiments, the investigators demonstrated that constitutive Ras potentiated MEK-ERK signaling in a canonical pattern through the WT RAF. The second mechanism involved overexpression of PDGFR$\beta$, activation downstream through WT RAF and MEK-ERK phosphorylation. While these mechanisms were found to be mutually exclusive, they both acted to promote resistance through WT Raf dimerization and the reactivation of MEK-ERK signals. Moreover, both molecular lesions were confirmed in patients that were either resistant or insensitive to PLX4032.

## Reactivation of MEK-ERK Signaling by COT

Overexpression of COT is a third resistance mechanism that has been identified. Importantly, COT overexpression leads to Raf-independent reactivation of MEK-ERK. Johannessen et al. [70] were the first to suggest this new paradigm. Using a cDNA library that encoded 75 % of the human kinome, the investigators introduced activated kinases into a V600E melanoma cell line and then analyzed for resistance to PLX4032. In addition to COT, Johannessen and colleagues identified C-Raf as a mediator for the resistance phenotype, thus providing an internal validation for their approach; however, the discovery that COT also mediated resistance was a breakthrough. COT, as reviewed earlier, activates MEK-ERK in a Raf-independent manner. Furthermore, its expression is inversely correlated to V600E expression, suggesting that mutant BRAF activation may downregulate COT. It is therefore feasible to hypothesize that when this pressure is removed (through pharmacological inhibition of V600E) COT can mediate resistance by directly activating MEK-ERK.

## Other Mechanisms of Resistance

In addition to the above described mechanisms of resistance, additional acquired resistance mechanisms have been reported; these include upregulation of insulin-like growth factor-1 receptor (IGF1R) signaling and acquisition of activating mutations in *MEK1* [71–73], high level expression of epidermal growth factor receptor (EGFR) [74] and activation of PI3 Kinase pathway as a result of loss of PTEN expression [72, 73].

Moreover, alternative splicing of BRAFV600E has also been described in a subset of melanoma tumor samples from patients who relapsed following BRAF inhibitor therapy [75]. The protein encoded by this alternatively spliced mRNA lacked the regulatory elements of BRAF, including the RAS binding site, resulting in a kinase that is capable of dimerizing and activating MEK even in the presence of a BRAF inhibitor. It is interesting to note that most of these resistance mechanisms involve the reactivation of the MEK/ERK pathway. However, increased insulin growth factor receptor 1 (IGF-1R) signaling or acquired loss of PTEN which correlated with increased activation of AKT [71–73] represent examples of MAPK-independent bypass mechanisms. Most of these mechanisms have been corroborated in small numbers of tumor samples from patient biopsies.

The discovery that RAS mutations promote resistance to RAF inhibitors led to a series of studies showing that RAF inhibitors often induce hyperactivation of MEK–ERK1/2 signaling in cells harboring N-RAS mutations [76]. Thus, ectopic expression of mutant N-RAS in a mutant B-RAF colorectal cancer cell line resulted in resistance to PLX4720 [77], a result that could also be seen in B-RAFV600E melanoma cells [69, 70]. As mentioned earlier, another possible mechanism to by-pass the activity of RAF inhibitors is through mutations in MEK1 and 2 which are downstream effectors of B-RAF. Through a random mutagenesis screen, P124L mutation in MEK1 was identified to confer cellular resistance to the MEK inhibitor, AZD624478. MEK1P124L expression also conferred resistance to PLX4720 in cell-based assays and was identified in a metastasis from a patient with acquired resistance to AZD6244, indicating the clinical relevance of this mutation [78]. Interestingly, the combination of PLX4720 plus AZD6244 overcame the resistance conferred by MEK1P124L expression. These and other data have prompted the use of MEK inhibitors in clinical trials enrolling mutant B-RAF melanoma patients who were previously treated with or without a BRAF inhibitor. Thus, phase I/II trials of the BRAF inhibitor dabrafenib in combination with the MEK inhibitor trametinib (GSK1120212, NCT01072175) and vemurafenib combined with the MEK inhibitor GDC-0973 (NCT01271803) are underway and results from the phase I/II trial of the trametinib-dabrafenib combination was reported to be associated with objective response rates of 77 % [79, 80]. Other clinical studies combining BRAF inhibitors with inhibitors of the mTOR/PI3K/AKT pathway are due to commence in the near future.

## Conclusions

The introduction of targeted therapies such as Gleevec and Vemurafinib represent a historical shift in our approach to cancer treatment. For the first time, serious malignancies have been attacked by therapeutic agents that account for our new and specialized understanding of the nature of cancer and the controls on cellular proliferation in normal cells. While the initial results in laboratory studies and in the clinic have proved promising, they have also provided cancer biologists and oncologists with a cautionary tale. Resistance to targeted therapy reveals a level of plasticity in the response of the cancer cell to the agents that seek its destruction. In some cases, that response is genetic selection, as is revealed by the allele specific suppressor mutations revealed in the BCR-Abl protein and the B-raf protein. In other cases, the response manifests itself by compensatory changes in the regulation of co-linear signaling pathways, such as growth factor receptor upregulation upstream of Raf, or constitutive activation of the MAP kinase cascade.

Strategies for the development of targeted therapies, such as Gleevec and Vemurafinib, rely on the identification of *druggable* targets. Those targets can be involved in what Weinstein and Joe described as *oncogene addiction* [81]. Oncogene addiction scenarios, such as the expression of the *bcr-abl* fusion gene, transform normal cells into cancer cells through creation of growth signaling dysequilibria. In the laboratory, restoration of the equilibrium in cultured cells through inactivation of these deleterious proteins results in the induction of apoptosis. This critical observation has provided the rationale for structure based rational drug design. Using molecular modeling and x-ray crystallography, teams of chemical biologists designed small molecules that inactivated the biochemical properties of druggable target proteins. As we have witnessed, the initial results in the clinic provided impressive results. Nevertheless, the difficulty of resistance remains. In the future, it may be possible to marshal the vast bioinformatics resources now available to design individualized therapies that account for signaling plasticity. Perhaps, using combinations of targeted therapies will limit the probability of resistance reactions.

A different therapy strategy may also prove more efficacious. At the present time, most targeted therapies rely on the deactivation of ATP-dependent reactions using ATP-mimetic small molecules that act as competitive inhibitors. Almost all of the available ATP mimetic compounds seek to inhibit kinase reactions in much the same manner that both Gleevec and Vemurafinib do. In the future, we may abandon structure-based rational drug design, and move to a development strategy in which cancer cell cytotoxicity takes precedence over affinity toward a target determined *a priori*. In that way, ATP mimetic scaffolds and their derivatives can be assayed in culture for both their ability to kill cultured tumor cell lines and their safety. Because of the highly conserved tertiary structure of kinase ATP-binding sites, these compounds will likely affect a combination of growth control kinases with varying affinity. In so doing, compounds developed by this method will likely affect multiple routes of aberrant signaling within cancer cells. By using cancer cell cytotoxicity as a first measure of efficacy, we should be able to accomplish two goals simultaneously: (1) the development of effective therapeutics and (2) the identification of novel targets or target combinations that limit the possibility of resistance.

The future of targeted therapy for cancer remains bright. Our continuing accumulation of knowledge from basic research into cellular growth control will serve to shed light on our already well-developed understanding of malignant transformation. In recognizing the phenomena of non-response and resistance, basic and applied researchers will learn and adapt to these challenges.

**No Conflict Statement** The authors declare no competing financial interests.

## References

1. Nordling CO. A new theory on cancer-inducing mechanism. Br J Cancer. 1953;7:68–72.
2. Knudson AG. Mutation and cancer: statistical study of retinoblastoma. Proc Natl Acad Sci U S A. 1971;68:820–3.
3. Druker BJ, Tamura S, Buchdunger E, Ohno S, Segal GM, Fanning S, Zimmermann J, Lydon NB. Effects of a selective inhibitor of the Abl tyrosine kinase on the growth of Bcr-Abl positive cells. Nat Med. 1996;2:561–6.
4. Heisterkamp N, Stam K, Groffen J, de Klein A, Grosveld G. Structural organization of the bcr gene and its role in the Ph' translocation. Nature. 1985;315:758–61.
5. Druker BJ, Sawyers CL, Kantarjian H, Resta DJ, Reese SF, Ford JM, Capdeville R, Talpaz M. Activity of a specific inhibitor of the BCR-ABL tyrosine kinase in the blast crisis of chronic myeloid leukemia and acute lymphoblastic leukemia with the Philadelphia chromosome. N Engl J Med. 2001;344:1038–42.
6. Hehlmann R, Current CML. Therapy: progress and dilemma. Leukemia. 2003;17:1010–2.
7. Shah NP, Sawyers CL. Mechanisms of resistance to STI571 in Philadelphia chromosome-associated leukemias. Oncogene. 2003;22:7389–95.
8. Tsao H, Chin L, Garraway LA, Fisher DE. Melanoma: from mutations to medicine. Genes Dev. 2012;26:1131–55.
9. Bar-Sagi D. A Ras by any other name. Mol Cell Biol. 2001;21:1441–3.
10. Weber CK, Slupsky JR, Herrmann C, Schuler M, Rapp UR, Block C. Mitogenic signaling of Ras is regulated by differential interaction with Raf isozymes. Oncogene. 2000;19:169–76.
11. Harvey JJ. An unidentified virus which causes the rapid production of tumours in mice. Nature. 1964;204:1104–5.
12. Kirsten WH, Mayer LA. Morphologic responses to a murine erythroblastosis virus. J Natl Cancer Inst. 1967;39:311–35.
13. Peters RL, Rabstein LS, Louise S, Van Vleck R, Kelloff GJ, Huebner RJ. Naturally occurring sarcoma virus of the BALB/cCr mouse. J Natl Cancer Inst. 1974;53:1725–9.
14. Dhar R, Ellis RW, Shih TY, Oroszlan S, Shapiro B, Maizel J, Lowy D, Scolnick E. Nucleotide sequence of the p21 transforming protein of Harvey murine sarcoma virus. Science. 1982;217:934–6.
15. Tsuchida N, Ryder T, Ohtsubo E. Nucleotide sequence of the oncogene encoding the p21 transforming protein of Kirsten murine sarcoma virus. Science. 1982;217:937–8.
16. Reddy EP, Lipman D, Anderson PR, Tronick SR, Aaronson SA. Nucleotide sequence analysis of BALB/c murine sarcoma virus transforming gene. J Virol. 1985;53:984–7.
17. Malumbres M, Barbacid M. RAS oncogenes: the first 30 years. Nat Rev Cancer. 2003;3:459–65.
18. Reddy EP, Reynolds RK, Santos E, Barbacid M. A point mutation is responsible for the acquisition of transforming properties by the T24 human bladder carcinoma oncogene. Nature (London). 1982;300:149–52.
19. Reddy EP. Nucleotide sequence analysis of the T24 human bladder carcinoma oncogene. Science. 1983;220:1061–3.
20. Tabin CJ, Bradley SM, Bargmann CI, Weinberg RA, Papageorge AG, Scolnick EM, Dhar R, Lowy DR, Chang EH. Mechanism of activation of a human oncogene. Nature (London). 1982;300:143–9.

21. Taparowsky E, Suard Y, Fasano O, Shimizu K, Goldfarb M, Wigler M. Activation of the T24 bladder carcinoma transforming gene is linked to a single amino acid change. Nature (London). 1982;300:762–5.
22. Yuasa Y, Srivastava SK, Dunn CY, Rhim JS, Reddy EP, Aaronson SA. Acquisition of transforming properties by alternative point mutations within c-bas/has human proto-oncogene. Nature (London). 1983;303:775–9.
23. Taparowsky E, Shimizu K, Goldfarb M, Wigler M. Structure and activation of the human N-ras gene. Cell. 1983;34:581–6.
24. Shimizu K, Birnbaum D, Ruley MA, Fasano O, Suard Y, Edlund L, Taparowsky E, Goldfarb M, Wigler M. Structure of the Ki-ras gene of the human lung carcinoma cell line Calu-1. Nature (London). 1983;304:497–500.
25. Bourne HR, Sanders DA, McCormick F. The GTPase superfamily: conserved structure and molecular mechanism. Nature. 1991;349:117–27.
26. Cox AD, Der CJ. Ras family signaling: therapeutic targeting. Cancer Biol Ther. 2002;1:599–606.
27. Downward J, Targeting RAS. signalling pathways in cancer therapy. Nat Rev Cancer. 2003;3:11–22.
28. Sebti SM, Hamilton AD. Design of growth factor antagonists with antiangiogenic and antitumor properties. Oncogene. 2000;19:6566–73.
29. Cox AD, Der CJ. Farnesyltransferase inhibitors: promises and realities. Curr Opin Pharmacol. 2002;2:388–93.
30. Rossi ED, Martini M, Capodimonti S, Lombardi CP, Pontecorvi A, Vellone VG, Zannoni GF, Larocca LM, Fadda G. BRAF (V600E) mutation analysis on liquid-based cytology-processed aspiration biopsies predicts bilaterality and lymph node involvement in papillary thyroid microcarcinoma. Cancer Cytopathol. 2012; Nov 28. doi:10.1002/cncy.21258.
31. Bösmüller H, Fischer A, Pham DL, Fehm T, Capper D, von Deimling A, Bonzheim I, Staebler A, Fend F. Detection of the BRAF V600E mutation in serous ovarian tumors: a comparative analysis of immunohistochemistry with a mutation-specific monoclonal antibody and allele-specific PCR. Hum Pathol. 2012; Oct 19. pii:S0046-8177(12)00266-3. doi:10.1016/j.humpath.2012.07.010.
32. Popovici V, Budinska E, Tejpar S, Weinrich S, Estrella H, Hodgson G, Van Cutsem E, Xie T, Bosman FT, Roth AD, Delorenzi M. Identification of a poor-prognosis BRAF-mutant-like population of patients with colon cancer. J Clin Oncol. 2012;30:1288–95.
33. Rapp UR, Goldsborough MD, Mark GE, Bonner TI, Groffen J, Reynolds FH Jr, Stephenson JR. Structure and biological activity of v-raf, a unique oncogene transduced by a retrovirus. Proc Natl Acad Sci U S A. 1983;80:4218–4222.
34. Schreck R, Rapp UR. Raf kinases: oncogenesis and drug discovery. Int J Cancer. 2006;119:2261–71.
35. Jansen HW, Patschinsky T, Bister K. Avian oncovirus MH2: molecular cloning of proviral DNA and structural analysis of viral RNA and protein. J Virol. 1983;48:61–73.
36. Sutrave P, Bonner TI, Rapp UR, Jansen HW, Patschinsky T, Bister K. Nucleotide sequence of avian retroviral oncogene v-mil: homologue of murine retroviral oncogene v-raf. Nature. 1984;309:85–8.
37. Moelling K, Heimann B, Rapp UR, Sander T. Serine- and threonine-specific protein kinase activities of purified gag-mil and gag-raf proteins. Nature. 1984;312:558–61.
38. Bonner TI, Kerby SB, Sutrave P, Gunnell MA, Mark G, Rapp UR. Structure and biological activity of human homologs of the raf/mil oncogene. Mol Cell Biol. 1985;5:1400–7.
39. Mark GE, MacIntyre RJ, Digan ME, Ambrosio L, Perrimon N. Drosophila melanogaster homologs of the raf oncogene. Mol Cell Biol. 1987;7:2134–40.
40. Han M, Golden A, Han Y, Sternberg PW. C. elegans lin-45 raf gene participates in let-60 ras-stimulated vulval differentiation. Nature. 1993;363:133–40.
41. Beck TW, Huleihel M, Gunnell M, Bonner TI, Rapp UR. The complete coding sequence of the human A-raf-1 oncogene and transforming activity of a human A-raf carrying retrovirus. Nucleic Acids Res. 1987;15:595–609.

42. Marx M, Eychène A, Laugier D, Béchade C, Crisanti P, Dezélée P, Pessac B, Calothy G. A novel oncogene related to c-mil is transduced in chicken neuroretina cells induced to proliferate by infection with an avian lymphomatosis virus. EMBO J. 1988;7:3369–73.
43. Barnier JV, Papin C, Eychène A, Lecoq O, Calothy G. The mouse B-raf gene encodes multiple protein isoforms with tissue-specific expression. J Biol Chem. 1995;270:23381–9.
44. Storm SM, Cleveland JL, Rapp UR. Expression of raf family proto-oncogenes in normal mouse tissues. Oncogene. 1990;5:345–51.
45. Wellbrock C, Karasarides M, Marais R. The RAF proteins take centre stage. Nat Rev Mol Cell Biol. 2004;5:875–85.
46. Ikawa S, Fukui M, Ueyama Y, Tamaoki N, Yamamoto T, Toyoshima K. B-raf, a new member of the raf family, is activated by DNA rearrangement. Mol Cell Biol. 1988;8:2651–2564.
47. Davies H, Bignell GR, Cox C, Stephens P, Edkins S, Clegg S, Futreal PA. Mutations of the BRAF gene in human cancer. Nature. 2002;417:949–54.
48. Dhillon AS, Kolch W. Oncogenic B-Raf mutations: crystal clear at last. Cancer Cell. 2004;5:303–4.
49. Zhang BH, Guan KL. Activation of B-Raf kinase requires phosphorylation of the conserved residues Thr598 and Ser601. EMBO J. 2000;19:5429–39.
50. Pollock PM, Harper UL, Hansen KS, Yudt LM, Stark M, Robbins CM, Moses TY, Hostetter G, Wagner U, Kakareka J, Salem G, Pohida T, Heenan P, Duray P, Kallioniemi O, Hayward NK, Trent JM. Meltzer PS High frequency of BRAF mutations in nevi. Nat Genet. 2003;33:19–20.
51. Sosman JA, Kim KB, Schuchter L, Gonzalez R, Pavlick AC, Weber JS, McArthur GA, Hutson TE, Moschos SJ, Flaherty KT, Hersey P, Kefford R, Lawrence D, Puzanov I, Lewis KD, Amaravadi RK, Chmielowski B, Lawrence HJ, Shyr Y, Ye F, Li J, Nolop KB, Lee RJ, Joe AK, Ribas A. Survival in BRAF V600-mutant advanced melanoma treated with vemurafenib. N Engl J Med. 2012;366:707–14.
52. Namba H, Nakashima M, Hayashi T, Hayashida N, Maeda S, Rogounovitch TI, Ohtsuru A, Saenko VA, Kanematsu T, Yamashita S. Clinical implication of hot spot BRAF mutation, V599E, in papillary thyroid cancers. J Clin Endocrinol Metab. 2003;88:4393–7.
53. Elisei R, Ugolini C, Viola D, Lupi C, Biagini A, Giannini R, Romei C, Miccoli P, Pinchera A, Basolo F. BRAF(V600E) mutation and outcome of patients with papillary thyroid carcinoma: a 15-year median follow-up study. J Clin Endocrinol Metab. 2008;93:3943–9.
54. Kobayashi M, Sonobe M, Takahashi T, Yoshizawa A, Ishikawa M, Kikuchi R, Okubo K, Huang CL, Date H. Clinical significance of BRAF gene mutations in patients with non-small cell lung cancer. Anticancer Res. 2011;31:4619–23.
55. Li WQ, Kawakami K, Ruszkiewicz A, Bennett G, Moore J, Iacopetta B. BRAF mutations are associated with distinctive clinical, pathological and molecular features of colorectal cancer independently of microsatellite instability status. Mol Cancer. 2006;5:2–8.
56. Tiacci E, Trifonov V, Schiavoni G, Holmes A, Kern W, Martelli MP, Pucciarini A, Bigerna B, Pacini R, Wells VA, Sportoletti P, Pettirossi V, Mannucci R, Elliott O, Liso A, Ambrosetti A, Pulsoni A, Forconi F, Trentin L, Semenzato G, Inghirami G, Capponi M, Di Raimondo F, Patti C, Arcaini L, Musto P, Pileri S, Haferlach C, Schnittger S, Pizzolo G, Foà R, Farinelli L, Haferlach T, Pasqualucci L, Rabadan R, Falini B. BRAF mutations in hairy-cell leukemia. N Engl J Med. 2011;364:2305–15.
57. Dumas J, Smith RA, Lowinger TB. Recent developments in the discovery of protein kinase inhibitors from the urea class. Curr Opin Drug Discov Devel. 2004;7:600–16.
58. Wilhelm SM, Carter C, Tang L, Wilkie D, McNabola A, Rong H, Chen C, Zhang X, Vincent P, McHugh M, Cao Y, Shujath J, Gawlak S, Eveleigh D, Rowley B, Liu L, Adnane L, Lynch M, Auclair D, Taylor I, Gedrich R, Voznesensky A, Riedl B, Post LE, Bollag G, Trail PA. BAY 43–9006 exhibits broad spectrum oral antitumor activity and targets the RAF/MEK/ERK pathway and receptor tyrosine kinases involved in tumor progression and angiogenesis. Cancer Res. 2004;64:7099–109.

59. Adnane L, Trail PA, Taylor I, Wilhelm SM. Sorafenib (BAY 43–9006, Nexavar), a dual-action inhibitor that targets RAF/MEK/ERK pathway in tumor cells and tyrosine kinases VEGFR/PDGFR in tumor vasculature. Methods Enzymol. 2006;407:597–612.
60. Yu C, Bruzek LM, Meng XW, Gores GJ, Carter CA, Kaufmann SH, Adjei AA. The role of Mcl-1 downregulation in the proapoptotic activity of the multikinase inhibitor BAY 43–9006. Oncogene. 2005;24:6861–9.
61. Hauschild A, Agarwala SS, Trefzer U, Hogg D, Robert C, Hersey P, Eggermont A, Grabbe S, Gonzalez R, Gille J, Peschel C, Schadendorf D, Garbe C, O'Day S, Daud A, White JM, Xia C, Patel K, Kirkwood JM, Keilholz U. Results of a phase III, randomized, placebo-controlled study of sorafenib in combination with carboplatin and paclitaxel as second-line treatment in patients with unresectable stage III or stage IV melanoma. J Clin Oncol. 2009;27:2823–30.
62. Shepherd C, Puzanov I, Sosman JA. B-RAF inhibitors: an evolving role in the therapy of malignant melanoma. Curr Oncol Rep. 2010;12:146–52.
63. Tsai J, Lee JT, Wang W, Zhang J, Cho H, Mamo S, Bremer R, Gillette S, Kong J, Haass NK, Sproesser K, Li L, Smalley KS, Fong D, Zhu YL, Marimuthu A, Nguyen H, Lam B, Liu J, Cheung I, Rice J, Suzuki Y, Luu C, Settachatgul C, Shellooe R, Cantwell J, Kim SH, Schlessinger J, Zhang KY, West BL, Powell B, Habets G, Zhang C, Ibrahim PN, Hirth P, Artis DR, Herlyn M, Bollag G. Discovery of a selective inhibitor of oncogenic B-Raf kinase with potent antimelanoma activity. Proc Natl Acad Sci U S A. 2008;105:3041–6.
64. Flaherty KT, Puzanov I, Kim KB, Ribas A, McArthur GA, Sosman JA, O'Dwyer PJ, Lee RJ, Grippo JF, Nolop K, Chapman PB. Inhibition of mutated, activated BRAF in metastatic melanoma. N Engl J Med. 2010;363:809–19.
65. Hauschild A, Grob JJ, Demidov LV, Jouary T, Gutzmer R, Millward M, Rutkowski P, Blank CU, Miller WH Jr, Kaempgen E, Martín-Algarra S, Karaszewska B, Mauch C, Chiarion-Sileni V, Martin AM, Swann S, Haney P, Mirakhur B, Guckert ME, Goodman V, Chapman PB. Dabrafenib in BRAF-mutated metastatic melanoma: a multicentre, open-label, phase 3 randomised controlled trial. Lancet. 2012;380:358–65.
66. Kim T, Kim J, Lee MG. Inhibition of mutated BRAF in melanoma. N Engl J Med. 2010;363:2261–2.
67. Solit DB, Garraway LA, Pratilas CA, Sawai A, Getz G, Basso A, Ye Q, Lobo JM, She Y, Osman I, Golub TR, Sebolt-Leopold J, Sellers WR, Rosen N. BRAF mutation predicts sensitivity to MEK inhibition. Nature. 2006;439:358–62.
68. Montagut C, Sharma SV, Shioda T, McDermott U, Ulman M, Ulkus LE, Dias-Santagata D, Stubbs H, Lee DY, Singh A, Drew L, Haber DA, Settleman J. Elevated CRAF as a potential mechanism of acquired resistance to BRAF inhibition in melanoma. Cancer Res. 2008;68:4853–61.
69. Nazarian R, Shi H, Wang Q, Kong X, Koya RC, Lee H, Chen Z, Lee MK, Attar N, Sazegar H, Chodon T, Nelson SF, McArthur G, Sosman JA, Ribas A, Lo RS. Melanomas acquire resistance to B-RAF(V600E) inhibition by RTK or N-RAS upregulation. Nature. 2010;468:973–7.
70. Johannessen CM, Boehm JS, Kim SY, Thomas SR, Wardwell L, Johnson LA, Emery CM, Stransky N, Cogdill AP, Barretina J, Caponigro G, Hieronymus H, Murray RR, Salehi-Ashtiani K, Hill DE, Vidal M, Zhao JJ, Yang X, Alkan O, Kim S, Harris JL, Wilson CJ, Myer VE, Finan PM, Root DE, Roberts TM, Golub T, Flaherty KT, Dummer R, Weber BL, Sellers WR, Schlegel R, Wargo JA, Hahn WC, Garraway LA. COT drives resistance to RAF inhibition through MAP kinase pathway reactivation. Nature. 2010;468:968–72.
71. Villanueva J, Vultur A, Lee JT, Somasundaram R, Fukunaga-Kalabis M, Cipolla AK, Wubbenhorst B, Xu X, Gimotty PA, Kee D, Santiago-Walker AE, Letrero R, D'Andrea K, Pushparajan A, Hayden JE, Brown KD, Laquerre S, McArthur GA, Sosman JA, Nathanson KL, Herlyn M. Acquired resistanceto BRAF inhibitors mediated by a RAF kinase switch in melanoma can be overcome by cotargeting MEK and IGF-1R/PI3K. Cancer Cell. 2010;18:683–95.
72. Paraiso KH, Xiang Y, Rebecca VW, Abel EV, Chen YA, Munko AC, Wood E, Fedorenko IV, Sondak VK, Anderson AR, Ribas A, Palma MD, Nathanson KL, Koomen JM, Messina

JL, Smalley KS. PTEN loss confers BRAF inhibitor resistance to melanoma cells through the suppression of BIM expression. Cancer Res. 2011;71:2750–60.
73. Fedorenko IV, Paraiso KH, Smalley KS. Acquired and intrinsic BRAF inhibitor resistance in BRAF V600E mutant melanoma. Biochem Pharmacol. 2011;82:201–9.
74. Prahallad A, Sun C, Huang S, Di Nicolantonio F, Salazar R, Zecchin D, Beijersbergen RL, Bardelli A, Bernards R. Unresponsiveness of colon cancer to BRAF(V600E) inhibition through feedback activation of EGFR. Nature. 2012;483:100–3.
75. Poulikakos PI, Persaud Y, Janakiraman M, Kong X, Ng C, Moriceau G, Shi H, Atefi M, Titz B, Gabay MT, Salton M, Dahlman KB, Tadi M, Wargo JA, Flaherty KT, Kelley MC, Misteli T, Chapman PB, Sosman JA, Graeber TG, Ribas A, Lo RS, Rosen N, Solit DB. RAF inhibitor resistance is mediated by dimerization of aberrantly splice BRAF(V600E). Nature. 2011;480:387–90.
76. Kaplan FM, Mastrangelo MJ, Aplin AE. The wrath of RAFs: rogue behavior of B-RAF kinase inhibitors. J Invest Dermatol. 2010;130:2669–71.
77. Poulikakos PI, Zhang C, Bollag G, Shokat KM, Rosen N. RAF inhibitors transactivate RAF dimers and ERK signalling in cells with wild-type BRAF. Nature. 2010;464:427–30.
78. Emery CM, Vijayendran KG, Zipser MC, Sawyer AM, Niu L, Kim JJ, Hatton C, Chopra R, Oberholzer PA, Karpova MB, MacConaill LE, Zhang J, Gray NS, Sellers WR, Dummer R, Garraway LA. MEK1 mutations confer resistance to MEK and B-RAF inhibition. Proc Natl Acad Sci USA. 2009;106:20411–6.
79. Infante JR, Falchook GS, Lawrence DP, Weber JS, Kefford RF, Bendell JC, Kurzrock R, Shapiro G, Kudchadkar RR, Long GV, Burris HA, Kim KB, Clements A, Peng S, Yi B, Allred AJ, Ouellet D, Patel K, Lebowitz PF, Flaherty KT. Phase I/II study of the oral MEK1/2 inhibitor GSK1120212 dosed in combination with the oral BRAF inhibitor GSK2118436. J Clin Oncol. 2011;29 18S:CRA8503.
80. Smalley KSM, McArthur GA. The current state of targeted therapy in melanoma: this time it's personal. Semin Oncol. 2012;39:204–14.
81. Weinstein IB, Joe A. Oncogene addiction. Cancer Res. 2008;68:3077–80.

# Role of $\beta 1$ Integrins in the Complication and Drug Resistance Against Lung Cancer: Targeting $\beta 1$ Integrins to Eradicate Lung Cancer

**Sirupa Mukhopadhyay, Parth Malik, Sunil Kumar Arora and Tapan K. Mukherjee**

**Abstract** Lung cancer is a major global concern of inevitable mortality and morbidity in modern times. In the context of lung carcinogenesis, bi-directional cell signaling events mediated by the interaction of transmembrane integrin receptors with the extracellular matrix (ECM) is highly critical. Structurally, the integrin superfamily comprises of a large family of receptors consisting of heterodimeric $\alpha$ and $\beta$ chains, noncovalently linked with each other. Eighteen different $\alpha$ subunits combine with 8 different $\beta$ subunits to emerge 24 known integrin family members. By far, $\beta 1$ integrins combine with their 12 different $\alpha$ subunits to yield the largest family members within the integrin superfamily. $\beta 1$ integrins are preferentially expressed in lung cancer cells and are significantly involved in lung carcinogenesis in terms of cell proliferation, survival, invasion and metastasis. Besides, the $\beta 1$ integrin family has a prime role in imparting chemoresistance to the lung cancer cells. The present discussion entitles a brief overview of the integrin superfamily, the cross-talk of $\beta 1$ integrins with ECM proteins, the importance of $\beta 1$ integrins as prognostic factors for lung cancer, the role of $\beta 1$ integrins in the complication of lung cancer and the associated drug resistance and, finally, targeting $\beta 1$ integrin functions to control lung cancer.

**Keywords** Non-small cell lung cancer (NSCLC) · Small cell lung cancer (SCLC) · $\beta 1$ integrins · Extracellular matrix (ECM) · Cell signaling · Drug resistance

---

S. Mukhopadhyay · S. K. Arora
Department of Immunopathology, Research block A, Post Graduate Institute of Medical Education and Research (PGIMER), Chandigarh, India

P. Malik · T. K. Mukherjee (✉)
Department of Biotechnology, Maharishi Markandeshwar University, Mullana, Ambala, India
e-mail: tapanmu@yahoo.com

T. K. Mukherjee
Department of Biology, IISER Mohali Sector-26, Chandigarh 160036, India

B. Bonavida (ed.), *Molecular Mechanisms of Tumor Cell Resistance to Chemotherapy*, Resistance to Targeted Anti-Cancer Therapeutics 1, DOI: 10.1007/978-1-4614-7070-0_5, 

**Abbreviations**

| | |
|---|---|
| BAD | $Bcl_2/Bclx_L$-associated death promoter |
| ECM | Extracellular matrix |
| EGFR TKIs | Epidermal growth factor receptor tyrosine kinase inhibitors |
| FAK | Focal adhesion kinases |
| mTOR | Mammalian target of rapamycin |
| NSCLC | Non-small cell lung cancer |
| PKB | Phosphatidyl inositol 3-kinase [PI(3)K]-protein kinase B |
| RGD | Arginine-(R)-glycine-(G)-aspartic acid-(D) |
| SCLC | Small cell lung cancer |
| ZO-1 | Zonula-occludensa |

## Introduction

Lung cancer is one of the prominent reasons of mortality in recent times owing to the cause of lung being the major vital organ that needs ultimate care and specialized supervision. There are mainly two types of lung cancer, namely, non-small cell lung cancer (NSCLC) and small cell lung cancer (SCLC). SCLC is further subdivided into adenocarcinoma and large cell differentiated carcinoma. In addition, carcinoid, Kaposi's sarcoma and melanoma are other types of lung cancers. The development of carcinogenesis in lung is a life long process, therefore, in all possibilities cancer cells from other organs may eventually metastasize and subsequently grow and develop in the lungs [1].

Normally, cells interact with their extracellular matrix (ECM) either by transmitting signals to the ECM (inside out signaling) or receive signals from ECM into the cell (outside in signaling) via cell surface integrins. Thus, integrins act as bidirectional signal transduction molecules [2, 3]. Biochemically, integrins are transmembrane glycoprotein receptors [4] that bind specifically with their corresponding ECM ligands, namely, fibronectins, collagens, vitronectins and laminins [5]. The integrin-ligand interaction and activation of bidirectional signal transduction are mediated through the activation of a number of so called proteins (e.g. talin) and the engagement of various kinases (e.g. focal adhesion kinases) (FAK) [6–11]. However, the integrins themselves lack any intrinsic kinase activity.

In practice, integrins promote cancer in two distinct ways; firstly, by preventing the apoptosis and, secondly, by imparting resistance of cancer cells to chemotherapeutic agents. Collectively, the inside out and outside in signaling mediated by integrins are key limiting steps in the regulation of antiapoptosis (pro-survival), growth, proliferation and the invasive properties of tumor cells into the surrounding tissues including blood vessels, and finally metastasizing to distant organs. Integrins are therefore regarded as the targets of cancer chemotherapy [12–14].

$\beta1$ integrins are the largest family among the integrin superfamily members [15]. Various members of the $\beta1$ family are preferentially expressed in the lung cancer tissues [16, 17]. The present review describes the central role of $\beta1$ integrins in the survival, proliferation, invasion, metastasis and drug resistance of SCLC and NSCLC. In addition, it also summarizes various drugs that target $\beta1$ integrins to combat various cancers including lung cancers.

## Expression of Various Integrin Superfamily Members in Lung Cancers

The integrin superfamily consists of 24 family members. These integrin family members are transmembrane receptors consisting of heterodimeric noncovalently linked different $\alpha$ and $\beta$ chains. Based on their specific functions, integrins can be commonly subdivided into three main types: cell–cell adhesion integrins, vascular integrins and cell-ECM adhesion integrins. Commonly, integrins having the $\beta1$ subunit constitute the majority of receptors for the ECM components and, therefore, hold an immense potential for cancer chemotherapy. The $\beta1$ integrin family is largely composed of 12 different $\alpha$ subunits members ($\alpha1$–$\alpha12$). Various $\alpha$ subunit members have the capacity to bind with the $\beta1$ subunit and, therefore, form 12 different $\beta1$ integrin subfamilies. Various subfamily members of the $\beta1$ integrins family are expressed in pulmonary tissues and, therefore, assume a very important role in the complication of lung cancer [15–18].

Basically, integrin expression is noticed in most of the nontransformed (normal) cells as well as cancer cells [18]. Comprehensive review articles by Gogali et al. [16] and Mizejewsk [17] described the distribution and action of various integrins in the lung and other cancer tissues. In general, SCLC cell lines fail to interact with the adhesive proteins in serum as well as with most ECM proteins, although, some SCLC cell lines can attach themselves onto the laminin substrates [19–21]. Feldman et al. [22] have shown that SCLC lung cancer cells viz, NCI-H187, H345, H146, H209, and N417 express sufficiently high levels of $\alpha3\beta1$. Similarly, Pellegrini et al. have demonstrated that the $\alpha6\beta1$ integrin is expressed by SCLC lung cancer cells [23]. In the following years, by using 19 human lung cancer cell lines, Hirasawa et al. have noticed that the main integrin expressed by SCLC cells is $\beta1$ and combined with $\alpha2$, $\alpha3$, $\alpha5$ and $\alpha6$, respectively [24]. In the same year, Falconi et al. have described that $\beta1$, $\beta3$, $\beta4$ and $\beta5$ integrins expressed by human lung carcinoma cells [25]. Using various SCLC cell lines, Sethi et al. have again reconfirmed that $\alpha2\beta1$, $\alpha3\beta1$, $\alpha6\beta1$ and $\alpha5\beta1$ are abundantly expressed in these cell lines [26]. Recently, Roman et al., by using mouse Lewis lung carcinoma cells, further demonstrated that the $\alpha5\beta1$ integrin is highly essential for adhesion with fibronectin and lack of the $\alpha5$ subunit of this integrin drastically affects cell proliferation, migration and colony formation [27]. Another recent study by

Dingemans et al. have indicated that the major integrin expressed by the NSCLC cells is $\alpha5\beta1$ [28]. Thus, the major $\beta1$ integrins expressed by various lung cancer cells/tissues are $\alpha5\beta1$, $\alpha3\beta1$, $\alpha2\beta1$, and $\alpha6\beta1$ integrins, respectively.

## Cross Talk of Integrins with Extracellular Matrix Proteins

ECM-cell interactions occur locally in specialized structural foci called focal adhesions (FA). FA are the specific focal contact zones between ECM and the cytoskeleton of the cell [8, 9], permitting the physical attachment of the integrins in membrane to their ECM ligands (e.g. fibronectins, vitronectins, laminins and collagen IV) and the intracellular transduction of downstream signals [29]. Other adhesion complexes mediated by integrins are focal complexes, fibrillar complexes, podosomes, etc. [30]. Critchley et al. have proved that the cytoplasmic domain of the $\beta$ subunit of the integrin is important in the localization of integrins to focal adhesions, and for integrin-mediated cell adhesion/spreading. The $\beta$-subunit of the integrins can be linked to F-actin via the actin-binding proteins talin, $\alpha$-actinin, filamin and vinculin [31]. Thus, these intracellular signaling proteins are highly essential for the connections of the integrins with the cytoskeleton.

The $\beta1$ integrins preferentially interact with various ECM proteins. For example, $\alpha1\beta1$ and $\alpha2\beta1$ integrins are major collagen receptors, whereas $\alpha4\beta1$ and $\alpha5\beta1$ integrins bind to fibronectins and $\alpha3\beta1$ and $\alpha6\beta1$ are receptors for laminins [3]. In SCLC lung cancer cells viz, NCI-H187, H345, H146, H209, and N417 cells, laminin is the major ligand which binds with the $\alpha3\beta1$ integrin [22]. Similarly, the $\alpha6\beta1$ integrin of the lung tissues binds with laminin [23]. Immunohistochemical analyses have confirmed the high level expression of collagen IV, which binds to $\alpha2\beta1$ or $\alpha3\beta1$ integrins and fibronectins bind to the $\alpha5\beta1$ integrin [32, 33] (Fig. 1 describe the various ligands of $\beta1$ integrins).

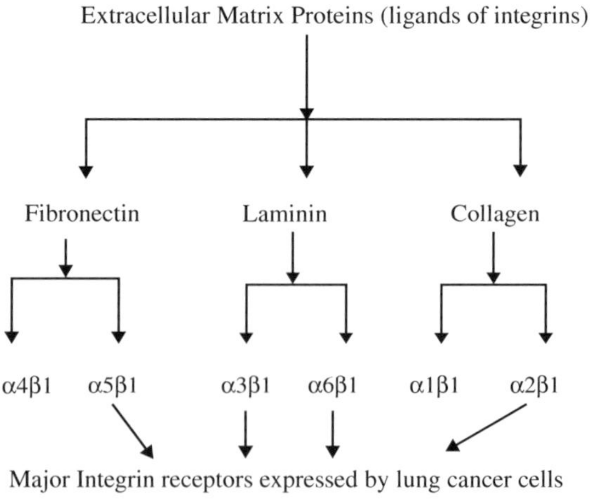


**Fig. 1** Cell surface integrin receptors binds with various extracellular matrix proteins namely fibronectins, laminins and collagen IV. $\alpha2\beta1$, $\alpha3\beta1$, $\alpha5\beta1$ and $\alpha6\beta1$ are the major integrins involved in lung cancer complication

Moreover, the SCLC cells express the functional CXCR4, the chemokine receptor for the stromal cell-derived factor 1 (SDF1). Expression of SDF-1 increases SCLC cell adhesion to collagen 1 and fibronectin via $\alpha 2\beta 1$ and $\alpha 5\beta 1$, respectively [34]. Of the various ECM molecules, the level of expression of fibronectins and collagens are unusually high in NSCLC cells and tissue samples [35, 36]. In A549 NSCLC cells, radiation-induced apoptosis or the cytotoxic drug ukrain-induced cell death are prevented by laminins and fibronectins, respectively [37]. In a comprehensive review article, Cacavari et al. have described the mechanism through which integrins-ECM interactions affects TGF$\beta$ and EGFR signaling in lung cancer cells [38]. Another study by Han et al. has demonstrated that the presence of fibronectin activates the survival of lung cancer cells by activating cyclooxyginase 2 (COX-2) through $\alpha 5\beta 1$ integrin [39].

## Interactions of Integrins with Extracellular Matrix Proteins

At the cellular level, several integrins bind with their corresponding ECM ligands (proteins), having a highly conserved amino acid sequence of the tripeptide arginine-(R)-glycine-(G)-aspartic acid-(D) [RGD]. $\alpha v\beta 3$, $\alpha 5\beta 1$ and $\alpha IIb\beta 3$ are the examples of this kind of integrins that recognize the RGD sequence of the ECM proteins. However, other integrins may recognize a different amino acid sequence of the ECM proteins. For example, the $\alpha 4\beta 1$ integrins recognize EILVD and REVD amino acid sequences in the alternatively spliced CS-1 fibronectin. In a comprehensive review article, Ruoslahti described RGD and other recognition sequences of various ECM proteins/ligands that interact with the integrins [40]. Mizejewsk [17] has tabulated the various ligands of integrins with respect to their structural motifs, recognition sites and functional activities. Identifying and understanding these unique binding sequences is important for targeting integrin-mediated therapy since the know-how of these amino acid sequences helps to develop therapeutic peptides or antibodies that could disrupt integrins-ECM interactions and regulate integrin-directed proliferation, growth, survival and metastasis of the cancer cells [41–44].

## $\beta 1$ Integrins as a Prognostic Factor in Small Cell Lung Cancer and Non Small Cell Lung Cancer

The level of expression of $\beta 1$ integrins is related to the prognosis of both SCLC and NSCLC. A clinical study of transbronchial biopsies revealed that the increased expression of $\beta 1$ integrins is directly correlated with chemoresistance and poor prognosis in SCLC [45]. Another study by the same group has shown that an association of $\beta 1$ integrin and oncogenic p53 gene expression together can act as a prognostic factor in SCLC [46]. In a recent in vivo study, by using samples from SCLC patients, Lawson et al. have shown that the levels of expressions of $\beta 1$ integrin and $Bcl_2$ are independent prognostic factors for the development of SCLC [47]. A number of other studies have also confirmed the elevated expression of $\beta 1$ integrins in SCLC cells but not of $\beta 3$, $\beta 4$, or $\beta 5$ integrins [48, 49]. However, in another very recent study using SCLC patients neither $\beta 1$ integrin nor its downstream signaling molecule rac1 are confirmed as a prognostic indicator [50].

In NSCLC, the expression of $\alpha 5$ integrins has been suggested to play a role in predicting the clinical course and prognosis of this disease [51, 52]. Okamura et al. [53] have shown that elevated $\beta 1$ integrin expression is a prognostic factor for overall survival in NSCLC patients. During early stages of the disease, the NSCLC patients suffer with a high risk of recurrence, therefore, determination of the level of $\alpha 5\beta 1$ expression may predict the prognosis of the disease in the individual patient [28]. A number of other studies also confirm the importance of $\alpha 5\beta 1$ as a prognostic factor against NSCLC. Thus, it is essential to check the level of expression of $\beta 1$ integrins and ECM proteins involved in various lung cancer cells.

## Role of $\beta 1$ Integrins in Prosurvival of Small Cell Lung Cancer Cells and Non Small Cell Lung Cancer: Implication on Anoikis

In normal cells, the interaction of cell surface integrins with various ECM proteins leads to pro-attachment signals. These pro-attachment signals generate prosurvival signals, whereas detachment from the ECM triggers many non-cancerous (normal) cell types to undergo "anoikis", a form of cell death due to cellular detachment [54, 55]. Understanding anoikis is very important for cancer cells since the acquisition of a resistance to anoikis constitutes a critical step in tumor progression and immortality, particularly in the emergence of invasive and metastatic cells [56]. Cancer cells have a unique potential of growth that enable them to have an anchorage-independent growth signaling which again signifies that cancer cells are not dependent on integrin-mediated cell-ECM interactions. Accordingly, cancer cells have a unique feature that allows them to escape and evade anoikis as they can survive without anchorage-dependent growth and, therefore, may rapidly metastasize to other body parts.

It must be mentioned that cancer cells exhibit major alterations in their repertoire of expressed integrins as well as in their surrounding ECM [57–59]. Moreover, metastatic cells display a marked resistance to anoikis [60–62]. For example, unlike normal cells, cell detachment-induced anoikis does not occur in cells expressing the oncogenes *Src* or *Ras*. These oncogenes must therefore express themselves properly with the active participation of specific integrins for the unlimited and infinite survival of associated tumor cells [63, 64].

Not only the enhanced expression of integrins and ECM proteins, the ability of integrins to prevent aniokis and to allow the cells to survive depends upon integrin-dependent activation of a number of prosurvival pathways [64]. However, unlike growth factors, integrins have no intrinsic kinase activity. Interestingly, integrins have the capacity to cocluster with kinases and adaptor proteins in focal complexes, thereby transmitting downstream prosurvival signals. A number of review articles describe in details integrin-dependent cell survival and inhibition of aniokis both in transformed cells and nontransformed cells [12]. At present, intensive attention has been given to FAK and Src in tumor progression [65–67], as well as two main downstream signaling pathways engaged by them, such as the PI(3)-K/AKT and MEK/ERK pathways, respectively [68, 69]. The most important signaling pathways that are involved in generating a prosurvival signal and eventually drug resistance in the lung cancer cells are discussed below.

SCLC is particularly the more aggressive form of lung cancer. In SCLC cells, the interaction of $\beta 1$ integrins with the ECM proteins activate the phosphatidyl inositol 3-kinase [PI(3)K]-protein kinase B (PKB) pathway, a prosurvival pathway and a key regulator of cell cycle progression [70, 71]. Of note, PKB is also known as AKT, a serine-threonine kinase. In SCLC cells, the PI(3)K-PKB prosurvival pathway thus inhibits the expression of p21 and p27, the critical inhibitors of cell cycle. It has also been observed that this prosurvival pathway is active even in the presence of DNA damage, indicating that DNA damage cannot prevent the cell cycle to proceed forward [72]. The integrins $\alpha 5\beta 1$ and $\alpha 5\beta 6$ are also involved in the elevated expression of the anti-apoptotic $Bcl_2$ protein [73]. Further studies showed that AKT activation actually leads to the generation of anti-apoptotic mechanisms such as the phosphorylated inactivation of the $Bcl_2/Bclx_L$-associated death promoter (BAD), a key negative regulator of cell survival that binds and blocks the anti-apoptotic $Bcl_2$ family proteins [74].

NSCLC is the most common type of lung cancer. Sometimes, the word NSCLC is also used as a generic term when a particular lung cancer cannot be properly classified. $\beta 1$ integrins play a very important role in NSCLC cells growth, angiogenesis, invasion, and metastasis to specific organs [75]. Han et al. have shown that in NSCLC cells, fibronectin activates the AKT-mammalian target of rapamycin (mTOR)-S6 kinase and inactivates the liver kinase B1 (LKB1)-5′-adenosine monophosphate-activated protein kinase (AMPK) pathway. The net effect of this dual action results in the enhanced proliferation of NSCLC cells [76]. In a recent study using NSCLC cells A549, $\beta 1$ integrin gene-silencing results in defective activity of the epidermal growth factor receptor (EGFR). This phenomenon leads to reduced in vitro proliferation, enhanced sensitivity towards the chemotherapeutic agents cisplatin and gefitinib and leading to impaired migration and invasive behavior of the A549 cells [77]. The integrin-linked kinase (ILK)-AKT pathway also provides a significant advantage to NSCLC cells for proliferation, survival, and invasion to distant organs (metastasis) by the protein kinase C$\varepsilon$ (PKC$\varepsilon$)-dependent signaling pathway [78].

## $\beta$1 Integrins in the Migration and Metastasis of Small Cell Lung Cancer Cells and Non Small Cell Lung Cancer Cells

The involvement of integrins in cancer cell invasion and migration are explained in a number of studies [79–81]. Normally, SCLC cells have a high tendency to metastasize in the bone marrow and bone marrow cells secrete a high amount of the chemokine stromal cell derived factor-12 (CXCL12) [82]. Of note, CXCL12 is a chemokine constitutively expressed by bone marrow stromal cells. CXCR4 is a lone receptor for the CXCL12 chemokine, which plays an important role in the housing of cancer cells within the bone marrow microenvironment [83–86]. The role of CXCL12-CXCR4 interaction is, thus, a very critical and an important driving factor for lung cancer onset as it drives the cytoskeletal organization of tumor cells involving the activation of vital cell surface receptors. This phenomenon also results in the attachment of cancer cells to the ECM via $\beta$1 integrins [86]. Therefore, the adhesion of SCLC cells through $\beta$1 integrins to fibronectin and collagen ligands is increased by the CXCL12 chemokine.

Likewise, studies using the NSCLC cell line NCI-H1299 have shown that integrin signaling induces various proteolytic enzymes (proteases) that are co-localized within the integrins regulating the interface between integrins and the intracellular cytoskeleton. In one study, urokinase, a protease, promotes the ligand-like binding of its receptor (urokinase receptor is a GPI linked membrane adaptor protein) to a set of $\beta$1 integrins. This binding, in turn, affects integrin-dependent cell invasion [87]. Similar findings by Takenaka et al. have shown that a highly metastatic variant of PC9 cells has increased $\alpha5\beta1$ integrin expression and that the metastatic potential of these cells can be reduced to more than half by treatment with an anti-$\beta$1 integrin antibody [88]. Interestingly, studies by Han et al. have demonstrated that increased expression of $\alpha5\beta1$ integrin is associated with lymph node metastasis in NSCLC patients [35]. An association between increased integrin $\alpha$5 or $\beta$1 expression and poor prognosis in NSCLC patients are reported previously [35].

A number of studies indicate the central role of PKCs in invasion and migration of NSCLC. For example, PKC$\varepsilon$ has been recently identified as a key regulator of a molecular network controlling the migration of NSCLC cells through the active coupling of zonula-occludensa-1 (ZO-1) proteins and $\alpha5\beta1$ integrin [78]. Protein kinase inhibitors prevent junction dissociation-induced by low extracellular calcium in MDCK epithelial cells [89]. Moreover, the clinical diagnosis of NSCLC cancer patients has revealed the presence of ZO-1-$\alpha5\beta1$ complex, being actively involved in the metastasis of the related disease. These clinical reports summarize the key signaling events brought about by the PKC$\varepsilon$-$\alpha5\beta1$-ZO-1 pathway suggesting its extremely sensitive role in the metastasis of NSCLC in humans. Although the molecular mechanism responsible for ILK over-expression in NSCLC cells is still not well understood, it is a highly likely perception that the integrin-ILK-AKT signaling pathway may provide a big boost for NSCLC cell proliferation, survival and invasion. So, inhibition of ILK expression to attenuate the onset of NSCLC presents an exciting approach to prevent the lung cancer occurrence. Some initial successes in this regard have been achieved following the use of KP-392 and cisplatin, a chemical inhibitor (drug) that prevents the growth and malignancy of NSCLC cells [90]. In a most recent study using A549 and H1299 cell lines, a novel histone deacetylase inhibitor HTPB significantly suppresses tumor metastasis, partly, through inhibition of the $\beta 1$ integrin-focal adhesion kinase (FAK)-matrix metallo proteinase (MMP)-RhoA-F-actin pathway [91]. These observations obviously lead to the conclusion that $\alpha 5\beta 1$ integrin is actively involved in promoting both NSCLC proliferation as well as its metastatic invasion.

## Role of Integrins Other than β1 in the Complication of Lung Cancer

While $\beta 1$ integrins are the major integrin subfamily involved in lung cancer survival, invasion, metastasis and drug resistance, a few other $\beta$ integrin family members are also involved in the complication of lung cancers. Two such integrin members of this type are $\alpha v\beta 5$ integrins and $\alpha v\beta 3$ integrins respectively, as discussed below.

### *Role of αvβ5 Integrins in Lung Cancer Complication*

In a significant attempt, Zhu et al. have investigated the combined role of the protease-activated receptor (PAR-1) and $\alpha v\beta 5$ integrin during the onset of thrombin-mediated lung cancer invasion, both in immobilized as well as in native state. They have found an active involvement of $\alpha v\beta 5$ integrins in both native as well as in the immobilized thrombin-mediated tumor cell invasion. $\alpha v\beta 5$ integrins promote lung cancer cell adhesion, migration and extracellular signal-regulated kinase (ERK) phosphorylation, in both native as well as in immobilized thrombin configuration. This study indicates that targeting of $\alpha v\beta 5$ integrins or the $\alpha v\beta 5$ integrin complex with PAR-1 can be an excellent therapeutic strategy in the treatment of lung cancer invasion [92].

### *Role of αvβ3 Integrins in Lung Cancer Complication*

Recently, Meng et al. have predicted that there is an active cooperative signaling between $\alpha v\beta 3$ integrins and the estrogen receptor alpha (ER$\alpha$), which in turn, is involved in the proliferation of lung cancer via thyroid hormone induction. In their study, Meng et al. have comprehensively analyzed the signaling mediated by ER$\alpha$-

positive human lung cancer cells and have successfully concluded that the proliferative action of the thyroid hormone is mediated by the corresponding activity of ERα. Their analysis has further shown that the lung cancer cell proliferation has an active dependence on the corresponding integrin receptors and that the thyroid hormone crosstalks with αvβ3 integrins and is an active endogenous factor supporting the proliferation of lung cancer [93].

Very recently two more studies have also described the importance of the contributory role of the αvβ3 integrin in the adhesion, invasion and metastasis of B6 melanoma cells in the lungs. In one of these studies, Lonsdorf et al. have observed a reduction in the hematogenous metastasis of B16 melanoma cells to the lungs following treatment with a monoclonal antibody aimed specifically at blocking the expression of α subunit of the αvβ3 integrin. This study further concludes that there is an active role of engagement of GPIIb/IIIa (platelet receptors) with αvβ3 integrin interaction that mediates the interaction of tumor cells with platelets and also explains how this interaction is involved in hematogenous tumor metastasis [94].

In another comprehensive study, Goldeman et al. have suggested crucial roles of β3 integrins in the progression and metastasis of lung cancer and have proposed these β3 integrins to be the potential targets in the treatment of lung cancer via therapeutic agents. They have used the RNA interference (siRNA) technology to study the consequence of β3 integrins impairment in the B16 murine melanoma cells. They have also screened their expression for tumor proliferation, metastasis, invasion, adhesion and migration. They have shown through their analysis that siRNA against β3 causes a significant reduction in the progression, metastasis, invasion and adhesive nature of the lung cancer. However, the proliferation rate of the lung cancer cells remains unaltered. These findings concluded that targeting the β3 integrins expression at the molecular level can be an excellent therapeutic approach in the treatment of lung cancer [95].

Interestingly, another study has demonstrated that α5β1 integrin is the major integrin expressed in the lung cancer cells and it regulates αvβ3-dependent angiogenesis by modulating the activity of the protein kinase A (PKA) [96]. Another study has also explained the signaling cooperation between β1 and β3 integrins [97]. Thus, it can be solely predicted that a combined inhibition of α5β1 and αvβ3 may be more effective to attenuate lung cancer angiogenesis instead of targeting α5β1or αvβ3 alone.

## Role of β1 Integrins in Drug Resistance in Small Cell Lung Cancer and Non-Small Cell Lung Cancer

Cancer cells are inherently different from their normal counterparts and have differential expression of integrins. While a particular cancer cell sufficiently expresses a few particular types of integrins, the same integrins may not be sufficiently expressed by some other cancer cells. Moreover, any particular cancer cell may express more than one type of integrins. Therefore, the task of understanding the role of integrins in the regulation of drug resistance of metastatic cancer cells seems to be a daunting subject. The rapid metastasis (for example to the bone marrow) and the tendency to develop resistance towards chemotherapeutic drugs are the two main reasons for the extreme malignancy of SCLC [98].

It is speculated that adhesion molecules are somehow related to drug resistance [99]. Interestingly, adhesion of SCLC lung carcinoma cells to fibronectins enhances tumorigenicity and confers resistance to apoptosis-induced by standard chemotherapeutic agents against cancers [33]. A similar study has reported that the adhesion of SCLC cells to fibronectin enhances their viability and cytoskeletal organization mainly by activating the PI(3)K-FAK pathway [100]. Kraus et al. [101] have demonstrated that chemoresistance of SCLC cells correlates with adhesion to ECM and constitutive activation of the Akt and MAPK-dependent prosurvival and proproliferative pathways. Moreover, Tsurutani et al. have shown that laminin-mediated activation of the PI(3)K-Akt pathway enables SCLC cells to escape from imatinib-induced apoptosis [102]. In a comprehensive work by Sethi et al., it is further discussed that adhesion of SCLC cells via $\beta1$ integrins to ECM components promotes cell survival and conferring resistance to chemotherapeutic agents. The chemotherapeutic resistance is possibly due to the activation of the PI(3)K-PKB prosurvival pathway which inhibits the expression of p21 and p27, the critical inhibitors of cell cycle. Interestingly, this prosurvival pathway is active even in the presence of DNA damage, indicating DNA damage cannot prevent the cell cycle to move forward [26]. Krystal et al. have described that inhibition of the PI(3)K-Akt pathway attenuates growth, promotes apoptosis and enhances sensitivity of SCLC cells to chemotherapy [70]. Similarly, Hartmann et al. have reported that in SCLC cells the interaction of $\alpha2$, $\alpha4$, $\alpha5$, and $\beta1$ integrins with the chemokine receptor CXCR4 induces adhesion and chemoresistance [33]. Contemporary studies have also reported that over expression of $\beta1$ integrins leads to chemoresistance, thus affecting the diagnosis of SCLC [45]. In another significant attempt, Ritzenthal et al. have demonstrated that there is an elevated expression of matrix glycoprotein fibronectins in most of the chronic lung disorders. They have explained that the increased expression of fibronectins in the lung cancer cells leads to an increased signaling with the cell surface integrins in these cells that further activate and stimulate the intercellular signals involved in the pathobiology of lung carcinogenesis and lung tumor chemoresistance including those of the MAPKs, GTPases and the PI3-kinase/Akt/mTOR pathways [103]. In yet another significant study, Kohmo et al. have explained the extreme aggressiveness of the malignant and metastatic nature of SCLC cells, citing chemoresistance as the primary cause after the very early initial trials. They have suggested that the signaling by ECM proteins is very acutely involved in imparting of the chemoresistance towards the mostly used cytotoxic drugs. They have also confirmed that there is an unusual and extraordinary subcellular signal transduction ability mediated via the activity of the tetraspanin family member CD9 which expresses itself in SCLC tumors and leading to the development of resistance against cisplatin and etoposide. They have further analyzed that the higher expression of CD-9 in SCLC cell lines continues even after 48 h of their exposure to the potential anticancerous drugs. In their study, Kohmo et al. have found that chemoresistant SCLC cell lines express the CD-9 receptor in an uncharacteristically higher concentration and are also bound to the fibronectins via $\beta 1$ integrins far more tightly than their non-chemoresistant counterparts. However, they are less metastatic due to lower motility. They have further analyzed that the secretion of cell surface CD-9 receptor on the chemoresistant cells are decreased significantly following treatment with the chemokine CXCL12 and this has also resulted in the restoration of the motility. To support their conclusive evidence, they have also found that selectively treating with either specific anti-CD-9 monoclonal antibody or the incorporation of CD-9 siRNA they resulted in the initiation of apoptosis in the chemoresistant SCLC cell lines [104]. Another attempt by Hoddkinson et al. have concluded that there is an extraordinary role of SCLC-ECM-proteins interactions via integrin activation and PI-3 kinase activity, which is responsible for the generation of a high degree of ECM protein mass around the SCLC cells. Therefore, ECM-integrin signaling becomes a hot target pathway for chemotherapy of lung cancers [72]. Thus, $\beta 1$ integrin expression is directly related to the metastatic potential and drug resistance of SCLC.

In the NSCLC cell line PC9/AB2, overexpression of $\beta 1$ integrin leads towards the resistance to epidermal growth factor receptor tyrosine kinase inhibitors (EGFR TKIs) such as gefitinib and erlotinib [92]. Notably, $\beta 1$ integrins control EGRF signaling and tumorigenic properties of lung cancer cells [77]. Altogether, these studies emphasize the critical role of $\beta 1$ integrins in both the malignancy and chemoresistance in lung cancer, in particular to SCLC cells.

## Modulation of Interactions Between $\beta 1$ Integrins and Extracellular Matrix Proteins

A number of molecules regulate the interactions between $\beta 1$ integrins and ECM proteins. Palecek et al. [105] have described in details the various molecules (e.g. COX-2) that may affect the integrin-ligand binding. Transforming growth factor beta (TGF$\beta$) is a potent regulator of integrin-ligand interactions [106]. In fact, the latent form of TGF$\beta$ acts as a ligand for $\alpha 5\beta 1$ integrins [107]. A recent study has shown that the chief tobacco alkaloid nicotine stimulates lung cancer growth via the elevated synthesis of the ECM protein fibronectin. Truly, $\alpha 5\beta 1$ integrins are the major receptors involved in the synthesis of the fibronectin ligand and its activation. Again, silencing the functional activity of $\alpha 5\beta 1$ integrins impairs the mitogenic effect of nicotine on lung cancer cells [108]. Thus, COX-2, TGF$\beta$ and nicotine are some of the molecules that affect the engagement of $\beta 1$ integrins with the ECM protein ligands and, thereby, regulate several cell signaling pathways both in SCLC and NSCLC cell types.

## $\beta$1 Integrins as a Drug Target Against Lung Cancer

Several studies have been tried in the last decade to establish the role of integrins as suitable therapeutic targets against various cancers. Some of these strategies are focused on the regulation of integrin-dependent angiogenesis. Notably, integrins have a very distinct role in cancer angiogenesis [109, 110]. In a comprehensive review, Brook has described various anti-angiogenic strategies for the treatment of lung cancer [111]. Similarly, Jin et al. have described the role of integrins in cancer development and as therapeutic targets [14]. They have described several clinical trials of various cancers by using anti-$\alpha5\beta1$ or anti-$\alpha v\beta3$ antibodies or peptides and concluded that these anti-integrin antibodies or peptides have promising anti-antiangiogenic effects against these cancers. Lately, Desgrosellier in a comprehensive review article has described various strategies utilized for anti-integrin therapy and thereby attenuating cancer complications [112]. In another comprehensive review, Chen has also described various synthetic peptides and proteins that targets the RGD sequence of integrin-ligands and integrin-targeted delivery of nanoparticles carrying therapeutic agents to regulate integrin ligand-integrin interactions [113]. All these above studies are limited to different phases of clinical trials. More importantly, in a study by Reynolds et al., it is shown that deletion of Itgb3 (which encodes $\alpha v\beta3$) or deletion of both Itgb3 and Itgb5 (which encodes $\alpha v\beta5$) failed to inhibit angiogenesis and actually potentiated angiogenesis [114]. Therefore, more studies are necessary with other integrins to conclusively determine the success of anti-integrin therapy to inhibit vascular angiogenesis.

Noteworthy, attention must also be given on several in vitro experiments using various lung cancer cell lines. For example, Takenake et al. have shown the involvement of the $\beta$1 subunit during attachment and this result is consistent with data showing reduced metastasis by a lung carcinoma cell line after treatment with blocking antibodies to the $\beta$1 integrin subunit [88]. In another attempt, the CXCL12-CXCR4 binding-axial orientation has been targeted with BKT140, an antagonist of the CXCR4 chemokine. This has helped in the inhibition of the growth of NSCLC cells and also helped in augmenting the remedial effects of chemotherapy and radiotherapy [115].

BRL 49653 and GW1929, the ligands of peroxisome proliferator-activated receptor gamma (PPAR$\gamma$), inhibit lung carcinoma cell growth by inhibiting the expression of $\alpha$5 subunit of $\alpha5\beta1$ integrins [116]. Tsai Y et al. have indicated that glabridin, a flavonoid, inhibits migration, invasion, and angiogenesis of in vitro cultured A549 cells by inhibiting the FAK/rho signaling pathway. Inhibition of this signaling pathway is possible due to proteasomal degradation of $\alpha v\beta3$ integrins [117]. The study by Aixinjueluo et al. has further elaborated the mechanism of anti-GD2-ganglioside-dependent apoptosis of the SCLC cell lines SK-LC-17, NCI-417, ACC-LC-171, and ACC-LC-96. This study also claimed that GD2$^+$ SCLC cells undergo anoikis through the conformational changes of integrin molecules and subsequent FAK dephosphorylation [118]. In SBC-5 and SBC-3 SCLC cell lines, siRNA-dependent down regulation of $\beta$3 integrins may prevent the metastatic potential of the cancer cells [119]. In a recent experimental metastasis study, a powerful subcellular in vivo imaging model is used to demonstrate how an anti-integrin antibody affects seeding and growth of osteosarcoma cells on the lungs. However, treatment with anti-$\beta$1 integrin monoclonal antibody, AIIB2, greatly inhibited the seeding of cancer cells on the lungs. This study has concluded that AIIB2 significantly inhibited the spontaneous lung metastasis but not the primary tumor growth [120]. In addition, in the human lung cancer cell line A549 and H1299, a novel histone deacetylase inhibitor, HTPB, significantly suppressed tumor metastasis partly through inhibition of the integrin-$\beta$1/FAK/MMP/RhoA/F-actin pathways [91].

## Conclusion and Future Perspective

The studies and findings collected in this review highlight the importance of the $\beta$1 integrin-ECM signaling in tumorigenesis and drug resistance, with particular reference to lung cancer. $\beta$1 integrins have been correctly implicated as very important mediators of the pathogenesis of lung cancer including proliferation, invasion and metastasis. Targeting integrins, particularly $\beta$1 integrins, could be one of the most crucial breakthroughs in the successful treatment of lung cancers. Studies aimed at inhibiting the expression of integrins and the weakening of their interaction with the ECM protein ligands are being investigated at the present time for the treatment of lung cancer in different parts of the world.

**Conflict of Interest** No conflict of interest.

## References

1. Pastorino U. Lung cancer screening. Br J Cancer. 2010;102:1681–6.
2. Hu P, Luo BH. Integrin bidirectional signaling across the plasma membrane. J Cell Physiol. 2013;228:306–12.
3. Hynes RO. Integrins: bidirectional, allosteric signaling machines. Cell. 2002;110:673–87.
4. Tamkun JW, DeSimone DW, Fonda D, Patel RS, Buck C, Horwitz AF, Hynes RO. Structure of integrin, a glycoprotein involved in the transmembrane linkage between fibronectin and actin. Cell. 1986;46:271–82.
5. Hynes RO. Integrins: versatility, modulation and signaling in cell adhesion. Cell. 1992;69:11–25.
6. Aplin AE, Howe A, Alahari SK, Juliano RL. Signal transduction and signal modulation by cell adhesion receptors: the role of integrins, cadherins, immunoglobulin-cell adhesion molecules, and selectins. Pharmacol Rev. 1998;50:197–263.
7. Jones JL, Walker RA. Integrins: a role as cell signaling molecules. J Clin Pathol: Mol Pathol. 1999;52:208–13.
8. Otey CA. pp 125FAK in the focal adhesion. Int Rev Cytol. 1996;167:161–83.
9. Guan JL. Role of focal adhesion kinase in integrin signaling. Int J Biochem Cell Biol. 1997;29:1085–96.

10. Anthis NJ, Campbell ID. The tail of integrin activation. Trends Biochem Sci. 2011;36:191–8.
11. Schwartz MA, Ginsberg MH. Networks and crosstalk: integrin signalling spreads. Nat Cell Biol. 2002;4:E65–8.
12. Aoudjit F, Vouri K. Integrin signaling in cancer cell survival and chemoresistance. Chemother Res Pract. 2012;2012:283181.
13. Cox D, Brennan B, Moran N. Integrins as therapeutic targets: lessons and opportunities. Nat Rev Drug Discov. 2010;9:804–20.
14. Jin H, Varner J. Integrins: roles in cancer development and as treatment targets. Br J Cancer. 2004;90:561–5.
15. Armulik A. Splice variants of human beta 1 integrins: origin, biosynthesis and functions. Front Biosci. 2002;7:d219–27.
16. Gogali A, Charalabopoulos K, Constantopoulos S. Integrin receptors in primary lung cancer. Exp Oncol. 2004;26:106–10.
17. Mizejewski GJ. Role of integrins in cancer: survey of expression patterns. Proc Soc Exp Biol Med. 1999;222:124–38.
18. Virtanen I, Korhonen M, Kariniemi AL, Gould VE, Laitinen L, Ylänne J. Integrins in human cells and tumors. Cell Differ Dev. 1990;32:215–27.
19. Fridman R, Giaccone G, Kanemoto T, Martin GR, Gazdar AF, Mulshine JL. Reconstituted basement membrane (matrigel) and laminin can enhance the tumorigenicity and the drug resistance of small cell lung cancer cell lines. Proc Natl Acad Sci USA. 1990;87:6698–702.
20. Tagliabue E, Martignone S, Mastroianni A, Menard S, Pellegrini R, Colnaghi MI. Laminin receptors on SCLC cells. Br J Cancer. 1991;Suppl. 14:83–5.
21. Giaccone G, Broers J, Jensen S, Fridman RI, Linnoila R, Gazdar AF. Increased expression of differentiation markers can accompany laminin-induced attachment of small cell lung cancer cells. Br J Cancer. 1992;66:488–95.
22. Feldman LE, Shin KC, Natale RB, Todd RF 3rd. Beta1 integrin expression on human small cell lung cancer cells. Cancer Res. 1991;51:1065–70.
23. Pellegrini R, Martignone S, Menard S, Colnaghi MI. Laminin receptor expression and function in small-cell lung carcinoma. Int J Cancer Suppl. 1994;8:116–20.
24. Hirasawa M, Shijubo N, Uede T, Abe S. Integrin expression and ability to adhere to extracellular matrix proteins and endothelial cells in human lung cancer lines. Br J Cancer. 1994;70:466–73.
25. Falconi R, Cimino L, Gentileschi M, D'Agnano I, Zupi G, Kennel SJ, Sacchi A. Expression of beta1, beta3, beta4 and beta5 integrins by human lung carcinoma cells of different histotypes. Exp Cell Res. 1994;210:113–22.
26. Sethi T, Rintoul RC, Moore SM, MacKinnon AC, Salter D, Choo C, Chilvers ER, Dransfield I, Donnelly SC, Strieter R, Haslett C. Extracellular matrix proteins protect small cell lung cancer cells against apoptosis: a mechanism for small cell lung cancer growth and drug resistance in vivo. Nat Med. 1999;5:662–8.
27. Roman J, Ritzenthaler JD, Roser-Page S, Sun X, Han S. Alpha5beta1-integrin expression is essential for tumor progression in experimental lung cancer. Am J Respir Cell Mol Biol. 2010;43:684–91.
28. Dingemans AMC, Boogaart VVD, Vosse BA, van Suylen RJ, Griffioen AW, Thijssen VL. Integrin expression profiling identifies integrin alpha5 and beta1 as prognostic factors in early stage non-small cell lung cancer. Mol Cancer. 2010;9:152.
29. Humphries JD, Byron A, Humphries MJ. Integrin ligands at a glance. J Cell Sci. 2006;119:3901–3.
30. Zaidel-Bar R, Itzkovitz S, Maayan A, Iyenger R, Geiger B. Functional atlas of the integrin adhesome. Nat Cell Biol. 2007;9:858–67.
31. Critchley DR, Holt MR, Barry ST, Priddle H, Hemmings L, Norman J. Integrin-mediated cell adhesion: the cytoskeletal connection. Biochem Soc Symp. 1999;65:79–99.
32. Buttery RC, Rintoul RC, Sethi T. Small cell lung cancer: the importance of the extracellular matrix. Int J Biochem Cell Biol. 2004;36:1154–60.

33. Rintoul RC, Sethi T. Extracellular matrix regulation of drug resistance in small-cell lung cancer. Clin Sci. 2002;102:417–24.
34. Hartmann TN, Burger JA, Glodek A, Fujii N, Burger M. CXCR4 chemokine receptor and integrin signaling co-operate in mediating adhesion and chemoresistance in small cell lung cancer (SCLC) cells. Oncogene. 2005;24:4462–71.
35. Han JY, Kim HS, Lee SH, Park WS, Lee JY, Yoo NJ. Immunohistochemical expression of integrins and extracellular matrix proteins in non-small cell lung cancer: correlation with lymph node metastasis. Lung Cancer. 2003;41:65–70.
36. Jakowlew SB, Mariano JM, You L, Mathias A. Differential regulation of protease and extra cellular matrix protein expression by transforming growth factor-beta 1 in non-small cell lung cancer cells and normal human bronchial epithelial cells. Biochim Biophys Acta. 1997;1353:157–70.
37. Cordes N, Blaese MA, Plasswilm L, Rodemann HP, Van Beuningen D. Fibronectin and laminin increase resistance to ionizing radiation and the cytotoxic drug Ukrain in human tumor and normal cells in vitro. Int J Radiat Biol. 2003;79:709–20.
38. Caccavari F, Valdembri D, Sandri C, Bussolino F, Serini G. Integrin signaling and lung cancer. Cell Adh Migr. 2010;4:124–9.
39. Han S, Sidell N, Roser-Page S, Roman J. Fibronectin stimulates human lung cancer cell growth by inducing cyclooxygenase-2 (COX-2) gene expression. Int J Cancer. 2004;111:322–31.
40. Ruoslahti E. RGD and other recognition sequences for integrins. Annu Rev Cell Dev Biol. 1996;12:697–715.
41. Humphries MJ, Olden K, Yamada KM. A synthetic peptide from fibronectin inhibits experimental metastasis of murine melanoma cells. Science. 1986;233:467–70.
42. Saiki I, Murata J, Iida J, Sakurai T, Nishi N, Matsuno K, Azuma IT. Antimetastatic effects of synthetic polypeptides containing repeated structures of the cell adhesive Arg-Gly-Asp (RGD) and Tyr-Ile-Gly-Ser-Arg (YIGSR) sequences. Br J Cancer. 1989;60:722–8.
43. Yamamura K, Kibbey MC, Jun SH, Kleinman HK. Effect of matrigel and laminin peptide YIGSR on tumor growth and metastasis. Semin Cancer Biol. 1993;4:259–65.
44. Vollmers HP, Imhof BA, Braun S, Waller CA, Schirrmacher V, Birchmeier W. Monoclonal antibodies which prevent experimental lung metastasis. Interference with the adhesion of tumor cells to laminin. FEBS Lett. 1984;172:17–20.
45. Oshita F, Kameda Y, Ikehara M, Tanaka G, Yamada K, Nomura I, Noda K, Shotsu A, Fujita A, Arai H, Ito H, Nakayama H, Mitsuda A. Increased expression of integrin $\beta 1$ is a poor prognostic factor in small-cell lung cancer. Anticancer Res. 2002;22:1065–70.
46. Oshita F, Kameda Y, Hamanaka N, Saito H, Yamada K, Noda K, Mitsuda A. High expression of integrin $\beta 1$ and p53 is a greater poor prognostic factor than clinical stage in small-cell lung cancer. Am J Clin Oncol. 2004;27:215–9.
47. Lawson MH, Cummings NM, Rassi DM, Vowler SL, Wickens M, Howat WJ, Brenton JD, Murphy G, Rintoul RC. Bcl2 and beta1 integrin predict survival in a tissue microarray of small cell lung cancer. Br J Cancer. 2010;103:1710–5.
48. Bartolazzi A, Cerboni C, Flamini G, Bigotti A, Lauriola L, Natali PG. Expression of $\alpha 3\beta 1$ integrin receptor and its ligands in human lung tumors. Int J Cancer. 1995;64:248–52.
49. Hibi K, Yamakawa K, Ueda R, Horio Y, Murata Y, Tamari M, Uchida K, Takahashi T, Nakamura Y, Takahashi T. Aberrant upregulation of a novel integrin $\alpha$ subunit gene at 3p21.3 in small cell lung cancer. Oncogene. 1994;9:611–9.
50. Chang MH, Lee K, Lee KY, Kim YS, Kim YK, Kang JH. Prognostic role of integrin $\beta 1$, E-cadherin, and rac1 expression in small cell lung cancer. APMIS: Acta Pathologica, Microbiologica et Immunologica. 2012;120:28–38.
51. Adachi M, Taki T, Higashiyama M, Kohno N, Inufusa H, Miyake M. Significance of integrin alpha5 gene expression as a prognostic factor in node-negative non-small cell lung cancer. Clin Cancer Res. 2000;6:96–101.

52. Adachi M, Taki T, Huang C, Higashiyama M, Doi O, Tsuji T, Miyake M. Reduced integrin alpha3 expression as a factor of poor prognosis of patients with adenocarcinoma of the lung. J Clin Oncol. 1998;16:1060–7.
53. Okamura M, Yamaji S, Nagashima Y, Nishikawa M, Yoshimoto N, Kido Y, Iemoto Y, Aoki I, Ishigatsubo Y. Prognostic value of integrin beta1-ILK-pAkt signaling pathway in non-small cell lung cancer. Hum Pathol. 2007;38:1081–91.
54. Chiarugi P, Giannoni E. Anoikis: a necessary death program for anchorage-dependent cells. Biochem Pharmacol. 2008;76:1352–64.
55. Kumar CC. Signaling by integrin receptors. Oncogene. 1998;17:1365–73.
56. Hehlgans S, Haase M, Cordes N. Signalling via integrins: implications for cell survival and anticancer strategies. Biochim Biophys Acta. 2007;1775:163–80.
57. Barkan D, Green JE, Chambers AF. Extracellular matrix: a gatekeeper in the transition from dormancy to metastatic growth. Eur J Cancer. 2010;46:1181–8.
58. Marastoni S, Ligresti G, Lorenzon E, Colombatti A, Mongiat M. Extracellular matrix: a matter of life and death. Connect Tissue Res. 2008;49:203–6.
59. Plantefaber LC, Hynes RO. Changes in integrin receptors on oncogenically transformed cells. Cell. 1989;56:281–90.
60. Nagaprashantha L, Vartak N, Awasthi S, Awasthi S, Singhal SS. Novel anti-cancer compounds for developing combinatorial therapies to target anoikis-resistant tumors. Pharm Res. 2012;29:621–36.
61. Taddei ML, Giannoni E, Fiaschi T, Chiarugi P. Anoikis: an emerging hallmark in health and diseases. J Pathol. 2012;226:380–93.
62. Zhong X, Rescorla FJ. Cell surface adhesion molecules and adhesion-initiated signaling: understanding of anoikis resistance mechanisms and therapeutic opportunities. Cell Signal. 2012;24:393–401.
63. Wary KK, Mainiero F, Isakoff SJ, Marcantonio EE, Giancotti FG. The adaptor protein Shc couples a class of integrins to the control of cell cycle progression. Cell. 1996;87:733–43.
64. Giancotti FG. Integrin signaling: specificity and control of cell survival and cell cycle progression. Curr Opin Cell Biol. 1997;9:691–700.
65. Parsons JT. Focal adhesion kinase: the first ten years. J Cell Sci. 2003;116:1409–16.
66. Provenzano PP, Keely PJ. The role of focal adhesion kinase in tumor initiation and progression. Cell Adh Migr. 2009;3:347–50.
67. Guarino M. Src signaling in cancer invasion. J Cell Physiol. 2010;223:14–26.
68. Osaki M, Oshimura M, Ito H. PI3K-Akt pathway: its functions and alterations in human cancer. Apoptosis. 2004;9:667–76.
69. Kim EK, Choi EJ. Pathological roles of MAPK signaling pathways in human diseases. Biochim Biophys Acta. 2010;1802:396–405.
70. Krystal GW, Sulanke G, Litz J. Inhibition of phosphatidylinositol 3-kinase-Akt signaling blocks growth, promotes apoptosis, and enhances sensitivity of small cell lung cancer cells to chemotherapy. Mol Cancer Ther. 2002;1:913–22.
71. Downward J. PI 3-kinase, Akt and cell survival. Semin Cell Dev Biol. 2004;15:177–82.
72. Hodkinson PS, Elliott T, Wong WS, Rintoul RC, Mackinnon AC, Haslett C, Sethi T. ECM overrides DNA damage-induced cell cycle arrest and apoptosis in small-cell lung cancer cells through $\beta 1$ integrin-dependent activation of PI3-kinase. Cell Death Differ. 2006;13:1776–88.
73. Matter ML, Ruoslahti E. A signaling pathway from the alpha5beta1 and alpha(v)beta3 integrins that elevates bcl2 transcription. J Biol Chem. 2001;276:27757–63.
74. Danial NN. BAD: undertaker by night, candyman by day. Oncogene. 2008;27(Suppl 1):S53–70.
75. Strieter RM, Belperio JA, Phillips RJ, Keane MP. Chemokines: angiogenesis and metastases in lung cancer. Novartis Found Symp. 2004;256:173–84.
76. Han S, Khuri FR, Roman J. Fibronectin stimulates non-small cell lung carcinoma cell growth through activation of Akt/mammalian target of rapamycin/S6 kinase and

inactivation of LKB1/AMP-activated protein kinase signal pathways. Cancer Res. 2006;66:315–23.

77. Morello V, Cabodi S, Sigismund S, Camacho-Leal MP, Repetto D, Volante M, Papotti M, Turco E, Defilippi P. $\beta 1$ integrin controls EGFR signaling and tumorigenic properties of lung cancer cells. Oncogene. 2011;30:4087–96.
78. Tuomi S, Mai A, Nevo J, Laine JO, Vilkki V, Ohman TJ, Gahmberg CG, Parker PJ, Ivaska J. PKCepsilon regulation of an alpha5 integrin-ZO-1 complex controls lamellae formation in migrating cancer cells. Sci Signal. 2009;2:ra32.
79. Rathinam R, Alahari SK. Important role of integrins in the cancer biology. Cancer Metastasis Rev. 2010;29:223–37.
80. Guo W, Giancotti FG. Integrin signalling during tumour progression. Nat Rev Mol Cell Biol. 2004;5:816–26.
81. Hood JD, Cheresh DA. Role of integrins in cell invasion and migration. Nature Rev Cancer. 2002;2:91–100.
82. Vokes EE, Gordon GS, Mauer AM, Rudin CM, Krauss SA, Szeto L, Golomb HM, Hoffman PC. A phase I study of STEALTH cisplatin (SPI-77) and vinorelbine in patients with advanced non small-cell lung cancer. Clin Lung Cancer. 2000;2:128–32.
83. Ganju RK, Brubaker SA, Meyer J, Dutt P, Yang Y, Qin S, Newman W, Groopman JE. The alpha-chemokine, stromal cell-derived factor-1alpha, binds to the transmembrane G-protein-coupled CXCR-4 receptor and activates multiple signal transduction pathways. J Biol Chem. 1998;273:23169–75.
84. Vicente-Manzanares M, Montoya MC, Mellado M, Frade JM, del Pozo MA, Nieto M, de Landazuri MO, Martínez-A C, Sánchez-Madrid F. The chemokine SDF-1alpha triggers a chemotactic response and induces cell polarization in human B lymphocytes. Eur J Immunol. 1998;28:2197–207.
85. Maulik G, Madhiwala P, Brooks S, Ma PC, Kijima T, Tibaldi EV, Schaefer E, Parmar K, Salgia R. Activated C-Met signals through PI3 K with dramatic effects on cytoskeletal functions in small cell lung cancer. J Cell Mol Med. 2002;6:539–53.
86. Burger M, Glodek A, Hartmann T, Schmitt-Gräff A, Silberstein LE, Fujii N, Kipps TJ, Burger JA. Functional expression of CXCR4 (CD184) on small-cell lung cancer cells mediates migration, integrin activation, and adhesion to stromal cells. Oncogene. 2003;22:8093–101.
87. Chapman HA, Wei Y, Simon DI, Waltz DA. Role of urokinase receptor and caveolin in regulation of integrin signaling. Thromb Haemost. 1999;82:291–7.
88. Takenaka K, Shibuya M, Takeda Y, Hibino S, Gemma A, Ono Y, Kudoh S. Altered expression and function of beta1 integrins in a highly metastatic human lung adenocarcinoma cell line. Int J Oncol. 2000;17:1187–94.
89. Citi S. Protein kinase inhibitors prevent junction dissociation induced by low extracellular calcium in MDCK epithelial cells. J Cell Biol. 1992;117:169–78.
90. Liu J, Costello PC, Pham NA, Pintillie M, Jabali M, Sanghera J, Tsao MS, Johnston MR. Integrin-linked kinase inhibitor KP-392 demonstrates clinical benefits in an orthotopic human non-small cell lung cancer model. J Thorac Oncol. 2006;1:771–9.
91. Shieh JM, Wei TT, Tang YA, Huang SM, Wen WL, Chen MY, Cheng HC, Salunke SB, Chen CS, Lin P, Chen CT, Wang YC. Mitochondrial apoptosis and FAK signaling disruption by a novel histone deacetylase inhibitor, HTPB, in antitumor and antimetastatic mouse models. PLoS ONE. 2012;7:e30240.
92. Ju L, Zhou C, Li W, Yan L. Integrin beta1 over-expression associates with resistance to tyrosine kinase inhibitor gefitinib in non-small cell lung cancer. J Cell Biochem. 2010;111:1565–74.
93. Meng R, Tang HY, Westfall J, London D, Cao JH, Mousa SA, Luidens M, Hercbergs A, Davis FB, Davis PJ, Lin HY. Crosstalk between integrin $\alpha v \beta 3$ and estrogen receptor-$\alpha$ is involved in thyroid hormone-induced proliferation in human lung carcinoma cells. PLoS ONE. 2011;6:e27547.

94. Lonsdorf AS, Krämer BF, Fahrleitner M, Schönberger T, Gnerlich S, Ring S, Gehring S, Schneider SW, Kruhlak MJ, Meuth SG, Nieswandt B, Gawaz M, Enk AH, Langer HF. Engagement of $\alpha$IIb$\beta$3 (GPIIb/IIIa) with $\alpha v\beta 3$ integrin mediates interaction of melanoma cells with platelets: a connection to hematogenous metastasis. J Biol Chem. 2012;287:2168–78.
95. Nasulewicz-Goldeman A, Uszczyńska B, Szczaurska-Nowak K, Wietrzyk J. siRNA-mediated silencing of integrin $\beta 3$ expression inhibits the metastatic potential of B16 melanoma cells. Oncol Rep. 2012;28:1567–73.
96. Kim S, Harris M, Varner JA. Regulation of integrin alpha vbeta3-mediated endothelial cell migration and angiogenesis by integrin alpha5beta1 and protein kinase A. J Biol Chem. 2000;275:33920–8.
97. Streuli CH, Akhtar N. Signal co-operation between integrins and other receptor systems. Biochem J. 2009;418:491–506.
98. Lassen U, Osterlind K, Hansen M, Dombernowsky P, Bergman B, Hansen HH. Long-term survival in small-cell lung cancer: post treatment characteristics in patients surviving 5 to 18+ years–an analysis of 1,714 consecutive patients. J Clin Oncol. 1995;13:1215–20.
99. Green SK, Frankel A, Kerbel RS. Adhesion-dependent multicellular drug resistance. Anticancer Drug Des. 1999;14:153–68.
100. Kijima T, Maulik G, Ma PC, Madhiwala P, Schaefer E, Salgia R. Fibronectin enhances viability and alters cytoskeletal functions (with effects on the phosphatidylinositol-3-kinase pathway) in small cell lung cancer. J Cell Mol Med. 2003;7:157–64.
101. Kraus AC, Ferber I, Bachmann SO, Specht H, Wimmel A, Gross MW, Schlegel J, Suske G, Schuermann M. In vitro chemo and radio-resistance in small cell lung cancer correlates with cell adhesion and constitutive activation of AKT and MAP kinase pathways. Oncogene. 2002;21:8683–95.
102. Tsurutani J, West KA, Sayyah J, Gills JJ, Dennis PA. Inhibition of the phosphatidylinositol 3-kinase/Akt/mammalian target of rapamycin pathway but not the MEK/ERK pathway attenuates laminin-mediated small cell lung cancer cellular survival and resistance to imatinib mesylate or chemotherapy. J Cancer Res. 2005;65:8423–32.
103. Ritzenthaler JD, Han S, Roman J. Stimulation of lung carcinoma cell growth by fibronectin-integrin signalling. Mol BioSyst. 2008;4:1160–9.
104. Kohmo S, Kijima T, Otani Y, Mori M, Minami T, Takahashi R, Nagatomo I, Takeda Y, Kida H, Goya S, Yoshida M, Kumagai T, Tachibana I, Yokota S, Kawase I. Cell surface tetraspanin CD9 mediates chemoresistance in small cell lung cancer. Cancer Res. 2010;70:8025–35.
105. Palecek SP, Loftus JC, Ginsburg MH, Lauffenburger DA, Horwitz AF. Integrin-ligand binding properties govern cell migration speed through cell-substratum adhesiveness. Nature. 1997;385:537–40.
106. Ignotz RA, Massaque J. Cell adhesion protein receptors as targets for transforming growth factor-$\beta$ action. Cell. 1987;51:189–97.
107. Munger JS, Harpel JG, Giancotti FG, Rifkin DB. Interactions between growth factors and integrins: latent forms of transforming growth factor $\beta$ are ligands for $\alpha$v$\beta$1. Mol Biol Cell. 1998;9:2627–38.
108. Zheng Y, Ritzenthaler JD, Roman J, Han S. Nicotine stimulates human lung cancer cell growth by inducing fibronectin expression. Am J Respir Cell Mol Biol. 2007;37:681–90.
109. Avraamides CJ, Garmy-Susini B, Varner JA. Integrins in angiogenesis and lymphangiogenesis. Nat Rev Cancer. 2008;8:604–17.
110. Silva R, D'Amico G, Hodivala-Dilke KM, Reynolds LE. Integrins: the keys to unlocking angiogenesis. Arteriosclerosis Thromb Vasc Biol. 2008;28:1703–13.
111. Brook CS, Lee SM. Anti-angiogenic strategies and vascular targeting in the treatment of lung cancer. Eur Respir J. 2002;19:557–70.
112. Desgrosellier JS, Cheresh DA. Integrins in cancers. Biological implications and therapeutic opportunities. Nat Rev Cancer. 2010;10:9–22.

113. Chen K, Chen X. Integrin targeted delivery of chemotherapeutics. Theranostics. 2011;1:189–200.
114. Reynolds LE, Wyder L, Lively JC, Taverna D, Robinson SD, Huang X, Sheppard D, Hynes RO, Hodivala-Dilke KM. Enhanced pathological angiogenesis in mice lacking $\beta 3$ integrin or $\beta 3$ and $\beta 5$ integrins. Nat Med. 2002;8:27–34.
115. Fahham D, Weiss ID, Abraham M, Beider K, Hanna W, Shlomai Z, Eizenberg O, Zamir G, Izhar U, Shapira OM, Peled A, Wald O. In vitro and in vivo therapeutic efficacy of CXCR4 antagonist BKT140 against human non-small cell lung cancer. J Thorac Cardiovasc Surg. 2012;144:1167–75.
116. Han S, Rivera HN, Roman J. Peroxisome proliferator-activated receptor-gamma ligands inhibit alpha5 integrin gene transcription in non-small cell lung carcinoma cells. Am J Respir Cell Mol Biol. 2005;32:350–9.
117. Tsai YM, Yang CJ, Hsu YL, Wu LY, Tsai YC, Hung JY, Lien CT, Huang MS, Kuo PL. Glabridin inhibits migration, invasion, and angiogenesis of human non-small cell lung cancer A549 cells by inhibiting the FAK/rho signaling pathway. Integr Cancer Ther. 2011;10:341–9.
118. Aixinjueluo W, Furukawa K, Zhang Q, Hamamura K, Tokuda N, Yoshida S, Ueda R, Furukawa K. Mechanisms for the apoptosis of small cell lung cancer cells induced by anti-GD2 monoclonal antibodies. J Biol Chem. 2005;280:298328–36.
119. Li N, Zhang JP, Guo S, Min J, Liu LL, Su HC, Feng YM, Zhang HL. Down-regulation of Î$^{2}$3-integrin inhibits bone metastasis of small cell lung cancer. Mol Biol Rep. 2012;39:3029–35.
120. Kimura H, Tome Y, Momiyama M, Hayashi K, Tsuchiya H, Bouvet M, Hoffman RM. Imaging the inhibition by anti-Î$^{2}$1 integrin antibody of lung seeding of single osteosarcoma cells in live mice. Int J Cancer. 2012;131:2027–33.

# Aldo-Keto Reductases as New Therapeutic Targets for Colon Cancer Chemoresistance

**Toshiyuki Matsunaga, Ossama El-Kabbani and Akira Hara**

**Abstract** The aldo-keto reductase (AKR) superfamily comprises NAD(P)(H)-dependent enzymes that catalyze the oxidoreduction of a variety of substrates, including prostaglandins, steroids, toxic aldehydes and drugs. Among members of this superfamily, AKR1B10, AKR1C1, AKR1C2 and/or AKR1C3 are overexpressed in several types of cancers. Out of the four AKRs, AKR1B10, AKR1C1 and AKR1C3 are also significantly up-regulated with acquisition of resistance to several anticancer drugs in colon cancer, although the up-regulated enzyme species differ among themselves depending on the drug types. Studies with cell-based experiments have proposed multiple mechanisms leading to the drug resistance through regulation of cell proliferation and detoxification of lipid-derived toxicants by the up-regulated enzymes. Thus, the three enzymes have been recognized not only as potential diagnostic and/or prognostic markers, but also as potential therapeutic targets for the prevention and treatment of the colon cancer chemoresistance. Recently, potent and selective inhibitors of AKR1B10, AKR1C1 and AKR1C3 have been reported, and experimentally used for reversal of the colon cancer chemoresistance. In this chapter, we describe the current literature focusing mainly on the expression profiles of the three AKRs in chemoresistance of colon cancer cells and availability of the inhibitors for overcoming the anticancer drug resistance.

**Keywords** Aldo-keto reductase · AKR1B10 · AKR1C1 · AKR1C3 · Colon cancers · Chemotherapy · Chemoresistance · Proliferation

---

T. Matsunaga · A. Hara (✉)
Laboratory of Biochemistry, Gifu Pharmaceutical University, 1-25-4 Daigaku-nishi, Gifu 501-1196, Japan
e-mail: hara@gifu-pu.ac.jp

O. El-Kabbani
Monash Institute of Pharmaceutical Sciences, Monash University, Parkville, VIC 3052, Australia

B. Bonavida (ed.), *Molecular Mechanisms of Tumor Cell Resistance to Chemotherapy*, Resistance to Targeted Anti-Cancer Therapeutics 1, DOI: 10.1007/978-1-4614-7070-0_6,

**Abbreviations**

| | |
|---|---|
| AKR | Aldo-keto reductase |
| ARE | Antioxidant response element |
| BPSA | 3-Bromo-5-phenylsalicylic acid |
| CDDP | Cisplatin |
| CPSA | 3-Chloro-5-phenyl salicylic acid |
| DOX | Doxorubicin |
| EGF | Epidermal growth factor |
| FAL | Farnesal |
| FOH | Farnesol |
| GGAL | Geranylgeranial |
| GGOH | Geranylgeraniol |
| HAHE | 3-(4-Hydroxy-2-methoxyphenyl)acrylic acid 3-(3-hydroxyphenyl) propyl ester |
| HNE | 4-Hydroxy-2-nonenal |
| HSD | Hydroxysteroid dehydrogenase |
| HT29/CDDP | HT29 phenotype resistant to cisplatin |
| Keap1 | Kelch-like ECH-associated protein 1 |
| LOHP | Oxaliplatin |
| MAPK | Mitogen-activated protein kinase |
| MMC | Mitomycin C |
| NSAID | Non-steroidal anti-inflammatory agent |
| NF$\kappa$B | Nuclear factor-$\kappa$B |
| Nrf2 | Nuclear factor-erythroid 2-related factor 2 |
| ONE | 4-Oxo-2-nonenal |
| PHPC | (*Z*)-2-(4-Methoxyphenylimino)-7-hydroxy-*N*-(pyridin-2-yl)-2*H*-chromene-3-carboxamide |
| PG | Prostaglandin |
| PPAR | Peroxisome proliferator-activated receptor |
| ROS | Reactive oxygen species |
| VEGF | Vascular endothelial growth factor |

## Introduction

Colon cancer is one of the most common cause of death from gastrointestinal malignancy and the first leading cause of cancer-related mortality in Japanese women [1]. A first-line of treatment for metastatic colon cancer is chemotherapy, in which chemotherapeutic drugs, such as 5-fluorouracil, leucovorin, oxaliplatin (LOHP), irinotecan, tegafur, doxifluridine, carmofur, capecitabine, doxorubicin (DOX) and mitomycin C (MMC), are clinically utilized for retracting the tumor prior to surgery or averting the recurrence after the surgery [2]. Since the cancer cells easily acquire tolerance toward the drugs during the long-term administration, most patients with metastatic colon cancer receive the combined chemotherapy and irradiation therapy [3] despite of the appearance of severe side effects including neutropenia and alopecia. Several different mechanisms underlying the resistance induction in colon cancer cells were previously reported. For example, one major mechanism underlying the chemotherapeutic therapy-induced resistance is the overexpression of P-glycoprotein [4], which secretes the chemotherapeutic agents from the cells before exerting the genotoxic efficacy. Other mechanisms of the resistance appear to be due to the reinforcements of the proliferative and tumorigenic potentials through hypersecretion of growth factors, e.g. epidermal growth factor (EGF) [5, 6] and vascular endothelial growth factor (VEGF) [6], and activation of the proteasome [7, 8], a multi-catalytic proteolytic apparatus for degrading damaged proteins, and an up-regulation of cell surface markers, CD133 [9, 10] and CD44 [6, 10], some of which could jointly lead to lowering the sensitivity to chemotherapy-based therapy. Some effective inhibitors for the above target molecules were developed [11–15], and have been in part approved for clinical use as the adjuvant drugs against the inducible chemoresistance of colon cancer.

The aldo-keto reductase (AKR) superfamily is a rapidly growing group of NAD(P)(H)-dependent oxidoreductases that metabolize carbohydrates, steroids, prostaglandins (PGs), and other endogenous aldehydes and ketones, as well as xenobiotic compounds [16–18]. Members of this superfamily are classified into 15 families and each individual family is further subdivided into several subfamilies based on their amino acid sequence similarities and substrate specificities. Structural and enzymatic features of all AKR members are collected and listed at an AKR superfamily homepage (http://www.med.upenn.edu/akr/). In humans, 15 genes for AKRs have been identified. Ten of the human AKRs belong to the AKR1 family, which include aldehyde reductase (AKR1A1), aldose reductase (AKR1B1) and its related enzymes (AKR1B10 and AKR1B15), and hydroxysteroid dehydrogenases (HSDs: AKR1C1, AKR1C2, AKR1C3 and AKR1C4). Among them, four AKR members (AKR1B10, AKR1C1, AKR1C2 and AKR1C3) have been reported to be up-regulated in several types of cancer and cultured cancer cells [19–21]. Therefore, the four AKRs act as potential markers for the corresponding overexpressing cancers. In addition, AKR1B10, AKR1C1 and AKR1C3 have been recently shown to be involved not only in facilitating proliferation of some cancer cells [16, 19, 21, 22], but also in developing chemoresistance of the cells [21, 23–25]. Thus, these AKRs have also received considerable attention as potential therapeutic targets for chemoresistance of colon cancer cells. Here, we provide an overview of the pathophysiological roles of the AKRs in the development of colon cancer cell resistance to anticancer drugs and the usefulness of their inhibitors in overcoming the chemoresistance.

## Possible Factors Involved in Colon Cancer Chemoresistance

The tumor microenvironment has been formerly accepted as a major mechanism that leads to the induction of drug resistance [4, 26, 27]. That is because the environment surrounding the tumor limits accessibility of antitumor drugs to the tumor cells and reduces the oxygen radicals generated during the treatment by the drugs [4, 26]. Recent reports provided substantial evidence indicating that the cancer chemoresistance is heterogenously developed by endogenous defensive responses to lethal stress, including sources derived from the microenvironment, and the possible mechanisms encompassing a variety of factors including the P-glycoprotein [4], EGF [5, 6], VEGF [6], CD133 [9, 10], CD44 [6, 10] and the proteasome [7, 8]. Here we outline the factors that are believed to be responsible for inducible chemoresistance of colon cancers, except for AKRs.

### *P-Glycoprotein*

The P-glycoprotein, encoded by the gene for human multidrug-resistance protein 1 [28], is a transporter ATP-binding cassette family protein, which is expressed in the apical membrane of epithelial cells in the small and large intestines and functions as a part of the gastrointestinal barrier for protecting against xenobiotics, bacterial toxicants and drugs [29, 30]. It also plays central roles in the absorption, distribution and excretion of chemotherapeutic drugs and actively extrudes xenobiotic substances including the cytotoxic drugs due to the overexpression in colon cancer cells, displaying drug resistance [4]. The aberrant cellular transportability by the overexpression is considered to be a major causal factor for mechanisms underlying drug resistance of colon cancer cells. Recently, glucosylceramide synthase, that converts ceramide into glucosylceramide and then reduces the ceramide-induced apoptosis, has been reported to be coincidently overexpressed with P-glycoprotein in drug-resistant cells [31, 32]. Based on these studies, the glucosylceramide synthase appears to modulate expression of the transporter protein through c-Src and $\beta$-catenin signaling pathways and confers the cell resistance to DOX, daunorubicin, and tumor necrosis factor-$\alpha$ [32]. Administration of zosuquidar, a potent and specific inhibitor of P-glycoprotein, reverses P-glycoprotein-mediated resistance in acute myeloid leukemia and breast cancer [11, 12], but little is known about the effect of the inhibitor on the colon cancer resistance.

## EGF Receptor and VEGF

The EGF receptor is expressed in a large proportion (30–85 %) of patients with colorectal cancer, and its signaling pathway is always activated in the cancer cells [5, 33]. Experiments show that persistent activation of the signaling through the EGF receptor promotes cell growth [34] and metastatic transformation [35], and inhibits apoptosis of the cells [36]. Upon EGF receptor stimulation by the binding of its ligand, including EGF and transforming growth factor, the ligand-receptor complex triggers a chain of the down-stream signaling pathways such as the mitogen-activated protein kinase (MAPK) and phosphoinositide 3-kinase [37]. Therefore, potent blocking agents of the EGF receptor are expected to suppress the colon cancer growth and resistance to certain drugs. Cetuximab is a monoclonal antibody that targets the extracellular domain of the EGF receptor and inhibits the ligand binding to the receptor [13]. Based on a randomized phase II clinical trial, the antibody was approved for use in combination with irinotecan or as mono-therapy in EGF receptor-positive irinotecan-refractory colorectal cancer [38]. Thus, the clinical use is limited for treatment of patients with overexpression of the EGF receptor.

Angiogenesis is an essential event for progression and metastasis of tumors [39]. The anti-angiogenic therapy interferes with the supply of oxygen and nutrients that are required for the tumor growth and is, therefore, utilized for the treatment of a variety of cancers as well as colon cancer [39]. A major target of the therapy is VEGF, which is known to promote several steps in angiogenesis including proliferation, migration and tube formation of endothelial cells. Bevacizumab, a monoclonal antibody that binds to VEGF, is approved for use in the treatment of metastatic colorectal cancer [14] and its application is expanded to non-small-cell lung carcinoma and breast cancer because of the clinical availability [40].

## CD133 and CD44

Significant increase in a cancer stem cell marker CD133-positive subpopulations of colorectal cancer was found in human colorectal cancer HT29 cells resistant to high doses of 5-fluorouracil [41]. A recent study also revealed that up-regulation of CD133 is associated with prognosis in patients bearing the drug-resistant colorectal cancer [42], suggesting that CD133 is a potential prognostic marker for the resistant cells. The drug tolerance seemed to be sensitized to 5-fluorouracil- or LOHP-mediated damage by using an anti-interleukin-4 neutralizing antibody [43]. In addition to the marker CD133, CD44 is recently proposed as the new potential marker, based on the data showing the positive relationship of its expression in colon SW620 cells with high potentials concerning colony formation, proliferation and chemoresistance development [44, 45].

### *Proteasome*

The proteasome is a large multi-subunit assembly that degrades unwanted proteins, which include oxidized and damaged proteins formed during exposure of cells to chemotherapeutic drugs, as well as proteins involved in the regulation of the cell cycle and transcription [46–48]. Compared to the normal cells, the proteolytic activity of the proteasome in colon cancer cells is considerably higher, but is lower than that in chemoresistant phenotypes of colon cancer cells [25]. The proteasome is considered an important potential therapeutic target for chemoresistance. One potential proteasome inhibitor that reverses the chemoresistance is bortezomib, and its availability is supported by the clinical trials [49] as well as data from the cell-based experiments [50]. The bortezomib-mediated sensitization of the cancer cells to drugs has been shown to be mediated by multiple regulations of cell cycle progression including the p21 cyclin-dependent kinase, checkpoint kinase and cyclins [51], and by controlling turnover of transcriptional factors including nuclear factor-$\kappa$B (NF$\kappa$B) and hypoxia-inducible factor-1 [15, 49]. Currently, bortezomib is approved for treating only refractory and recurrent multiple myeloma in Japan and has been tested against other cancers such as lymphoma [52], lung cancer [53] and colorectal cancer [54].

## Up-Regulation of AKRs Related to Development of Cancer and Chemoresistance

### *High Expression of AKRs in Tumors and Carcinomas Cells*

AKR1B10 is a 36-kDa cytosolic NADPH-dependent reductase that is originally identified as aldose reductase-like 1 or small intestine aldose reductase [55]. The enzyme is expressed in human gastrointestinal tract and adrenal gland, but its expression levels in other tissues including lung and liver are low [56, 57]. Clinical investigation exhibited that high expression of AKR1B10 is found in airway of healthy smokers compared to those of healthy non-smokers [57–62], suggesting that the enzyme is up-regulated due to components of tobacco smoke. The significant overexpression is also reported in the malignant tumors such as squamous cell lung carcinomas [63–66], hepatocellular carcinomas [67, 68], uterine carcinomas [69], cholangiocarcinomas [67], and gastric [70] and esophageal tumors [71], albeit at different expressions depending on their stage and sites examined. Thus, AKR1B10 is the tumor marker, and is suggested that its overexpression is implicated in carcinogenesis of the above types of cancers.

Four human members (AKR1C1, AKR1C2, AKR1C3 and AKR1C4) in the AKR1C subfamily are cytosolic HSDs with different positional and stereochemical preferences for steroid substrates [16]. AKR1C1 (20$\alpha$-HSD), AKR1C2 (type 3 3$\alpha$-HSD), AKR1C3 (type 2 3$\alpha$-HSD and type 5 17$\beta$-HSD) and AKR1C4 (type 1 3$\alpha$-HSD) share >84 % amino acid sequence identity, and collectively called AKR1C isoforms. With the exception of liver-specific AKR1C4, the other three isoforms are expressed in many human tissues, in which the extents of their expressions are different from each other [16–18]. The three isoforms are overexpressed in cancer tissues, but their expression levels are different depending on types of cancer. For example, AKR1C1 is overexpressed in lung [63, 72, 73], uterine cervix [74] and colon cancers [23, 48]. Expression of AKR1C2 is elevated in lung [72], bladder [75] and esophageal cancers [71]. In contrast, AKR1C3 is up-regulated in leukemia [76], lung cancers [77], squamous cell carcinomas of head and neck [78], and carcinomas of hormone-sensitive tissues such as the prostate [79] and the mammary gland [80]. The differences in the expression levels of AKR1C1, AKR1C2 and AKR1C3 may reflect their various roles in carcinogenesis and/or cancer cell proliferation [19–22].

## *Alteration in the AKR Expression in Colon Cancer Chemoresistance*

Resistance to chemotherapy is a principal problem in treating most common solid tumors. Like the solid tumors, colon cancers are frequently resistant to the chemotherapy and irradiation. One of the major problems is that the resistance leads to malignant transformation and metastasis of tumors [2, 3]. Despite progressive improvement of the therapeutic approach for colorectal cancer, a high proportion of patients carrying the tumors eventually succumb to the metastatic transformation resulting from the resistance. Therefore, it is of great importance to discover new cardinal factors that are involved in the development of colon cancer chemoresistance, besides the above factors mentioned in section Possible Factors Involved in Colon Cancer Chemoresistance.

There is growing evidence that some members of the AKR superfamily are induced during the development of chemoresistance in a variety of cancers. The alteration in expression of the AKR members (AKR1B10, AKR1C1, AKR1C2 and AKR1C3) in various cancer cells resistant to anticancer drugs has been reviewed [21]. In colon cancers, there are four papers that describe the changes in expression of the AKRs caused by chemoresistance [21, 23–25], which are summarized in Table 1. Although other AKR species were not examined, AKR1C1 is up-regulated in methotrexate-resistant HT29 cells, and is suggested to abolish the cell cycle arrest at S-phase and apoptotic response provoked by the drug [23]. In HT29 phenotypes resistant to MMC [24] and 2 μM LOHP (low-grade) [21], the AKRs, other than AKR1C2, are overexpressed, and the level is the highest in AKR1B10, followed by AKR1C1 and AKR1C3. However, the expressions of the four enzymes are decreased in the HT29 phenotype resistant to 20 μM LOHP (high-grade) [21]. Table 1 includes our unpublished data on the expression levels of AKR1B10, AKR1C1, AKR1C2 and AKR1C3 in the HT29 phenotypes resistant to

**Table 1** Expression of AKRs in chemoresistant colon cancer cells

| Cell | Drug[a] | Concentration[b] (μM) | Expression of AKR[c] | | | | Ref. |
|---|---|---|---|---|---|---|---|
| | | | 1B10 | 1C1 | 1C2 | 1C3 | |
| HT29 | MTX | 10 | nd | ↑ | nd | nd | [23] |
| | MMC | 0.5 | ↑↑ | ↑ | → | ↑ | [24] |
| | LOHP | 2 (low-grade) | ↑↑ | ↑ | → | ↑ | [21] |
| | | 20 (high-grade) | ↓ | ↓ | ↓ | ↓ | [21] |
| | DOX | 0.5 | ↑↑ | ↑ | → | ↑ | Unpublished |
| | CDDP | 10 | ↓ | ↓ | ↓ | ↓ | Unpublished |
| HCT15 | CDDP | 10 | ne | ↑↑ | → | ↑↑ | [25] |
| RKO | CDDP | 5 | ne | ↑↑ | → | ↑↑ | Unpublished |

[a] *MTX* methotrexate; *MMC* mitomycin C; *LOHP* oxaliplatin; *DOX* doxorubicin; and *CDDP* cisplatin

[b] Drug concentration that is used to establish the resistant cells

[c] Change compared to the parental cells: ↑↑ Highly up-regulation; ↑ up-regulation; → no change; ↓ down-regulation. *nd* not determined. *ne* no expression is observed

DOX and CDDP (HT29/CDDP). Gain of DOX resistance increases the expression of AKR1B10, AKR1C1 and AKR1C3, and the expression patterns of the four AKRs are similar to those in the MMC- and low-grade LOHP-resistant cells. In contrast, the CDDP resistance resulted in significant decreases in the expression levels of the four AKRs, which are similar to those in the high-grade LOHP-resistant cells. The diversity in the AKR expression may be due to different toxic mechanisms toward HT29 cells among the two platin-based drugs (CDDP and LOHP), DOX and MMC. Presumably, the difference is also explained by the results that the HT29/CDDP phenotype had high cross-resistance to toxicity induced by LOHP, but the cross-resistance to DOX or MMC was low (unpublished data). On the other hand, our recent study revealed that significant overexpressions of AKR1C1 and AKR1C3 result from acquisition of CDDP resistance of other colon cancer HCT15 and RKO cells, in which the mRNA for AKR1B10 is hardly expressed [25]. Thus, alteration in the AKR expression due to the chemoresistance is affected by both the enzyme distribution and cell types employed. Considering that the four AKR members (AKR1C1, AKR1C2, AKR1C3 and AKR1B10) are up-regulated with the development of chemoresistance of cancer cells derived from other tissues [19–21], these enzymes are suggested to be closely involved in cancer cell chemoresistance, except in some cases using the two platin drugs.

Nuclear factor-erythroid 2-related factor 2 (Nrf2) is a key transcriptional regulator of many enzymes including AKR1B10, AKR1C1, AKR1C2 and AKR1C3 (Fig. 1) [61, 81–83]. Under normal conditions, Kelch-like ECH-associated protein 1 (Keap1) recruits Nrf2 into the Cul3 containing ubiquitin ligase E3 complex for ubiquitin conjugation and subsequent proteasomal degradation. When cells receive various stimuli including reactive oxygen species (ROS) [84] and electrophiles [85], Nrf2 is dissociated from Keap1 for translocation into the nuclei, and binds with Maf to genes of antioxidant response element (ARE) to transcriptionally induce proteins including the above AKRs. Ciaccio et al. showed that the

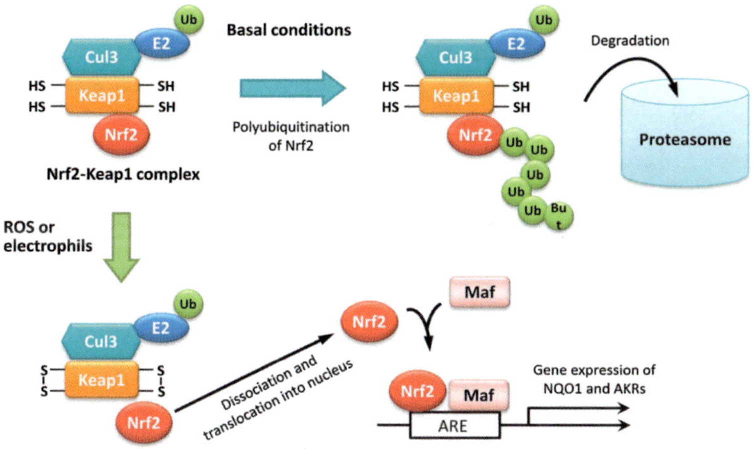


**Fig. 1** Induction of AKRs by Nrf2-dependent mechanism. Under basal conditions, Nrf2 complexed with Keap1-Cul3-ubiquitin (Ub) ligase E3 is subjected to polyubiquitination, and the modified proteins are degraded in the Ub-dependent proteasome system. Upon stimulation with oxidants such as ROS and electrophiles, Nrf2 is dissociated from Keap1 through its structural modification by formation of disulfide bonds. Nrf2 is translocated to the nucleus, forms a heterodimer with its obligatory partner Maf, and ultimately induces the expression of ARE-dependent proteins such as NADPH: quinone oxidoreductase (NQO) 1 and AKRs

acquisition of resistance to ethacrynic acid, a potent AKR inducer and Nrf2 activator, up-regulates the expression of AKRs in HT29 cells, and the resistant cells also exhibit cross-resistance to adriamycin and MMC [86, 87]. These findings apparently indicate that the AKR overexpression in cancer showing chemoresistance is conceivably due to the Nrf2 activation. Induction of chemoresistance in cancer cells frequently requires some somatic mutations [88, 89], which disrupt the interaction between Nrf2 and Keap1. Therefore, constitutive up-regulation of these AKRs may be triggered by an abnormality of the Nrf2-Keap1 system that occurred during acquisition of chemoresistance. The mechanistic base for hyperactivation of Nrf2 in the chemotherapy-resistant cancer cells is also explained by down-regulation of Cul3 ubiquitin E3 ligase [90]. Thus, the constitutive activation of Nrf2 is considered as a crucial event in the development of chemoresistance.

## Possible Roles of AKRs in Colon Cancer Chemoresistance

### *Cell Proliferation*

Retinoids function as major regulators for proliferation, differentiation and morphogenesis of various types of cells (Fig. 2) [91]. Retinols are oxidized into retinoic acids via retinals, and the resulting retinoic acids (all-*trans*-retinal, 9-*cis*-retinal and

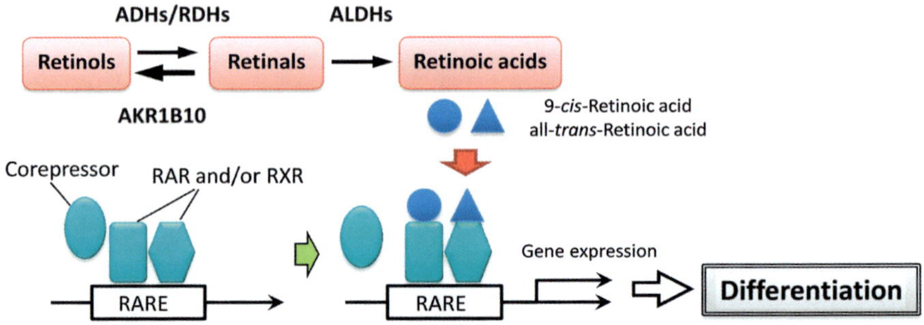


**Fig. 2** Retinoid metabolism and cell proliferation. Retinols are oxidized to retinoic acids via retinals by alcohol dehydrogenases (ADHs) or retinol dehydrogenases (RDHs), and aldehyde dehydrogenase (ALDHs). 9-*cis*- and all-*trans*-Retinoic acids bind to nuclear retinoic acid receptor heterodimers composed of RAR and/or RXR, and promotes the expression of genes involved in cell differentiation through binding of the heterodimer to retinoic acid response element (RARE) and consequently dissociating a corepressor protein. AKR1B10 efficiently reduces retinals back to retinols, leading to cell proliferation through cellular low concentrations of the active retinoic acids

13-*cis*-retinal) promote cell differentiation through bindings to retinoic acid receptors or retinoid X receptors. In cancer cells, retinoic acids are maintained at low concentrations, resulting in the high growth potentials [91]. AKR1B10 efficiently reduces the retinals to their corresponding retinols, as its catalytic efficiency is much higher than those of other enzymes including AKR1B1, AKR1C1 and AKR1C3 [92,93]. AKR1B10 overexpressed in cancer cells is thought to contribute to cancer development by lowering the cellular levels of retinoic acids by reducing retinals back to retinols [92–94].

Three AKR1C isoforms (AKR1C1, AKR1C2 and AKR1C3) are involved in the synthesis of active androgen and estrogen, whose hyperproduction is recently considered to promote proliferation of prostate and breast cancer cells [16, 19]. In addition, the metabolism of PGs by the AKR1C isoforms is likely to be one potential mechanism of cellular proliferation (Fig. 3). AKR1C1 and AKR1C2 convert $PGD_2$ into its 11$\beta$-OH form, 9$\alpha$,11$\beta$-$PGF_2$, but have low $PGF_{2\alpha}$ synthase activity [95]. As AKR1C3 was originally identified as a PGF synthase, it efficiently catalyzes both the formation of $PGF_{2\alpha}$ from $PGH_2$ and the reduction of $PGD_2$ into 9$\alpha$,11$\beta$-$PGF_2$ [96]. The produced PGs ($PGF_{2\alpha}$ and 9$\alpha$,11$\beta$-$PGF_2$) are ligands for the PGF receptor, and their binding to the receptor activates signaling pathways including MAPK and NF$\kappa$B, finally promoting cellular proliferation [97, 98]. At the same time, the up-regulation of the enzymes decreases the formation of 15-deoxy-$PGJ_2$, an endogenous ligand for peroxisome proliferator-activated receptor (PPAR) $\gamma$, resulting in inactivation of the pro-apoptotic signaling pathway [95].

In addition to the PG-dependent signaling, we have recently proposed the contribution of two isoprene derivatives to the proliferative potential of cells (Fig. 4) [21, 99]. Isoprenyl aldehydes, farnesal (FAL) and geranylgeranial (GGAL), are the intermediates in the metabolism of farnesol (FOH) and geranylgeraniol (GGOH),

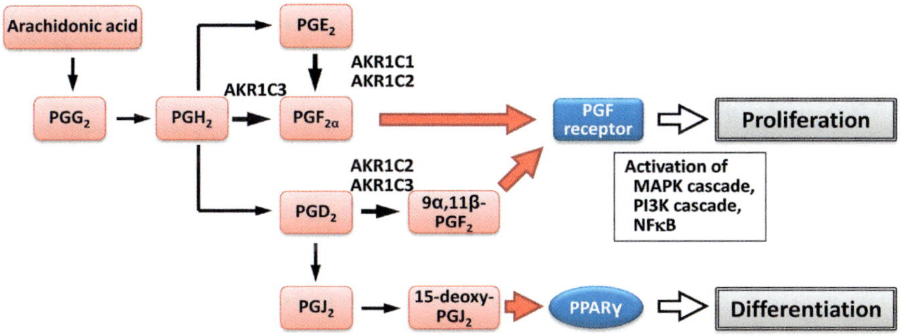


**Fig. 3** PG metabolism by AKRs related to cell proliferation. $PGF_{2\alpha}$ and $9\alpha,11\beta$-$PGF_2$ are formed from $PGH_2$ and $PGD_2$, receptivity, by AKR1C1, AKR1C2 and/or AKR1C3, and are ligands for PGF receptor, leading to cell proliferation through activation of signal cascades including MAPK, PI3K and/or NF$\kappa$B. $PGD_2$ is also metabolized to 15-deoxy-$PGJ_2$, which is assumed to link to the initiation of differentiative process via its binding to PPAR$\gamma$ and/or inactivation of NF$\kappa$B. The reduction of $PGD_2$ by the AKRs decreases the cellular level of 15-deoxy-$PGJ_2$, and the low level of 15-deoxy-$PGJ_2$ also promotes cell proliferation

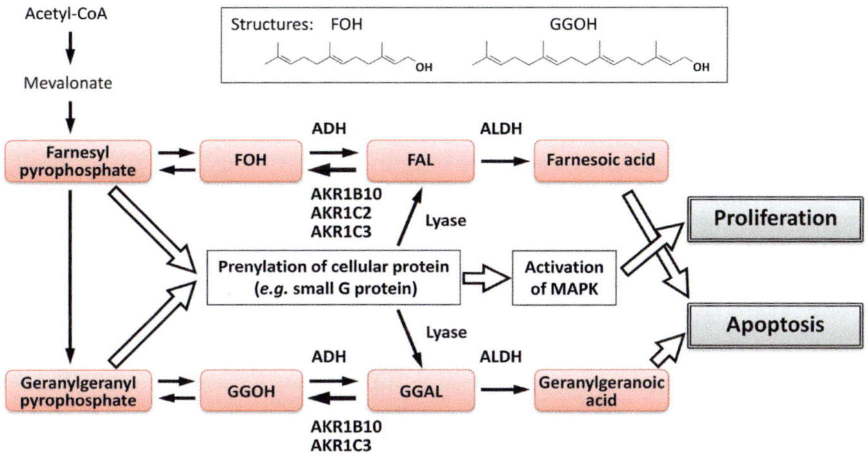


**Fig. 4** Metabolic pathway of isoprenoids. Two prenyl pyrophosphates (farnesyl and geranylgeranyl pyrophosphates), intermediates of the mevalonate pathway, are dephosphorylated to the corresponding alcohols (FOH and GGOH), which are oxidized to the carboxylic acids by alcohol dehydrogenase (ADH) and aldehyde dehydrogenase (ALDH) via the aldehydes (FAL and GAL). FAL and GAL are also released from prenylated proteins by lysosomal prenylcysteine lyase (Lyase). The carboxylic acid metabolites possibly lead to activation of apoptotic signaling pathway. Like the retinoid metabolism, the reduction of FAL and GAL by AKRs decreases the cellular concentrations of the apoptotic carboxylic acid metabolites, and increases those of FOH and GGOH. FOH and GGOH are converted to their pyrophosphates, which are thought to provoke both prenylation of cellular proteins including Ras and activation of MAPK cascade, resulting in cell proliferation

respectively, into their carboxylic acids (farnesoic acid and geranylgeranoic acid) by the catalysis of dehydrogenases. The two isoprenyl alcohols (FOH and GGOH) are phosphorylated to their pyrophosphates, which are rendered to prenylation of cellular proteins including small G-proteins and activation of the downstream MAPK signaling, leading to promotion of cell proliferation [100]. AKR1B10 efficiently reduces FAL and GGAL into FOH and GGOH, respectively [99]. Comparison of the kinetic constants ($k_{cat}/K_m$ values) involved in reduction of the aldehydes among seven AKR members (AKR1A1, AKR1B1, AKR1B10, AKR1C1, AKR1C2, AKR1C3 and AKR1C4) reveals that AKR1B10 and AKR1C3 are predominant enzymes involved in the reduction of FAL and GGAL, respectively [99]. Consistent with the enzymatic activities, enhancement of growth potential by forced overexpressions of AKR1B10 and AKR1C3 was observed in human leukemia U937 [101] and prostate PC3 cells [102], and is in accordance with data in similar experiments using other cell lines [103–105]. In addition, it should be stated that the proliferation rate in colon cancer HT29 cells is positively correlated with the AKR1B10 expression, and enhanced by the addition of low concentrations of FAL [21]. These findings clearly indicate that the AKR1B10-mediated reduction of FAL into FOH promotes the cell proliferation. When compared with the proliferation rates between AKR1B10- and AKR1C3-overexpressing U937 cells prepared under the same transfection conditions, the overexpression of AKR1B10 resulted in a 1.5-fold higher growth rate than that of AKR1C3 (unpublished data). Although the precise mechanisms underlying the difference remain uncertain, elevation of the proliferative capacity by the overexpression of AKR1B10 may be due to promotion of fatty acid and lipid synthesis through acetyl-CoA carboxylase $\alpha$ [106, 107], in addition to the above mentioned metabolisms of retinoids and isoprenoids.

## *Antioxidant Capacity*

Exposure of some chemotherapeutic agents (DOX, MMC and CDDP) to cancer cells generates ROS. The accumulation of ROS into the cells evokes the oxidative damage of biomolecules, such as proteins, lipids and nucleic acids, and consequently forms highly reactive and toxic carbonyl compounds including lipid peroxidation-derived 4-hydroxy-2-nonenal (HNE) and 4-oxo-2-nonenal (ONE). It has been previously proposed that glutathione *S*-transferase [108], AKRs (1B1, 1B10 and 1C1) [107, 109–111] and aldehyde dehydrogenases (ALDH) [112] are the major detoxifying enzymes of the lipid peroxidation-derived aldehydes and their glutathione-conjugated derivatives. Among them, one of the most effective enzymes is AKR1B10, because overexpression of AKR1B10 lowers damage induced by lipid aldehydes including HNE, ONE and acrolein in non-resistant and resistant cancer cells [24, 103]. In addition, knockdown of *AKR1B10* gene by small interference RNAs sensitizes colon cancer HCT-8 cells to acrolein and crotonaldehyde [103]. Thus, the detoxification of lipid peroxidation-derived carbonyl compounds is a role of AKR1B10 in tumor development. In colon cancer showing chemoresistance, up-regulation of AKR1B10 is responsible for inducing MMC resistance in HT29 cells, and the major molecular basis is suggested to be due to the enzyme-mediated detoxification of the lipid aldehydes resulting from the MMC treatment [24]. In HCT15 cells that do not express AKR1B10, AKR1C1 and AKR1C3 are overexpressed and detoxify the aldehydes in the CDDP resistant cells [25]. This study also shows that the AKR-mediated detoxification of the toxic aldehydes and proteolytic activity of the proteasome toward the oxidatively damaged proteins formed during the CDDP treatment collaborate to confer the colon cancer chemoresistance. Since the treatment with proteasome inhibitors bortezomib and MG132 enhances the expression levels of AKR1C1, AKR1C3 and AKR1B10 in four colon cancer cells (SW480, HT29, Caco2 and HRT-18) [47, 48], it is very intriguing to evaluate the relationship between AKR expression and proteasomal activation.

### *Metabolism of Anticancer Drugs*

AKR1B10 and AKR1C3 are suggested to be involved in the metabolic inactivation of anticancer drugs, such as daunorubicin, idarubicin, epirubicin, oracin and DOX, by reducing the carbonyl groups of the drugs into their less active alcohol metabolites [113–119]. Recently, we have also found high NADPH-linked reductase activity towards MMC in the drug-resistant HT29 cells (unpublished data), in which AKR1B10 is significantly overexpressed compared to the parental cells [24]. Indeed, AKR1B10 exhibits the MMC reductase activity, as it efficiently reduces several other *p*-quinones [120]. In addition to the above two mechanisms of AKR1B10-overexpression in the development of MMC resistance, the ability of AKR1B10 to metabolize this drug may be in part responsible for the drug resistance [24].

## Reversal of Colon Cancer Chemoresistance Due to Inhibitors Specific to Individual AKRs

As mentioned above, AKR1B10, AKR1C1 and AKR1C3 are predominant factors for the gain of colon cancer chemoresistance, and their specific inhibitors are therefore expected to be potential adjuvants to suppress the cancer tolerance to anticancer drugs, leading to the persistence of the therapeutic efficacy. Since the enzymes are recognized to be targets for the prevention and treatment of several types of cancers described in the Introduction, various kinds of their inhibitors have been reported [20, 21, 121, 122]. However, there are few studies that show the selectivity of the inhibitors to the target enzyme in comparison with their inhibitory potencies for the structurally similar enzymes with distinct functions. The amino acid sequence identity between AKR1B10 and AKR1B1 is 73 %, and that among the four AKR1C isoforms (1C1-1C4) is >84 %. In particular, AKR1C1 and AKR1C2 differ only by seven amino acids. Recently, several specific inhibitors of the respective enzymes have been found by molecular docking and virtual screening based on the crystal structures of the enzymes and databases of compounds that may show anti-tumor properties. In this section, we briefly review recent progress on specific inhibitors of AKR1B10, AKR1C1 and AKR1C3, in relation to prevention of resistance to anticancer drugs.

## *AKR1B10 Inhibitors*

Various synthetic and natural compounds show inhibitory effects on AKR1B10, as reviewed previously [21]. Table 2 summarizes the inhibitory potency and selectivity (ratio of AKR1B10 to AKR1B1) of the inhibitors including those reported recently. Among the natural compounds, γ-mangostin is the most potent inhibitor of AKR1B10, and oleanoic acid, followed by bisdemethoxycurcumin, show high selectivity. The natural compounds listed in Table 2 have been described as being effective in the chemoprevention of cancer or having potential antitumor actions, and their inhibitory activities of AKR1B10 are believed to be responsible for their anti-cancer properties [101, 123–125]. In the synthetic inhibitors, (*Z*)-2-(4-methoxyphenylimino)-7-hydroxy-*N*-(pyridine-2-yl)-*2H*-chromene-3-carboxamide (PHPC) and 3-(4-hydroxy-2-methoxyphenyl)acrylic acid 3-(3-hydroxyphenyl)propyl ester (HAHE) are the most potent inhibitors. HAHE is highly selective, because it is designed based on molecular docking of caffeic acid phenthyl ester in AKR1B10, for synthesis of potent and selective inhibitors [101]. PHPC and 9-methyl-2,3,7-trihydroxy-6-fluorone were found by virtual screening of chemical databases based on the crystal structure of AKR1B10 [126, 127], but like an AKR1B1 inhibitor tolrestat, they lack the selectivity to AKR1B10.

**Table 2** Natural and synthetic inhibitors of AKR1B10

| Inhibitor | AKR1B10 | | AKR1B1 | $IC_{50}$ ratio | Ref. |
|---|---|---|---|---|---|
| | $IC_{50}$ (μM) | $K_i$ (nM) | $IC_{50}$ (μM) | (1B1/1B10) | |
| *Natural compounds* | | | | | |
| γ-Mangostin | 0.018 | 5.6 | 0.29 | 16 | [123] |
| Bisdemethoxycurcumin | 0.060 | 22 | 5.1 | 85 | [124] |
| CAPE[a] | 0.080 | 46 | 0.57 | 7 | [101] |
| Oleanolic acid | 0.090 | 72 | 124 | 1,370 | [125] |
| *Synthetic compounds* | | | | | |
| PHPC | 0.0060 | 2.7 | 0.011 | 2 | [126] |
| HAHE | 0.0062 | 2.6 | 4.9 | 790 | [101] |
| Tolrestat | 0.054 | 46 | 0.014 | 0.3 | [128] |
| MTF[b] | 0.40 | 200 | 1.6 | 4 | [127] |

[a] *CAPE* caffeic acid phenthyl ester; [b] *MTF* 9-methyl-2,3,7-trihydroxy-6-fluorone

The efficacies of bisdemethoxycurcumin [124], oleanolic acid [125], PHPC [126], HAHE [101] and tolrestat [128] as the inhibitors were demonstrated in cellular AKR1B10-mediated metabolism. In addition, PHPC [21] and HAHE [101] suppress the proliferative potential in AKR1B10-overexpressing U937 cells. The proliferation capacity of HT29 cells resistant to MMC and oxaliplatin is also decreased by the addition of oleanolic acid [125] and PHPC [21], respectively. Furthermore, an endogenous AKR1B10 inhibitor, $PGA_1$, exhibits anti-tumoral activity, and is also suggested to counteract the DOX-induced resistance of lung cancer A549 cells through the inhibition of DOX metabolism by this enzyme [129].

## *Inhibitors of AKR1C1 and AKR1C3*

In 2011, three elegant reviews on the inhibitors have been published [20, 121, 122]. Recently, new inhibitors with high potency and/or selectivity have been reported, and are focused in this section. The potent inhibitor of AKR1C1 is 3-chloro-5-phenyl salicylic acid (CPSA), which shows an inhibition constant ($K_i$) value of 0.86 nM [130]. The inhibitory potency and selectivity are compared with those of several known inhibitors [20, 131–135] in Table 3, in which the $IC_{50}$ values are determined under the same assay conditions using *S*-1-tetralol as the substrate. CPSA shows the highest selectivity to AKR1C1 over other AKR1C isoforms, but its selectivity versus AKR1C2 is still low (only 30-fold). This low selectivity is due to the structural similarity between AKR1C1 and AKR1C2, which differs only by one active-site residue (Leu54 versus Val54). Table 3 also includes two flavonoids, naringenin and 7-hydroxyflavone, which are reported to inhibit AKR1C1 potently and selectively [134, 135]. The inhibitory potency and selectivity of the two flavonoids are much lower than those of the synthetic inhibitors, CPSA and its bromo derivative, 3-bromo-5-phenylsalicylic acid (BPSA).

AKR1C3 is inhibited by several structurally different compounds, which are divided into steroidal and non-steroidal inhibitors, as previously reviewed [121, 122]. Representative steroidal inhibitors are medroxyprogesterone acetate and steroidal

**Table 3** Inhibitory potency and selectivity of AKR1C1 inhibitors

| Inhibitor | $IC_{50}$ (μM)[a] | | | | $IC_{50}$ ratio (1C2/1C1) | Ref. |
|---|---|---|---|---|---|---|
| | AKR1C1 | AKR1C2 | AKR1C3 | AKR1C4 | | |
| CPSA | 0.0037 | 0.11 | >50 | >50 | 30 | [130] |
| BPSA | 0.018 | 0.38 | >50 | 33 | 21 | [131] |
| 3,5-Dichlorosalicylic acid | 0.044 | 0.39 | >50 | >50 | 9 | [132] |
| Benzbromarone | 0.048 | 0.35 | 2.2 | 0.73 | 7 | [133] |
| Naringenin | 0.31 | 4.0 | 0.91 | 0.62 | 13 | [134] |
| 7-Hydroxyflavone | 0.74 | 2.5 | 1.0 | 12 | 3 | [135] |

[a] The values are determined with substrate (*S*-1-tetralol) concentrations of 0.1 mM (for AKR1C1) and 1.0 mM (for other enzymes) in the presence of 0.25 mM $NADP^+$ at pH 7.4

lactones, and the non-steroidal inhibitors are non-steroidal anti-inflammatory agents (NSAIDs) and their analogues, benzodiazepines, prostaglandin analogues, flavonoids, cyclopentane derivatives, and cinnamic acids. These inhibitors lack AKR1C3 selectivity or their inhibitory effects on other AKR1C isoforms have not been tested. Among the previously known inhibitors, a highly selective inhibitor is an indomethacin analogue, N-(4-chlorobenzoyl)-melatonin, but its $IC_{50}$ value is 7.8 μM [136]. Recently, more potent and selective AKR1C3 inhibitors have been reported and are listed in Table 4. Two NSAIDs, tolfenamic acid and *R*-flurbiprofen, are more potent and/or selective inhibitors of AKR1C3 than N-(4-chlorobenzoyl)-melatonin [99]. Using a NSAID, flufenamic acid, as lead compound, several highly potent and selective inhibitors of AKR1C3 have been synthesized [137]. The selectivity of a representative inhibitor, compound 40, is shown in Table 4. The remarkable feature of this inhibitor is the lack of cyclooxygenase inhibition, which is an undesirable action for AKR1C3 inhibitors derived from NSAIDs. In addition, more selective inhibitors that are structurally distinct from NSAIDs have been found by biological evaluation of natural compounds and virtual screening of chemical libraries. Baccharin [3-prenyl-4-(dihydrocinnamoyloxy)cinnamic acid], a component of honeybee propolis, is a potent competitive inhibitor ($K_i$ 56 nM) with high selectivity to AKR1C3 [102]. Compounds, 38 and 26, are actually specific inhibitors of AKR1C3 [138, 139], and compound 38 shows the highest inhibitory potency, and inhibits cellular AKR1C3-mediated metabolism at low nM concentration [138].

**Table 4** Selective AKR1C3 inhibitors

| Inhibitor | $IC_{50}$ (μM) | | | | $IC_{50}$ ratio | | | Ref. |
|---|---|---|---|---|---|---|---|---|
| | AKR1C3 | AKR1C1 | AKR1C2 | AKR1C4 | 1C1/1C3 | 1C2/1C3 | 1C4/1C3 | |
| Compound 38[a] | 0.0061 | >30 | >30 | >30 | >4,918 | >4,918 | >4,918 | [138] |
| Compound 40[a] | 0.062 | 22.7 | 15.4 | 62.7 | 368 | 249 | 1,015 | [137] |
| Tolfenamic acid | 0.017 | 0.71 | 0.57 | 53 | 42 | 34 | 3,118 | [99] |
| Baccharin | 0.11 | >100 | >100 | >100 | >909 | >909 | >909 | [102] |
| Compound 26[a] | 0.213 | >300 | 206 | >300 | >1,408 | 967 | >1,408 | [139] |
| *R*-Flurbiprofen | 1.5 | >50 | >50 | act | >30 | >30 | – | [99] |
| CBM[b] | 7.8 | >100 | >100 | nd | >13 | >13 | – | [136] |

[a] Structures

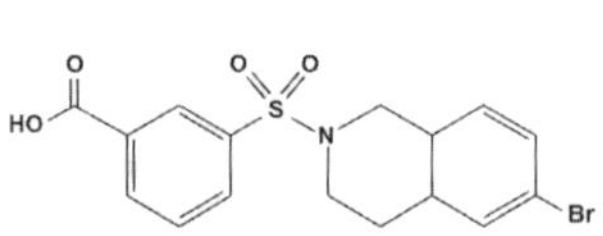


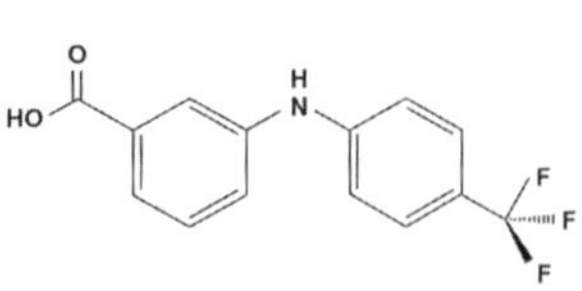


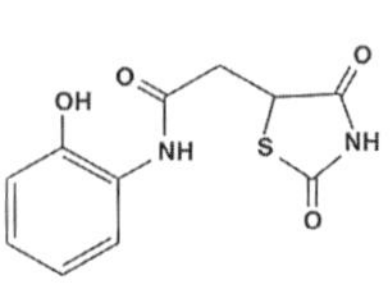


Compound 38 Compound 40 Compound 26

[b] Abbreviations *CBM* N-(4-chlorobenzoyl)-melatonin; *act* activation; and *nd* not determined

In the cell-based experiments, the AKR1C1 inhibitors BPSA [131] and CPSA [130] effectively inhibit cellular progesterone metabolism by AKR1C1, showing low $IC_{50}$ values of 0.46 and 0.10 μM, respectively. The AKR1C3 inhibitors, tolfenamic acid [99] and baccharin [102], are capable of mitigating facilitation of proliferation potential of U937 and PC3 cells, respectively, by overexpression of AKR1C3, in addition to almost completely inhibiting the cellular metabolism of the preferred substrates (FAL and 17-ketosteroids). Treatment with tolfenamic acid is also found to promote U937 cell differentiation through conceivably lowering the proliferative effect of AKR1C3 [140]. Moreover, our recent study shows that combination of BPSA and tolfenamic acid increases the sensitivity of CDDP-resistant HCT15 cells to the anti-cancer drug [25]. This result may confer direct evidence that AKR1C1 and AKR1C3 are key enzymes responsible for CDDP resistance of the colon cancer cells. However, there is limited evidence in the literature regarding the suppressive effects of the inhibitors on colon cancer chemoresistance. It is therefore necessary to further determine efficacy of the adjuvant therapy using the inhibitors in various malignant stages, such as invasion, metastasis and tumorigenesis of cancer cells, caused with the chemoresistance.

## Conclusion

Treatment with chemotherapeutic agents against colon cancer is currently limited by appearance of the therapy-resistant cancer cells. Previous studies for evaluating the molecular basis have proposed some candidates of the adjuvant therapy targets, which include P-glycoprotein, growth factors, cell membrane antigens and proteasomes. Besides them, other rationale candidates are three AKRs (AKR1B10, AKR1C1 and AKR1C3), because they are up-regulated and play roles in not only developing carcinogenesis of cells, but also acquiring colon cancer chemoresistance. Therefore, accumulation of such experimental data using the inhibitors specific to the enzymes would lead to the development of a new strategy, which makes combination of anticancer drugs and the inhibitors applicable to colon cancer chemotherapy. Since the cancer tolerance to chemotherapy is the most significant problem for other carcinomas including lung cancer, we believe that exploration of potent and specific inhibitors of the respective AKRs directly leads to the development of adjuvant drugs useful for the treatments of various kinds of cancers.

**No Conflict Statement** No potential conflicts of interest were disclosed.

## References


1. Pham NM, Mizoue T, Tanaka K, Tsuji I, Tamakoshi A, Matsuo K, Ito H, Wakai K, Nagata C, Sasazuki S, Inoue M, Tsugane S. Research group for the development and evaluation of cancer prevention strategies in Japan. Physical activity and colorectal cancer risk: an evaluation based on a systematic review of epidemiologic evidence among the Japanese population. Jpn J Clin Oncol. 2012;42:2–13.
2. Kim R, Yamaguchi Y, Toge T. Adjuvant therapy for colorectal carcinoma. Anticancer Res. 2002;22:2413–8.
3. Saltz LB. Adjuvant therapy for colon cancer. Surg Oncol Clin N Am. 2010;19:819–27.
4. Dietel M. Molecular mechanisms and possibilities of overcoming drug resistance in gastrointestinal tumors. Recent Results Cancer Res. 1996;142:89–101.
5. Modjtahedi H, Essapen S. Epidermal growth factor receptor inhibitors in cancer treatment: advances challenges and opportunities. Anticancer Drugs. 2009;20:851–5.
6. Ahmed FE. Molecular markers that predict response to colon cancer therapy. Expert Rev Mol Diagn. 2005;5:353–75.
7. Ogiso Y, Tomida A, Lei S, Omura S, Tsuruo T. Proteasome inhibition circumvents solid tumor resistance to topoisomerase II-directed drugs. Cancer Res. 2000;60:2429–34.
8. Van Geelen CM, de Vries EG, de Jong S. Lessons from TRAIL-resistance mechanisms in colorectal cancer cells: paving the road to patient-tailored therapy. Drug Resist Updat. 2004;7:345–58.
9. Todaro M, Perez Alea M, Scopelliti A, Medema JP, Stassi G. IL-4-mediated drug resistance in colon cancer stem cells. Cell Cycle. 2008;7:309–313.
10. Todaro M, Francipane MG, Medema JP, Stassi G. Colon cancer stem cells: promise of targeted therapy. Gastroenterology. 2010;138:2151–62.
11. Lehne G, De Angelis P, den Boer M, Rugstad HE. Growth inhibition, cytokinesis failure and apoptosis of multidrug-resistant leukemia cells after treatment with P-glycoprotein inhibitory agents. Leukemia. 1999;13:768–78.
12. Fracasso PM, Goldstein LJ, de Alwis DP, Rader JS, Arquette MA, Goodner SA, Wright LP, Fears CL, Gazak RJ, Andre VA, Burgess MF, Slapak CA, Schellens JH. Phase I study of docetaxel in combination with the P-glycoprotein inhibitor, zosuquidar, in resistant malignancies. Clin Cancer Res. 2004;10:7220–8.
13. Enrique AA, Gema PC, Jeronimo JC, Auxiliadora GE. Role of anti-EGFR target therapy in colorectal carcinoma. Front Biosci. 2012;4:12–22.
14. Ferrarotto R, Hoff PM. Antiangiogenic drugs for colorectal cancer: exploring new possibilities. Clin Colorectal Cancer. 2013;12:1–7.
15. Hochster HS. Opportunities for newer agents in combination with oxaliplatin. Semin Oncol. 2003;30:62–7.
16. Bauman DR, Steckelbroeck S, Penning TM. The roles of aldo-keto reductases in steroid hormone action. Drug News Perspect. 2004;17:563–78.
17. Jin Y, Penning TM. Aldo-keto reductases and bioactivation/detoxication. Annu Rev Pharmacol Toxicol. 2007;47:263–92.
18. Barski OA, Tipparaju SM, Bhatnagar A. The aldo-keto reductase superfamily and its role in drug metabolism and detoxification. Drug Metab Rev. 2008;40:553–624.
19. Penning TM, Byrns MC. Steroid hormone transforming aldo-keto reductases and cancer. Ann N Y Acad Sci. 2009;1155:33–42.
20. El-Kabbani O, Dhagat U, Hara A. Inhibitors of human 20$\alpha$-hydroxysteroid dehydrogenase (AKR1C1). J Steroid Biochem Mol Biol. 2011;125:105–11.
21. Matsunaga T, Wada Y, Endo S, Soda M, El-Kabbani O, Hara A. Aldo-keto reductase 1B10 and its role in proliferation capacity of drug-resistant cancers. Front Pharmacol. 2012;3:5.
22. Rižner TL. Enzymes of the AKR1B and AKR1C subfamilies and uterine diseases. Front Pharmacol. 2012;3:34.

23. Selga E, Noé V, Ciudad CJ. Transcriptional regulation of aldo-keto reductase 1C1 in HT29 human colon cancer cells resistant to methotrexate: role in the cell cycle and apoptosis. Biochem Pharmacol. 2008;75:414–26.
24. Matsunaga T, Yamane Y, Iida K, Endo S, Banno Y, El-Kabbani O, Hara A. Involvement of the aldo-keto reductase, AKR1B10, in mitomycin-c resistance through reactive oxygen species-dependent mechanisms. Anticancer Drugs. 2011;22:402–8.
25. Matsunaga T, Hojo A, Yamane Y, Endo S, El-Kabbani O, Hara A. Pathophysiological roles of aldo-keto reductases (AKR1C1 and AKR1C3) in development of cisplatin resistance in human colon cancers. Chem Biol Interact; 2013;202:234–42.
26. Mekhail-Ishak K, Hudson N, Tsao MS, Batist G. Implications for therapy of drug-metabolizing enzymes in human colon cancer. Cancer Res. 1989;49:4866–9.
27. Lotz C, Kelleher DK, Gassner B, Gekle M, Vaupel P, Thews O. Role of the tumor microenvironment in the activity and expression of the P-glycoprotein in human colon carcinoma cells. Oncol Rep. 2007;17:239–44.
28. Ueda K, Pastan I, Gottesman MM. Isolation and sequence of the promoter region of the human multidrug-resistance (P-glycoprotein) gene. J Biol Chem. 1987;262:17432–6.
29. Thiebaut F, Tsuruo T, Hamada H, Gottesman MM, Pastan I, Willingham MC. Cellular localization of the multidrug-resistance gene product P-glycoprotein in normal human tissues. Proc Natl Acad Sci USA. 1987;84:7735–8.
30. Cordon-Cardo C, O'Brien JP, Boccia J, Casals D, Bertino JR, Melamed MR. Expression of the multidrug resistance gene product (P-glycoprotein) in human normal and tumor tissues. J Histochem Cytochem. 1990;38:1277–87.
31. Klappe K, Hinrichs JW, Kroesen BJ, Sietsma H, Kok JW. MRP1 and glucosylceramide are coordinately over expressed and enriched in rafts during multidrug resistance acquisition in colon cancer cells. Int J Cancer. 2004;110:511–22.
32. Liu YY, Gupta V, Patwardhan GA, Bhinge K, Zhao Y, Bao J, Mehendale H, Cabot MC, Li YT, Jazwinski SM. Glucosylceramide synthase upregulates *MDR1* expression in the regulation of cancer drug resistance through *cSrc* and *β*-catenin signaling. Mol Cancer. 2010;9:145.
33. Mendelsohn J, Baselga J. Epidermal growth factor receptor targeting in cancer. Semin Oncol. 2006;33:369–85.
34. Huang SA, Lin PF, Fan D, Price JE, Trujillo JM, Chakrabarty S. Growth modulation by epidermal growth factor (EGF) in human colonic carcinoma cells: constitutive expression of the human EGF gene. J Cell Physiol. 1991;148:220–7.
35. Radinsky R, Risin S, Fan D, Dong Z, Bielenberg D, Bucana CD, Fidler IJ. Level and function of epidermal growth factor receptor predict the metastatic potential of human colon carcinoma cells. Clin Cancer Res. 1995;1:19–31.
36. Karnes WE Jr, Weller SG, Adjei PN, Kottke TJ, Glenn KS, Gores GJ, Kaufmann SH. Inhibition of epidermal growth factor receptor kinase induces protease-dependent apoptosis in human colon cancer cells. Gastroenterology. 1998;114:930–9.
37. Jones MK, Tomikawa M, Mohajer B, Tarnawski AS. Gastrointestinal mucosal regeneration: role of growth factors. Front Biosci. 1999;4:D303–9.
38. Vincenzi B, Santini D, Rabitti C, Coppola R, Beomonte Zobel B, Trodella L, Tonini G. Cetuximab and irinotecan as third-line therapy in advanced colorectal cancer patients: a single centre phase II trial. Br J Cancer. 2006;94:792–797.
39. Stoeltzing O, Liu W, Reinmuth N, Parikh A, Ahmad SA, Jung YD, Fan F, Ellis LM. Angiogenesis and antiangiogenic therapy of colon cancer liver metastasis. Ann Surg Oncol. 2003;10:722–33.
40. Braghiroli MI, Sabbaga J, Hoff PM. Bevacizumab: overview of the literature. Expert Rev Anticancer Ther. 2012;12:567–80.
41. Dallas NA, Xia L, Fan F, Gray MJ, Gaur P, van Buren G 2nd, Samuel S, Kim MP, Lim SJ, Ellis LM. Chemoresistant colorectal cancer cells, the cancer stem cell phenotype, and increased sensitivity to insulin-like growth factor-I receptor inhibition. Cancer Res. 2009;69:1951–1957.

42. Reggiani Bonetti L, Migaldi M, Caredda E, Boninsegna A, Ponz De Leon M, Di Gregorio C, Barresi V, Scannone D, Danese S, Cittadini A, Sgambato A. Increased expression of CD133 is a strong predictor of poor outcome in stage I colorectal cancer patients. Scand J Gastroenterol. 2012;47:1211–1217.
43. Todaro M, Alea MP, Di Stefano AB, Cammareri P, Vermeulen L, Iovino F, Tripodo C, Russo A, Gulotta G, Medema JP, Stassi G. Colon cancer stem cells dictate tumor growth and resist cell death by production of interleukin-4. Cell Stem Cell. 2007;1:389–402.
44. Lakshman M, Subramaniam V, Rubenthiran U, Jothy S. CD44 promotes resistance to apoptosis in human colon cancer cells. Exp Mol Pathol. 2004;77:18–25.
45. Wang C, Xie J, Guo J, Manning HC, Gore JC, Guo N. Evaluation of CD44 and CD133 as cancer stem cell markers for colorectal cancer. Oncol Rep. 2012;28:1301–8.
46. Cory AH, Cory JG. Lactacystin, a proteasome inhibitor, potentiates the apoptotic effect of parthenolide, an inhibitor of NF$\kappa$B activation, on drug-resistant mouse leukemia L1210 cells. Anticancer Res. 2002;22:3805–9.
47. Loeffler-Ragg J, Mueller D, Gamerith G, Auer T, Skvortsov S, Sarg B, Skvortsova I, Schmitz KJ, Martin HJ, Krugmann J, Alakus H, Maser E, Menzel J, Hilbe W, Lindner H, Schmid KW, Zwierzina H. Proteomic identification of aldo-keto reductase AKR1B10 induction after treatment of colorectal cancer cells with the proteasome inhibitor bortezomib. Mol Cancer Ther. 2009;8:1995–2006.
48. Ebert B, Kisiela M, Wsól V, Maser E. Proteasome inhibitors MG-132 and Bortezomib induce AKR1C1, AKR1C3, AKR1B1, and AKR1B10 in human colon cancer cell lines SW-480 and HT-29. Chem Biol Interact. 2011;191:239–49.
49. Mackay H, Hedley D, Major P, Townsley C, Mackenzie M, Vincent M, Degendorfer P, Tsao MS, Nicklee T, Birle D, Wright J, Siu L, Moore M, Oza A. A phase II trial with pharmacodynamic endpoints of the Proteasome inhibitor Bortezomib in patients with metastatic colorectal cancer. Clin Cancer Res. 2005;11:5526–33.
50. Boccadoro M, Morgan G, Cavenagh J. Preclinical evaluation of the proteasome inhibitor bortezomib in cancer therapy. Cancer Cell Int. 2005;5:18.
51. Pitts TM, Morrow M, Kaufman SA, Tentler JJ, Eckhardt SG. Vorinostat and Bortezomib exert synergistic antiproliferative and proapoptotic effects in colon cancer cell models. Mol Cancer Ther. 2009;8:342–9.
52. Iwata S, Yano S, Ito Y, Ushijima Y, Gotoh K, Kawada J, Fujiwara S, Sugimoto K, Isobe Y, Nishiyama Y, Kimura H. Bortezomib induces apoptosis in T lymphoma cells and natural killer lymphoma cells independent of Epstein–Barr virus infection. Int J Cancer. 2011;129:2263–73.
53. Kusumoto S, Sugiyama T, Ando K, Hosaka T, Ishida H, Shirai T, Yamaoka T, Okuda K, Hirose T, Ohnishi T, Inoue F, Kanome T, Kadofuku T, Saijo N, Adachii M, Ohmori T. Combination effect between bortezomib and tumor necrosis factor alpha on gefitinib-resistant non-small cell lung cancer cell lines. Anticancer Res. 2009;29:2315–22.
54. Yanaba K, Asano Y, Tada Y, Sugaya M, Kadono T, Sato S. Proteasome inhibitor bortezomib ameliorates intestinal injury in mice. PLoS One. 2012;7:e34587.
55. Cao D, Fan ST, Chung SS. Identification and characterization of a novel human aldose reductase-like gene. J Biol Chem. 1998;273:11429–35.
56. Hyndman DJ, Flynn TG. Sequence and expression levels in human tissues of a new member of the aldo-keto reductase family. Biochim Biophys Acta. 1998;1399:198–202.
57. Fukumoto S, Yamauchi N, Moriguchi H, Hippo Y, Watanabe A, Shibahara J, Taniguchi H, Ishikawa S, Ito H, Yamamoto S, Iwanari H, Hironaka M, Ishikawa Y, Niki T, Sohara Y, Kodama T, Nishimura M, Fukayama M, Dosaka-Akita H, Aburatani H. Overexpression of the aldo-keto reductase family protein AKR1B10 is highly correlated with smokers' non-small cell lung carcinomas. Clin Cancer Res. 2005;11:1776–85.
58. Nagaraj NS, Beckers S, Mensah JK, Waigel S, Vigneswaran N, Zacharias W. Cigarette smoke condensate induces cytochromes P450 and aldo-keto reductases in oral cancer cells. Toxicol Lett. 2006;165:182–94.

59. Pierrou S, Broberg P, O'Donnell RA, Pawłowski K, Virtala R, Lindqvist E, Richter A, Wilson SJ, Angco G, Möller S, Bergstrand H, Koopmann W, Wieslander E, Strömstedt PE, Holgate ST, Davies DE, Lund J, Djukanovic R. Expression of genes involved in oxidative stress responses in airway epithelial cells of smokers with chronic obstructive pulmonary disease. Am J Respir Crit Care Med. 2007;175:577–86.
60. Zhang L, Lee JJ, Tang H, Fan YH, Xiao L, Ren H, Kurie J, Morice RC, Hong WK, Mao L. Impact of smoking cessation on global gene expression in the bronchial epithelium of chronic smokers. Cancer Prev Res (Phila). 2008;1:112–8.
61. Penning TM, Lerman C. Genomics of smoking exposure and cessation: lessons for cancer prevention and treatment. Cancer Prev Res (Phila). 2008;1:80–3.
62. Wang R, Wang G, Ricard MJ, Ferris B, Strulovici-Barel Y, Salit J, Hackett NR, Gudas LJ, Crystal RG. Smoking-induced upregulation of AKR1B10 expression in the airway epithelium of healthy individuals. Chest. 2010;138:1402–10.
63. Woenckhaus M, Klein-Hitpass L, Grepmeier U, Merk J, Pfeifer M, Wild P, Bettstetter M, Wuensch P, Blaszyk H, Hartmann A, Hofstaedter F, Dietmaier W. Smoking and cancer-related gene expression in bronchial epithelium and non-small-cell lung cancers. J Pathol. 2006;210:192–204.
64. Kim B, Lee HJ, Choi HY, Shin Y, Nam S, Seo G, Son DS, Jo J, Kim J, Lee J, Kim J, Kim K, Lee S. Clinical validity of the lung cancer biomarkers identified by bioinformatics analysis of public expression data. Cancer Res. 2007;67:7431–8.
65. Li CP, Goto A, Watanabe A, Murata K, Ota S, Niki T, Aburatani H, Fukayama M. AKR1B10 in usual interstitial pneumonia: expression in squamous metaplasia in association with smoking and lung cancer. Pathol Res Pract. 2008;204:295–304.
66. Kang MW, Lee ES, Yoon SY, Jo J, Lee J, Kim HK, Choi YS, Kim K, Shim YM, Kim J, Kim H. AKR1B10 is associated with smoking and smoking-related non-small-cell lung cancer. J Int Med Res. 2011;39:78–85.
67. Heringlake S, Hofdmann M, Fiebeler A, Manns MP, Schmiegel W, Tannapfel A. Identification and expression analysis of the aldo-ketoreductase 1–B10 gene in primary malignant liver tumours. J Hepatol. 2010;52:220–7.
68. Schmitz KJ, Sotiropoulos GC, Baba HA, Schmid KW, Müller D, Paul A, Auer T, Gamerith G, Loeffler-Ragg J. AKR1B10 expression is associated with less aggressive hepatocellular carcinoma: a clinicopathological study of 168 cases. Liver Int. 2011;31:810–6.
69. Yoshitake H, Takahashi M, Ishikawa H, Nojima M, Iwanari H, Watanabe A, Aburatani H, Yoshida K, Ishi K, Takamori K, Ogawa H, Hamakubo T, Kodama T, Araki Y. Aldo-keto reductase family 1, member B10 in uterine carcinomas: a potential risk factor of recurrence after surgical therapy in cervical cancer. Int J Gynecol Cancer. 2007;17:1300–6.
70. Lee HJ, Nam KT, Park HS, Kim MA, Lafleur BJ, Aburatani H, Yang HK, Kim WH, Goldenring JR. Gene expression profiling of metaplastic lineages identifies CDH17 as a prognostic marker in early stage gastric cancer. Gastroenterology. 2010;139:213–25.
71. Breton J, Gage MC, Hay AW, Keen JN, Wild CP, Donnellan C, Findlay JB, Hardie LJ. Proteomic screening of a cell line model of esophageal carcinogenesis identifies cathepsin d and aldo-keto reductase 1C2 and 1B10 dysregulation in Barrett's esophagus and esophageal adenocarcinoma. J Proteome Res. 2008;7:1953–62.
72. Palackal NT, Lee SH, Harvey RG, Blair IA, Penning TM. Activation of polycyclic aromatic hydrocarbon *trans*-dihydrodiol proximate carcinogens by human aldo-keto reductase (AKR1C) enzymes and their functional overexpression in human lung carcinoma (A549) cells. J Biol Chem. 2002;277:24799–808.
73. Wang HW, Lin CP, Chiu JH, Chow KC, Kuo KT, Lin CS, Wang LS. Reversal of inflammation-associated dihydrodiol dehydrogenases (AKR1C1 and AKR1C2) overexpression and drug resistance in nonsmall cell lung cancer cells by Wogonin and Chrysin. Int J Cancer. 2007;120:2019–27.
74. Rižner TL, Smuc T, Rupreht R, Sinkovec J, Penning TM. AKR1C1 and AKR1C3 may determine progesterone and estrogen ratios in endometrial cancer. Mol Cell Endocrinol. 2006;248:126–35.

75. Tai HL, Lin TS, Huang HH, Lin TY, Chou MC, Chiou SH, Chow KC. Overexpression of aldo-keto reductase 1C2 as a high-risk factor in bladder cancer. Oncol Rep. 2007;17:305–11.
76. Desmond JC, Mountford JC, Drayson MT, Walker EA, Hewison M, Ride JP, Luong QT, Hayden RE, Vanin EF, Bunce CM. The aldo-keto reductase AKR1C3 is a novel suppressor of cell differentiation that provides a plausible target for the non-cyclooxygenase-dependent antineoplastic actions of nonsteroidal anti-inflammatory drugs. Cancer Res. 2003;63:505–12.
77. Miller VL, Lin HK, Murugan P, Fan M, Penning TM, Brame LS, Yang Q, Fung KM. Aldo-keto reductase family 1 member C3 (AKR1C3) is expressed in adenocarcinoma and squamous cell carcinoma but not small cell carcinoma. Int J Clin Exp Pathol. 2012;5:278–89.
78. Martinez I, Wang J, Hobson KF, Ferris RL, Khan SA. Identification of differentially expressed genes in HPV-positive and HPV-negative oropharyngeal squamous cell carcinomas. Eur J Cancer. 2007;43:415–32.
79. Stanbrough M, Bubley GJ, Ross K, Golub TR, Rubin MA, Penning TM, Febbo PG, Balk SP. Increased expression of genes converting adrenal androgens to testosterone in androgen-independent prostate cancer. Cancer Res. 2006;66:2815–25.
80. Lewis MJ, Wiebe JP, Heathcote JG. Expression of progesterone metabolizing enzyme genes (AKR1C1, AKR1C2, AKR1C3, SRD5A1, SRD5A2) is altered in human breast carcinoma. BMC Cancer. 2004;4:27.
81. Lau A, Villeneuve NF, Sun Z, Wong PK, Zhang DD. Dual roles of Nrf2 in cancer. Pharmacol Res. 2008;58:262–70.
82. MacLeod AK, McMahon M, Plummer SM, Higgins LG, Penning TM, Igarashi K, Hayes JD. Characterization of the cancer chemopreventive NRF2-dependent gene battery in human keratinocytes: demonstration that the KEAP1-NRF2 pathway, and not the BACH1-NRF2 pathway, controls cytoprotection against electrophiles as well as redox-cycling compounds. Carcinogenesis. 2009;30:1571–80.
83. Nishinaka T, Miura T, Okumura M, Nakao F, Nakamura H, Terada T. Regulation of aldo-keto reductase AKR1B10 gene expression: involvement of transcription factor Nrf2. Chem Biol Interact. 2011;191:185–91.
84. De Vries HE, Witte M, Hondius D, Rozemuller AJ, Drukarch B, Hoozemans J, van Horssen J. Nrf2-induced antioxidant protection: a promising target to counteract ROS-mediated damage in neurodegenerative disease? Free Radic Biol Med. 2008;45:1375–83.
85. Levonen AL, Landar A, Ramachandran A, Ceaser EK, Dickinson DA, Zanoni G, Morrow JD, Darley-Usmar VM. Cellular mechanisms of redox cell signalling: role of cysteine modification in controlling antioxidant defences in response to electrophilic lipid oxidation products. Biochem J. 2004;378:373–82.
86. Ciaccio PJ, Stuart JE, Tew KD. Overproduction of a 37.5 kDa cytosolic protein structurally related to prostaglandin F synthase in ethacrynic acid-resistant human colon cells. Mol Pharmacol. 1993;43:845–53.
87. Ciaccio PJ, Jaiswal AK, Tew KD. Regulation of human dihydrodiol dehydrogenase by michael acceptor xenobiotics. J Biol Chem. 1994;269:15558–62.
88. Singh A, Misra V, Thimmulappa RK, Lee H, Ames S, Hoque MO, Herman JG, Baylin SB, Sidransky D, Gabrielson E, Brock MV, Biswal S. Dysfunctional KEAP1-NRF2 interaction in non-small-cell lung cancer. PLoS Med. 2006;3:e420.
89. Shibata T, Kokubu A, Gotoh M, Ojima H, Ohta T, Yamamoto M, Hirohashi S. Genetic alteration of Keap1 confers constitutive Nrf2 activation and resistance to chemotherapy in gallbladder cancer. Gastroenterology. 2008;135:1358–68.
90. Loignon M, Miao W, Hu L, Bier A, Bismar TA, Scrivens PJ, Mann K, Basik M, Bouchard A, Fiset PO, Batist Z, Batist G. Cul3 overexpression depletes Nrf2 in breast cancer and is associated with sensitivity to carcinogens, to oxidative stress, and to chemotherapy. Mol Cancer Ther. 2009;8:2432–40.

91. Tang XH, Gudas LJ. Retinoids, retinoic acid receptors, and cancer. Annu Rev Pathol. 2011;6:345–64.
92. Crosas B, Hyndman DJ, Gallego O, Martras S, Parés X, Flynn TG, Farrés J. Human aldose reductase and human small intestine aldose reductase are efficient retinal reductases: consequences for retinoid metabolism. Biochem J. 2003;373:973–9.
93. Gallego O, Ruiz FX, Ardèvol A, Domínguez M, Alvarez R, de Lera AR, Rovira C, Farrés J, Fita I, Parés X. Structural basis for the high all-trans-retinaldehyde reductase activity of the tumor marker AKR1B10. Proc Natl Acad Sci USA. 2007;104:20764–9.
94. Ruiz FX, Gallego O, Ardèvol A, Moro A, Domínguez M, Alvarez S, Alvarez R, de Lera AR, Rovira C, Fita I, Parés X, Farrés J. Aldo-keto reductases from the AKR1B subfamily: retinoid specificity and control of cellular retinoic acid levels. Chem Biol Interact. 2009;178:171–7.
95. Matsuura K, Shiraishi H, Hara A, Sato K, Deyashiki Y, Ninomiya M, Sakai S. Identification of a principal mRNA species for human 3$\alpha$-hydroxysteroid dehydrogenase isoform (AKR1C3) that exhibits high prostaglandin $D_2$ 11-ketoreductase activity. J Biochem. 1998;124:940–6.
96. Suzuki-Yamamoto T, Nishizawa M, Fukui M, Okuda-Ashitaka E, Nakajima T, Ito S, Watanabe K. cDNA cloning, expression and characterization of human prostaglandin F synthase. FEBS Lett. 1999;462:335–40.
97. Adams JW, Sah VP, Henderson SA, Brown JH. Tyrosine kinase and c-Jun $NH_2$-terminal kinase mediate hypertrophic responses to prostaglandin $F_{2\alpha}$ in cultured neonatal rat ventricular myocytes. Circ Res. 1998;83:167–78.
98. Zaragoza DB, Wilson RR, Mitchell BF, Olson DM. The interleukin 1$\beta$-induced expression of human prostaglandin $F_{2\alpha}$ receptor messenger RNA in human myometrial-derived ULTR cells requires the transcription factor NF$\kappa$B. Biol Reprod. 2006;75:697–704.
99. Endo S, Matsunaga T, Ohta C, Soda M, Kanamori A, Kitade Y, Ohno S, Tajima K, El-Kabbani O, Hara A. Roles of rat and human aldo-keto reductases in metabolism of farnesol and geranylgeraniol. Chem Biol Interact. 2011;191:261–8.
100. Winter-Vann AM, Casey PJ. Post-prenylation-processing enzymes as new targets in oncogenesis. Nat Rev Cancer. 2005;5:405–12.
101. Soda M, Hu D, Endo S, Takemura M, Li J, Wada R, Ifuku S, Zhao HT, El-Kabbani O, Ohta S, Yamamura K, Toyooka N, Hara A, Matsunaga T. Design, synthesis and evaluation of caffeic acid phenethyl ester-based inhibitors targeting a selectivity pocket in the active site of human aldo-keto reductase 1B10. Eur J Med Chem. 2012;48:321–9.
102. Endo S, Matsunaga T, Kanamori A, Otsuji Y, Nagai H, Sundaram K, El-Kabbani O, Toyooka N, Ohta S, Hara A. Selective inhibition of human type-5 17$\beta$-hydroxysteroid dehydrogenase (AKR1C3) by baccharin, a component of Brazilian propolis. J Nat Prod. 2012;75:716–21.
103. Yan R, Zu X, Ma J, Liu Z, Adeyanju M, Cao D. Aldo-keto reductase family 1B10 gene silencing results in growth inhibition of colorectal cancer cells: implication for cancer intervention. Int J Cancer. 2007;121:2301–6.
104. Zu X, Yan R, Ma J, Liao D, Cao D. AKR1B10: a potential target for cancer therapy. Biosci Hypothesis. 2009;2:31–3.
105. Byrns MC, Mindnich R, Duan L, Penning TM. Overexpression of aldo-keto reductase 1C3 (AKR1C3) in LNCaP cells diverts androgen metabolism towards testosterone resulting in resistance to the 5$\alpha$-reductase inhibitor finasteride. J Steroid Biochem Mol Biol. 2012;130:7–15.
106. Ma J, Yan R, Zu X, Cheng JM, Rao K, Liao DF, Cao D. Aldo-keto reductase family 1B10 affects fatty acid synthesis by regulating the stability of acetyl-CoA carboxylase-$\alpha$ in breast cancer cells. J Biol Chem. 2008;283:3418–23.
107. Wang C, Yan R, Luo D, Watabe K, Liao DF, Cao D. Aldo-keto reductase family 1 member B10 promotes cell survival by regulating lipid synthesis and eliminating carbonyls. J Biol Chem. 2009;284:26742–8.

108. Lesgards JF, Gauthier C, Iovanna J, Vidal N, Dolla A, Stocker P. Effect of reactive oxygen and carbonyl species on crucial cellular antioxidant enzymes. Chem Biol Interact. 2011;190:28–34.
109. Burczynski ME, Sridhar GR, Palackal NT, Penning TM. The reactive oxygen species- and michael acceptor-inducible human aldo-keto reductase AKR1C1 reduces the α, β-unsaturated aldehyde 4-hydroxy-2-nonenal to 1,4-dihydroxy-2-nonene. J Biol Chem. 2001;276:2890–7.
110. Martin HJ, Maser E. Role of human aldo-keto-reductase AKR1B10 in the protection against toxic aldehydes. Chem Biol Interact. 2009;178:145–50.
111. Shen Y, Zhong L, Johnson S, Cao D. Human aldo-keto reductases 1B1 and 1B10: a comparative study on their enzyme activity toward electrophilic carbonyl compounds. Chem Biol Interact. 2011;191:192–8.
112. Hartley DP, Ruth JA, Petersen DR. The hepatocellular metabolism of 4-hydroxynonenal by alcohol dehydrogenase, aldehyde dehydrogenase, and glutathione *S*-transferase. Arch Biochem Biophys. 1995;316:197–205.
113. Martin HJ, Breyer-Pfaff U, Wsol V, Venz S, Block S, Maser E. Purification and characterization of AKR1B10 from human liver: role in carbonyl reduction of xenobiotics. Drug Metab Dispos. 2006;34:464–70.
114. Novotna R, Wsol V, Xiong G, Maser E. Inactivation of the anticancer drugs doxorubicin and oracin by aldo-keto reductase (AKR) 1C3. Toxicol Lett. 2008;181(1):1–6.
115. Kassner N, Huse K, Martin HJ, Gödtel-Armbrust U, Metzger A, Meineke I, Brockmöller J, Klein K, Zanger UM, Maser E, Wojnowski L. Carbonyl reductase 1 is a predominant doxorubicin reductase in the human liver. Drug Metab Dispos. 2008;36:2113–20.
116. Balendiran GK. Fibrates in the chemical action of daunorubicin. Curr Cancer Drug Targets. 2009;9:366–9.
117. Balendiran GK, Martin HJ, El-Hawari Y, Maser E. Cancer biomarker AKR1B10 and carbonyl metabolism. Chem Biol Interact. 2009;178:134–7.
118. Bains OS, Grigliatti TA, Reid RE, Riggs KW. Naturally occurring variants of human aldo-keto reductases with reduced in vitro metabolism of daunorubicin and doxorubicin. J Pharmacol Exp Ther. 2010;335:533–45.
119. Zhong L, Shen H, Huang C, Jing H, Cao D. AKR1B10 induces cell resistance to daunorubicin and idarubicin by reducing C13 ketonic group. Toxicol Appl Pharmacol. 2011;255:40–7.
120. Matsunaga T, Endo S, Takemura M, Soda M, Yamamura K, Tajima K, Miura T, Terada T, El-Kabbani O, Hara A. Reduction of cytotoxic *p*-quinone metabolites of *tert*-butylhydroquinone by human aldo-keto reductase (AKR) 1B10. Drug Metab Pharmacokinet. 2012;27:553–8.
121. Byrns MC, Jin Y, Penning TM. Inhibitors of type 5 17β-hydroxysteroid dehydrogenase (AKR1C3): overview and structural insights. J Steroid Biochem Mol Biol. 2011;125:95–104.
122. Brožič P, Turk S, Rižner TL, Gobec S. Inhibitors of aldo-keto reductases AKR1C1-AKR1C4. Curr Med Chem. 2011;18:2554–65.
123. Soda M, Endo S, Matsunaga T, Zhao HT, El-Kabbani O, Iinuma M, Yamamura K, Hara A. Inhibition of human aldose reductase-like protein (AKR1B10) by α- and γ-mangostins, major components of pericarps of mangosteen. Biol Pharm Bull. 2012;35:2075–80.
124. Matsunaga T, Endo S, Soda M, Zhao HT, El-Kabbani O, Tajima K, Hara A. Potent and selective inhibition of the tumor marker AKR1B10 by bisdemethoxycurcumin: probing the active site of the enzyme with molecular modeling and site-directed mutagenesis. Biochem Biophys Res Commun. 2009;389:128–32.
125. Takemura M, Endo S, Matsunaga T, Soda M, Zhao HT, El-Kabbani O, Tajima K, Iinuma M, Hara A. Selective inhibition of the tumor marker aldo-keto reductase family member 1B10 by oleanolic acid. J Nat Prod. 2011;74:1201–6.

126. Endo S, Matsunaga T, Kuwata K, Zhao HT, El-Kabbani O, Kitade Y, Hara A. Chromene-3-carboxamide derivatives discovered from virtual screening as potent inhibitors of the tumour maker, AKR1B10. Bioorg Med Chem. 2010;18:2485–90.
127. Zhao HT, Soda M, Endo S, Hara A, El-Kabbani O. Selectivity determinants of inhibitor binding to the tumour marker human aldose reductase-like protein (AKR1B10) discovered from molecular docking and database screening. Eur J Med Chem. 2010;45:4354–7.
128. Endo S, Matsunaga T, Mamiya H, Ohta C, Soda M, Kitade Y, Tajima K, Zhao HT, El-Kabbani O, Hara A. Kinetic studies of AKR1B10, human aldose reductase-like protein: endogenous substrates and inhibition by steroids. Arch Biochem Biophys. 2009;487:1–9.
129. Díez-Dacal B, Gayarre J, Gharbi S, Timms JF, Coderch C, Gago F, Pérez-Sala D. Identification of aldo-keto reductase AKR1B10 as a selective target for modification and inhibition by prostaglandin A(1): implications for antitumoral activity. Cancer Res. 2011;71:4161–71.
130. El-Kabbani O, Scammells PJ, Day T, Dhagat U, Endo S, Matsunaga T, Soda M, Hara A. Structure-Based optimization and biological evaluation of human 20α-hydroxysteroid dehydrogenase (AKR1C1) salicylic acid-based inhibitors. Eur J Med Chem. 2010;45:5309–17.
131. El-Kabbani O, Scammells PJ, Gosling J, Dhagat U, Endo S, Matsunaga T, Soda M, Hara A. Structure-guided design, synthesis, and evaluation of salicylic acid-based inhibitors targeting a selectivity pocket in the active site of human 20α-hydroxysteroid dehydrogenase (AKR1C1). J Med Chem. 2009;52:3259–64.
132. Dhagat U, Endo S, Sumii R, Hara A, El-Kabbani O. Selectivity determinants of inhibitor binding to human 20α-hydroxysteroid dehydrogenase: crystal structure of the enzyme in ternary complex with coenzyme and the potent inhibitor 3,5-dichlorosalicylic acid. J Med Chem. 2008;51:4844–8.
133. Higaki Y, Usami N, Shintani S, Ishikura S, El-Kabbani O, Hara A. Selective and potent inhibitors of human 20α-hydroxysteroid dehydrogenase (AKR1C1) that metabolizes neurosteroids derived from progesterone. Chem Biol Interact. 2003;143–144:503–13.
134. Brozic P, Smuc T, Gobec S, Rizner TL. Phytoestrogens as inhibitors of the human progesterone metabolizing enzyme AKR1C1. Mol Cell Endocrinol. 2006;259:30–42.
135. Skarydová L, Zivná L, Xiong G, Maser E, Wsól V. AKR1C3 as a potential target for the inhibitory effect of dietary flavonoids. Chem Biol Interact. 2009;178:138–44.
136. Byrns MC, Steckelbroeck S, Penning TM. An indomethacin analogue, N-(4-chlorobenzoyl)-melatonin, is a selective inhibitor of aldo-keto reductase 1C3 (type 2 3α-HSD, type 5 17β-HSD, and prostaglandin F synthase), a potential target for the treatment of hormone dependent and hormone independent malignancies. Biochem Pharmacol. 2008;75:484–93.
137. Adeniji AO, Twenter BM, Byrns MC, Jin Y, Chen M, Winkler JD, Penning TM. Development of potent and selective inhibitors of aldo-keto reductase 1C3 (Type 5 17β-hydroxysteroid dehydrogenase) based on N-phenyl-aminobenzoates and their structure-activity relationships. J Med Chem. 2012;55:2311–23.
138. Jamieson SM, Brooke DG, Heinrich D, Atwell GJ, Silva S, Hamilton EJ, Turnbull AP, Rigoreau LJ, Trivier E, Soudy C, Samlal SS, Owen PJ, Schroeder E, Raynham T, Flanagan JU, Denny WA. 3-(3,4-dihydroisoquinolin-2(1H)-ylsulfonyl)benzoic acids: highly potent and selective inhibitors of the type 5 17β-hydroxysteroid dehydrogenase AKR1C3. J Med Chem. 2012;55:7746–58.
139. Brožič P, Turk S, Adeniji AO, Konc J, Janežič D, Penning TM, Lanišnik Rižner T, Gobec S. Selective inhibitors of aldo-keto reductases AKR1C1 and AKR1C3 discovered by virtual screening of a fragment library. J Med Chem. 2012;55:7417–24.
140. Matsunaga T, Hosogai M, Arakaki M, Endo S, El-Kabbani O, Hara A. 9,10-Phenanthrenequinone induces monocytic differentiation of U937 cells through regulating expression of aldo-keto reductase 1C3. Biol Pharm Bull. 2012;35:1598–602.

# Overcoming Drug Resistance Through Elevation of ROS in Cancer

**Amit K. Maiti**

**Abstract** Drug resistance is the most devastating problem in treating cancer and drug-resistant cells are harder to kill with the same drug. The mechanism of drug resistance is believed to differ in various cancers for different anticancer drugs. Most of the anticancer agents induce the generation of Reactive Oxygen Species (ROS) to kill cancer cells by apoptosis. However, prolonged treatment with the same drug reduces the ROS level and this reduced ROS level causes drug-sensitive cancer cells to become drug-resistant. Exogenous ROS in conjunction with the same drug resensitizes these drug-resistant cells. Thus, the apoptosis of cancer cells by inducing ROS generation or drug resistance by lack of ROS could be the principle mechanisms of drug sensitivity or drug resistance in various cancer cells. The genetic mechanism of ROS-induced drug sensitivity and ROS depletion leading to drug resistance in various cancer cells with most of the anticancer drugs could involve 'common molecular pathways'. Understanding the molecular mechanism of ROS generation and maintenance could identify distinct targets for subsequent manipulation to elevate ROS levels in cancer cells. Thus, a 'combinational chemotherapy' could be designed using an anticancer drug while maintaining an elevated level of ROS in the cell during the drug treatment for developing a successful chemotherapy.

**Keywords** Apoptosis • ROS • Drug sensitivity • Drug resistance • Chemotherapy • Gene network

**Abbreviations**

| | |
|---|---|
| ABCB1 | ATP-binding cassette sub-family B member 1 |
| APEX1 | APEX nuclease (multifunctional DNA repair enzyme) 1 |
| CASPs | Caspases |

A. K. Maiti (✉)
Department of Arthritis and Clinical Immunology, Oklahoma Medical Research Foundation, Oklahoma city, OK 73104, USA
e-mail: amit-maiti@omrf.orgakmit123@yahoo.com

B. Bonavida (ed.), *Molecular Mechanisms of Tumor Cell Resistance to Chemotherapy*, Resistance to Targeted Anti-Cancer Therapeutics 1, DOI: 10.1007/978-1-4614-7070-0_7,

| | |
|---|---|
| CAT | Catalase |
| CDDO-ME | C-28 methyl ester derivative methyl-2-cyano-3,12-dioxooleana-1,9(11)-dien-28-oate |
| ERK1/2 | Extracellular-signal-regulated kinases |
| IPA | Ingenuity Pathway Analysis |
| GST | Glutathione-S-transferase |
| HIF 2-alpha | Hypoxia-inducible factor 2, alpha subunit |
| hTERT | Human telomerase reverse transcriptase |
| JUN | Jun proto-oncogene |
| KEAP1 | Kelch-like ECH-associated protein 1 |
| MAPK8 | Mitogen-activated protein kinase 8 |
| MCTS | Multi-cellular tumor spheroids |
| MDR1 | Multi drug resistance protein 1 |
| NFE2L2 | Nuclear factor (erythroid-derived 2)-like 2 |
| NF-$\kappa$B | Nuclear factor kappa-light-chain-enhancer of activated B cells |
| PEITC | Phynyl isothiocyanate |
| ROS | Reactive Oxygen Species |
| SOD | Superoxide dismutase |

## Introduction

Cancer, a complex disease, arises as a result of a progressive accumulation of genetic and epigenetic changes that lead to the breakdown of normal cell division checkpoints. The World Health Organization (WHO 2008) estimated that 10 million patients worldwide die of this disease every year [1]. The remarkable fact is that most deaths occur after subsequent medical intervention with anticancer therapy that includes radio, chemo, targeted, immune and gene therapy. Despite initial high response rates, a large proportion of patients relapse, resulting in a therapeutic challenge. Many of these patients are not curable with current chemotherapeutic strategies, so the goal of therapy is only to improve the quality and length of life [2]. Consequently, more focus is needed on understanding drug resistance in chemotherapy.

Chemotherapy is the most commonly used therapy that randomly kills rapidly growing cancer cells. These chemotherapeutic drugs are classified in different families such as DNA damaging agents (cisplatin, carboplatin), alkylating agents (chlorambucil, cyclophosphamide, temozolomide, carmustine), topoisomerase inhibitors (irinotecan, etoposide,), anti-metabolites (5-fluorouracil, methotrexate, capecitabine) and oncoprotein targeting antibodies and small molecule inhibitors [3]. These anticancer agents generally affect various cellular functions leading to rapid cell death. Although anticancer agents kill cancer cells, biological barriers make it difficult for anticancer drugs to access specific targets in cancer cells. The major problems are that efflux pumps (p-glycoprotein, encoded by ABCB1 or MDR1) that expel the drugs from these cells into the blood may lead to adequate levels of the drug in the bloodstream with only small concentrations actually reaching the targeted area and cells. Sometimes, drugs accumulate in isolated cellular areas, known as reservoirs, and never reach the targeted cells [4].

Therefore, improvements in the successful therapeutic management of cancer require an understanding of multi drug resistance (MDR) mechanisms and the identification of the underlying drug resistant pathways to successfully manipulate genes for all classes of tumors. This process is believed to be dependent on the cellular and microenvironmental context that include oncogene expression, apoptosis mechanisms, cell cycle control and regulation, DNA repair and mutation, vascularization, and many others.

The cellular signaling leading to these above mentioned molecular processes depends on redox signaling in the cell. Altered redox signaling affects the stability and integrity of cellular proteins, DNA, RNA and other small molecules and modifies normal cellular pathways. Thus, understanding the redox regulation and its role in developing drug resistance in cancer cells is immensely important in overcoming chemotherapeutic challenges.

## How ROS Changes Cellular Pathways

Reactive oxygen species (ROS) with a highly reactive oxygen atom (Fig. 1) reacts with DNA, amino acids of proteins and unsaturated fatty acids leading to oxidation of these biomolecules. ROS can lead to oxidation of amino acid residues of side chains, formation of protein–protein cross-linkages and oxidation of the protein backbone resulting in protein fragmentation [5]. These oxidized proteins, in turn,

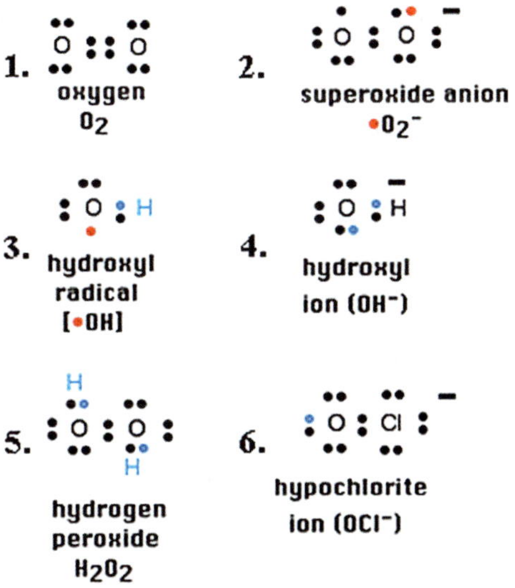


**Fig. 1** Most common reactive oxygen species (ROS). *Red* is the unpaired electron which makes an extremely unstable configuration and reacts with other molecules or radicals to achieve stable configuration. The superoxide anion, which is both ion (*2*) and radical (*1*). Hydroxyl radical (*3*) is the most reactive of all radicals. It differs from the hydroxyl ion (*4*) and hydrogen peroxide (*5*). Ions like the hypochlorite ion (*6*) is also very reactive than other ions

modify normal protein functions, and have profound effects in cellular signaling and create oxidative stresses in the cell, thus, compelling cells to adopt altered molecular pathways.

## How ROS is Generated or Maintained in Cancer Cells

ROS is generally produced at the electron transport chain of the mitochondria as a byproduct during ATP generation and at the plasma membrane to kill cellular microorganisms such as bacteria and viruses. Increasing evidence suggests that mitochondrial ROS generation is the principal contributor in the growth of cancer cells [6–8]. The Warburg effect states that cancer originates from a shift towards excess glycolysis due to defective mitochondrial oxidative phosphorylation that promotes differentiated cells to become undifferentiated cells [9]. Mutations in the TCA cycle enzymes, indeed, show mitochondrial dysfunction with excess glycolysis in cancer cells [10] and oncogene activation increases the chance of mitochondrial dysfunction [11].

External agents such as chemicals, radiation, and viruses induce mutations in cancer-causing genes and proteins that lead to cancer. Hypoxia (low glucose leading to low oxygen supply) increases HIF2-alpha expression leading to suppression of the DNA mismatch repair system that continually increases mitochondrial ROS generation and induces oxidative damage in the mitochondrial and nuclear genome. This view is widely supported as tumor cells have often found to harbor numerous mutations in the mitochondrial genome [12, 13] leading to abnormal metabolism and increased glycolysis [14] (Fig. 2). Thus, in cancer cells, a cycle is maintained when excess ROS induces mutations in the mitochondrial genome and excess mitochondrial genome mutations produce abnormal enzymes that induce ROS generation .

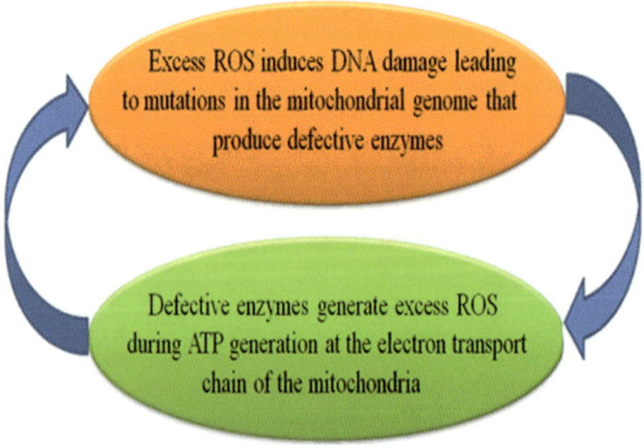


**Fig. 2** *ROS generation in cancer cells.* Excess ROS in cancer cells induces mitochondrial DNA damage leading to secondary mutations that virtually produce nonfunctional enzymes and, in turn, generates more ROS through aberrant respiration. The excess ROS production at the mitochondria elevates the ROS level in cancer cells

Apart from ROS generation, cancer cells have distinct mechanisms for ROS maintenance. Mitochondrial oxidative damage also triggers redox-sensitive transcription to regulate ROS levels by eliminating ROS-scavenging systems such as NFE2L2 (NRF2)-KEAP1, catalase, superoxide dismutases, glutathione peroxidases, peroxiredoxins, glutaredoxins and thioredoxins [15]. A general mechanism of ROS maintenance includes NF2EL2-KEAP1 systems (Fig. 3). NF2EL2 remains bound with KEAP1 in the cytoplasm. However, excess ROS in the cell releases KEAP1 from NF2EL2 and free NF2EL2 is phosphorylated by kinases and enters the nucleus. Phosphorylated NFE2L2 binds to Antioxidant Responsive Element (ARE) of the antioxidant genes such as catalase (*CAT*) or glutathione S-transferase (*GST*) [16] and facilitates transcription. The minimum sequence requirement for ARE is 5′-gagTcACaGTgAGtCggCAaaatt-3′ or TMAnnRTGAYnnnGCRwwww or TGA(C/T)nnnGCA [17, 18] and the presence of two or more copies of the ARE in close proximity to each other often serves as a bona fide ARE [19]. As a result of NFE2L2 binding at the promoter, increased transcription of antioxidant genes leads to increased protein synthesis that reduces the ROS level in the cell. However, all

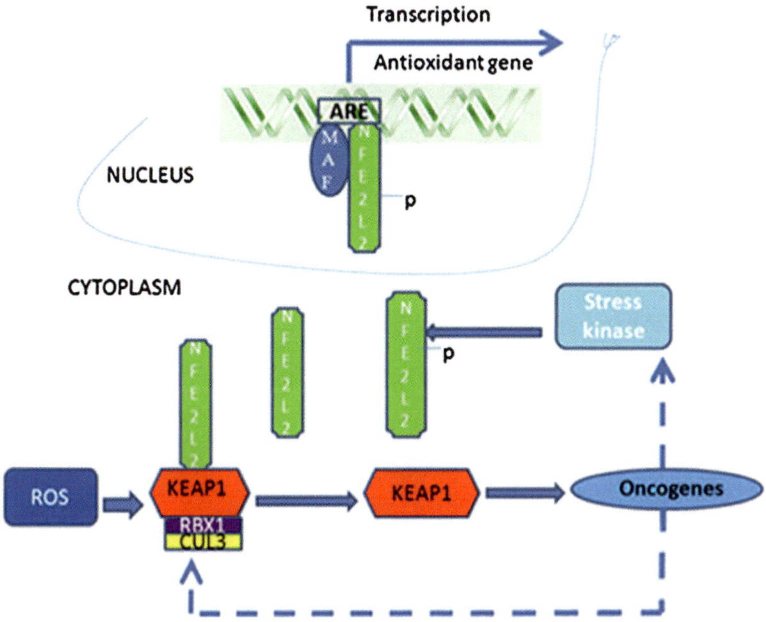


**Fig. 3** NFE2L2—KEAP1 *mechanism for maintaining the ROS* ***level in cancer cells***. NFE2L2 remains bound with KEAP1, RBX1 and CUL3 at the cytoplasm. An excess ROS release KEAP1, RBX1 and CUL3 from NFE2L2 and free NFE2L2 is phosphorylated in the cytoplasm. Phosphorylated NFE2L2 travels to the nucleus and binds at the promoter of *ARE* sequence carrying antioxidant genes to facilitate transcription leading to increased antioxidant enzymes production that reduce excess ROS. P, the phosphate group, MAF-oncoprotein v-MAF family members

antioxidant genes do not possess ARE sequences at their promoters such as SOD1, SOD2, SOD3 etc. and must be activated through an unknown mechanism.

## The Role of ROS in Tumor Development

ROS concentrations act as a double-edged sword for tumor progression as high concentrations of ROS are needed for cancer progression but are toxic to normal cells. The positive ROS signaling is a necessary prerequisite for the development of tumors [15, 20] and cancer cells have higher ROS content than normal cells [14, 21]. ROS acts as a second messenger and its signaling is necessary for cancer metastasis, cellular adhesion and spreading [22, 23]. Recent studies also strongly support the important relationship between the extracellular redox state and cancer cell aggressiveness [24]. ROS compartmentalization and distribution within cancer cells also plays a critical role in tumor progression [25]. ROS oxidizes and inhibits MAPK phosphatases resulting in enhanced cancer cell proliferation and survival. ROS also affects transcription through phosphorylation, activation, oxidation of transcription factors such as APEX1, NF-$\kappa$B complex, p53, and HIF-2 alpha leading to changes in targeted gene expression [26–29].

An opposite view is also speculated by the observation of the fact that the increase of some of the oncogenes such as *K-Ras, B-Raf and Myc* induce *NFE2L2* expression that detoxify excess ROS in the cell and induce tumorigenesis in mice [30]. This hypothesis considers that low level of ROS promotes tumorigenesis. Several factors should be considered before considering this apparent conflict between the ROS reduction and generation-inducing tumorigenesis. The effect of only three oncogenes on an increase of *NFE2L2* expression may not be sufficient to assume that overall reduction of the ROS level induces tumor development. In human lung cancer, several somatic mutations in *KEAP1* are identified and although the functions of these individual mutations are not studied [31], it is believed that these mutations abolish KEAP1 binding with NFE2L2. Therefore, free NFE2L2 could activate the antioxidant system to decrease the ROS level in lung cancer but the direct evidence for such an assumption is not verified. Unfortunately, Keap1$^{-/-}$ mice expressed *NFE2L2* consistently but did not live more than 21 days, making it impossible to assess its direct role in tumorigenesis [32]. In contrast to oncogenes, tumor suppressor genes, such as *p53* down regulation induces ROS generation during tumorigenesis [33] and have also been shown to harbor numerous mutations in many cancers [34, 35]. Thus, *p53* mutation supports the view that positive ROS signaling is necessary for tumorigenesis.

In addition, no germline mutations in *NFE2L2* or *KEAP1* have been identified in any cancers that would justify the notion that an increase in antioxidants facilitates tumor development. It is believed that cancer stem cells or normal cells when becoming cancerous do need increased ROS signaling [20, 23], but mature tumor cells cannot sustain excessive ROS generation due to abnormal mitochondrial mutations in cancer cells [6, 14, 36]. Any cell, including a cancer cell is vulnerable to excess ROS, thus somatic mutations in *NFE2L2* or *KEAP1* genes observed in cancer cells could activate antioxidant systems to reduce the ROS level that actually helps tumor progression, but may not initiate tumor development. Importantly, several *NFE2L2* mutations are also observed in lung cancer patients (11/103 patients) and are believed to help tumor progression [31]. It has also been proposed that excess ROS, generated by cancer cells, itself drives cancer cells from their primary site towards the bloodstream and is a molecular basis for metastasis [37]. This assumes that the adherence properties of cancer cells are reduced in the presence of excessive ROS signaling and ROS-mediated physiological changes help to prepare them to move from primary sites. Although extensive research is needed to establish this, it appears that tumorigenesis depends on a critical level of ROS in a dose-dependent manner but not solely on a decrease or increase of ROS level in the cell [38, 39]. By combining both views it is also possible that before transformation of normal cells to tumor cells, the ROS level is low but it increases in the tumor cells due to secondary mutations in the mitochondrial genome and then reaches a tolerable level of cancer cells before metastasis.

## Anticancer Drugs Induce the Generation of ROS to Kill Cancer Cells Through Apoptosis

Excessive ROS kills cancer cells through induction of apoptosis [40]. Most of the anticancer agents or drugs initially induce ROS generation to kill cancer cells through apoptosis [41, 42]. Numerous anticancer agents including rotenone, thermoquinone, sulforaphane, menadione, rosmarinic acid, tanshinone, chlorambucil, cisplatin, chrysophenol, dithiophene and others initially induce the generation of ROS in various cancer cells [43]. However, it is unknown how these anticancer agents influence cellular mechanisms to generate ROS. Direct relationships between anticancer drugs and antioxidant modulating genes or proteins are yet to be established. It is not known whether ROS generation is the only way to induce apoptosis, or if other mechanisms also play an important role in inducing apoptosis. It is evident that various anticancer agents induce ROS in various cancers by activating different genes and these genes activate the same set of master genes that could induce apoptosis. For example, in breast cancer, rotenone activates ERK1/2, JUN and MAPK8 [44], but sulforaphane is known to inhibit hTERT functions [45]. Dithiophene induces IL-24 in pancreatic cancer cells to activate apoptosis [46]. Similarly, in ovarian cancer, chlorambucil and cisplatin activate NF-$\kappa$B and p53 [47] whereas CDDO-ME (C-28 methyl ester derivative methyl-2-cyano-3,12-dioxooleana-1,9(11)-dien-28-oate) downregulates NFE2L2 [48]. *APEX1*, a DNA transcription factor and DNA repair gene, confers drug resistance through ROS generation and MDR activation [29]. Recently, Li et al. [49] showed that APEX1 regulates mitochondrial membrane potential and ROS production after photodynamic therapy of

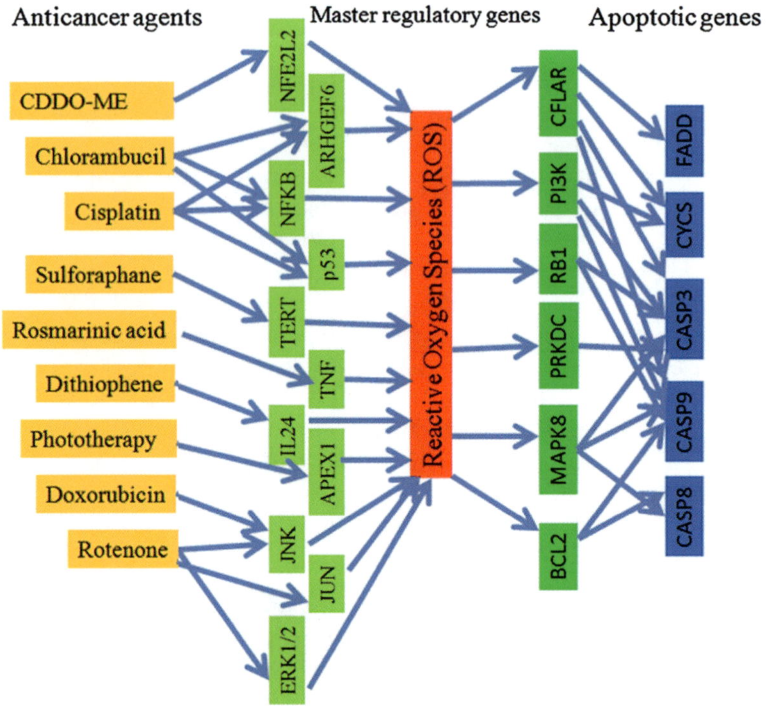


**Fig. 4** *ROS-mediated apoptosis through common molecular pathways*. Various anticancer drugs activate different genes to induce ROS generation. However, these genes act on some master regulatory genes that act on apoptotic genes leading to the induction of apoptosis. Thus, anticancer agents share 'common molecular pathways' for ROS-mediated apoptosis

lung cancer cells and induces apoptosis. Again, ROS activates a master set of genes such as *CFLAR, MAPK8, PRKDC, p53*, and *RB1* that act on apoptosis-inducing genes leading to apoptosis [43, 50, 51]. IPA analysis (Ingenuity Pathway Analysis; www.ingenuity.com) has shown the interacting pathways of these genes (Fig. 4). These master regulatory genes could activate Caspases *(CASPs)*, *FADD/MORT* and cytochrome C to activate apoptosis. Thus, the ROS-apoptosis model demonstrates that various anticancer agents in most of the cancer cells induce apoptosis through 'common molecular pathways' [52].

## ROS Reduction is the Key Underlying Mechanism of Drug Resistance

Although most of these anticancer drugs induce apoptosis through ROS generation, prolonged treatment with the same drug reduces the ROS level in ovarian cancer cells [47]. Thus, drug resistant cells have a lower ROS content than the drug sensitive cells. Addition of exogenous ROS in conjunction with the drug resensitizes drug resistant cells to drug sensitive cells. The specific mechanism for this observed phenomenon is unknown. Also, it is not clear whether a reduced ROS level makes the sensitive cancer cells become resistant or if resistant cells reduce ROS levels in the cell for their survival. More clearly, upon prolonged drug treatment, it is unclear whether ROS reduction is a primary mechanism for drug sensitive cancer cells to become drug resistant cells or a secondary consequence of resistant cells. Nevertheless, based on these observations, an ROS management cycle in cancer cells could be established, which demonstrates these events (Fig. 5).

In support of this view, evidence is accumulating that the reduced ROS level could be the primary reason for acquired drug resistance in various cancers. It is observed that drug resistant cells have a higher expression of catalase at the plasma membrane which could reduce some of the ROS level [53] in the cells derived from human ovarian, gastric and cervical carcinomas. In fibrosarcoma cells, p21-mediated apoptosis could be blocked by overexpression of catalase at the mitochondria of these cells [54]. Overexpression of human catalase in transgenic mice breast cancer model shows resistance to $H_2O_2$ associated oxidative stress [55]. Recent observations also indicate that overexpression of *NFE2L2*, that mechanistically should reduce ROS, actually confers resistance to lung and ovarian epithelial cancers for platinum based drugs [56, 57]. Depletion of glutathione S transferase (GST), an

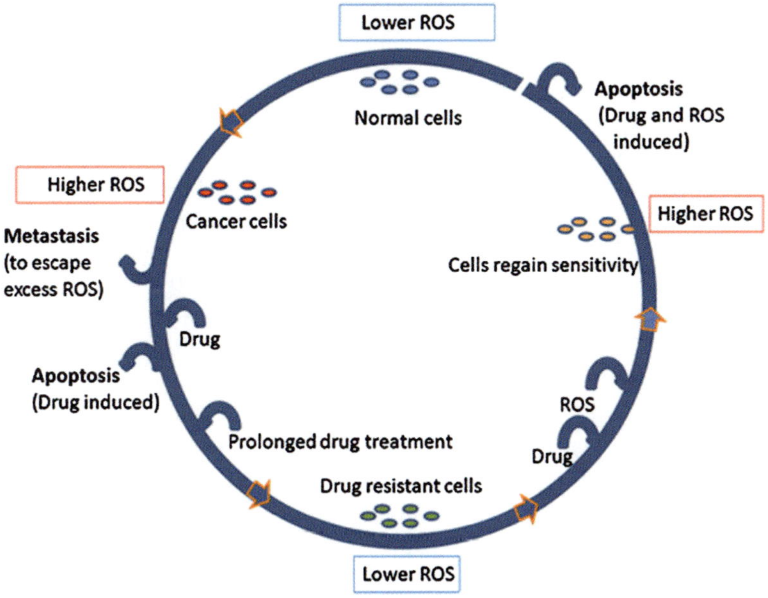


**Fig. 5** *ROS cycle of drug resistance*. A flow cycle of ROS-mediated drug resistance is shown where anticancer drugs initially induce the generation of ROS that kills cancer cells by apoptosis. However, prolonged treatment of the same drug reduces the ROS level and transforms drug-sensitive cancer cells to drug-resistant cells. Exogenous ROS in combination with the drug sensitizes drug-resistance cells

antioxidant producing enzyme, induces apoptosis through Phenyl isothiocyanate (PEITC) in MCF7 breast cancer cells [58]. However, it remains to be seen whether prolong treated anticancer drug-mediated ROS reduction is a general mechanism of various cancer cells for most of the anticancer drugs.

## Drug Resistance Could be Overcome Through Modulation of ROS Along with Drug Treatment as a 'Combinational Chemotherapy'

Elevation of the ROS level in combination with a drug is a useful strategy for 'combinational chemotherapy'. Along with drug treatment, ROS level in cancer cells could be elevated by modulating specific ROS level-elevating genes or proteins. It could be achieved by (1) inactivating ROS generating and maintenance genes or proteins, such as *CAT, SODs, GSTPs* or (2) modulating ROS regulatory master genes, such as, *ARHGEF6, NFE2L2, KEAP1, p53, APEX1* etc. or (3) direct delivery of ROS at the tumor site during chemotherapy.

(1) *Inactivating ROS generating and maintenance genes or proteins.*

Manipulation of drug-resistant mechanisms by redox modulation could have significant therapeutic implications by inhibiting the antioxidative enzyme systems of tumor cells [15, 53, 59]. These are superoxide dismutases (SOD1, SOD2 and SOD3), glutathione peroxidases, peroxiredoxins, glutaredoxins, thioredoxins and catalases. Using optimal concentrations of the catalase inhibitor [3-aminotriazole (3-AT)], or site-specific generation of hydroxyl radicals at the cell membrane of the tumor could prove useful for maintaining excess ROS levels in cancer cells [53]. The function of these target proteins could also be impaired by screening small molecules or siRNA or triplex oligo-mediated gene silencing.

Drugs such as mitocans are being developed that selectively target the mitochondria of malignant cells without adversely affecting mitochondria of normal cells. Mitocans selectively interfere with the bioenergetic functions of cancer cell mitochondria, causing major disruptions that lead to increased ROS production and induction of intrinsic apoptotic pathways [6].

(2) *Modulating ROS regulatory master genes.*

Modulation of ROS regulatory genes should have a significant impact to maintain the elevated level of ROS in the cell. As most cancers share common pathways for antioxidant regulation to induce apoptosis, these strategies could be useful for developing advanced chemotherapy for many cancers. siRNA silencing, triplex-mediated oligo or regulating gene specific microRNA (miRNA) could be used for elevation of ROS in cancer cells. Systematic knockdown of *NFE2L2* or *KEAP1* using siRNA confers resistance to carboplatin-treated epithelial ovarian cancer cells [57]. However, it is not known whether *KEAP1* downregulation overcomes complete resistance in these cells. Experiments with combined manipulation of several genes would be necessary for overcoming complete resistance in drug-treated cancer cells. Cancer cell specific delivery of inhibitors of these genes could be achieved using cancer specific biomarkers attached to suitable nanoparticles [60, 61]. In these cases, a tumor specific attached biomarker would guide the nanoparticles specifically to the tumor cells and the inhibitors directed to specific ROS-reducing genes, such as *NFE2L2 or ARHGEF6,* could be inhibited leading to excess ROS generation. Efficient delivery of oligo to cancer cells through lipoplexes and polyplexes has also been recently developed as a useful tool for efficient gene silencing [62].

(3) *Direct delivery of ROS at the tumor site during chemotherapy.*

Finally, the most important aspect is to explore the direct delivery of ROS into tumor cells along with drug treatment. Designing suitable vehicles to efficiently deliver or generate ROS in conjunction with specific drugs could help to overcome drug resistance in cancer cells in vivo. Among the emerging techniques, liposome-polycation-DNA (LPDI and LPDII) nanoparticles conjugated with an ROS-generating agent, is a promising discovery for treating cancer cells. LPD nanoparticles with a guanidinium containing cationic lipid such as DSAA [***N***,***N***-distearyl-***N***-methyl-***N***-2-(***N***′-arginyl) aminoethyl ammonium chloride] can induce ROS, down regulate MDR transporter expression (Pgp-glycoprotein mediated drug efflux), increase doxorubixin uptake and also shows a significant improvement in tumor growth inhibition [63]. Magnetically engineered nanoparticles also have advantages of inducing ROS in cancer cells [64]. In another approach, multicellular tumor spheroids (MCTS) offer an excellent in vitro system that mimics the endogenous oxidative stress often observed in tumors and could potentially be exploited for ROS generation in vivo [65].

## Conclusion

Increased generation of ROS with an altered redox status might not only be exploited for treatment of primary tumors, but has the potential to be used for sensitizing drug-resistant cancer. Numerous antitumor agents induce ROS production and activate ROS-dependent apoptotic pathways. As observed in ovarian carcinoma cells, anticancer drugs initially increase ROS levels, but prolonged treatment with the same drugs reduces ROS levels in cancer cells resulting in drug resistance. Thus, constant maintenance of higher ROS levels in cancer cells may be necessary for better drug efficacy. A rigorous investigation could be initiated to assess whether drug resistance due to ROS reduction is a general mechanism for most cancer types with most of the anticancer drugs. Nevertheless, using a ‘combinational chemotherapy’ with anticancer drug and elevation of ROS could be an efficient alternative to increase the efficacy of drug treatments, thereby precluding drug resistance.

**Acknowledgements** I thank Ms B. Mikkola for critically reading this manuscript.

**Conflicts of Interest** No potential conflicts of interest were disclosed.

## References

1. Higginson IJ, Costantini M. Dying with cancer, living well with advanced cancer. Eur J Cancer. 2008;44:1414–24.
2. Fung-Kee-Fung M, Oliver T, Elit L, Oza A, Hirte HW, Bryson P. Optimal chemotherapy treatment for women with recurrent ovarian cancer. Curr Oncol. 2007;14:195–208.
3. Lai D, Visser-Grieve S, Yang X. Tumour suppressor genes in chemotherapeutic drug response. Biosci Rep. 2012;32:361–74.
4. Donnenberg VS, Donnenberg AD. Multiple drug resistance in cancer revisited: the cancer stem cell hypothesis. J Clin Pharmacol. 2005;45:872–7.
5. Berlett BS, Stadtman ER. Protein oxidation in aging, disease, and oxidative stress. J Biol Chem. 1997;272:20313–6.
6. Ralph SJ, Rodriguez-Enriquez S, Neuzil J, Moreno-Sanchez R. Bioenergetic pathways in tumor mitochondria as targets for cancer therapy and the importance of the ROS-induced apoptotic trigger. Mol Aspects Med. 2010;31:29–59.
7. Weinberg F, Hamanaka R, Wheaton WW, Weinberg S, Joseph J, Lopez M, Kalyanaraman B, Mutlu GM, Budinger GR, Chandel NS. Mitochondrial metabolism and ROS generation are essential for Kras-mediated tumorigenicity. Proc Natl Acad Sci USA. 2010;107:8788–93.
8. Wallace DC. Mitochondria and cancer. Warburg addressed. Cold Spring Harb Symp Quant Biol. 2005;70:363–74.
9. Warburg O. On the origin of cancer cells. Science. 1956;123:309–14.
10. Gottlieb E, Tomlinson IP. Mitochondrial tumour suppressors. A genetic and biochemical update. Nat Rev Cancer. 2005;5:857–66.
11. Funes JM, Quintero M, Henderson S, Martinez D, Qureshi U, Westwood C, Clements MO, Bourboulia D, Pedley RB, Moncada S, Boshoff C. Transformation of human mesenchymal stem cells increases their dependency on oxidative phosphorylation for energy production. Proc Natl Acad Sci USA. 2007;104:6223–8.
12. Singh KK, Kulawiec M. Mitochondrial DNA polymorphism and risk of cancer. Methods Mol Biol. 2009;471:291–303.
13. Taylor RW, Turnbull DM. Mitochondrial DNA mutations in human disease. Nat Rev Genet. 2005;6:389–402.
14. Ralph SJ, Rodriguez-Enriquez S, Neuzil J, Saavedra E, Moreno-Sanchez R. The causes of cancer revisited: "mitochondrial malignancy" and ROS-induced oncogenic transformation—why mitochondria are targets for cancer therapy. Mol Aspects Med. 2010;31:145–70.
15. Trachootham D, Alexandre J, Huang P. Targeting cancer cells by ROS-mediated mechanisms: a radical therapeutic approach? Nat Rev Drug Discov. 2009;8:579–91.
16. Rushmore TH, Morton MR, Pickett CB. The antioxidant responsive element. Activation by oxidative stress and identification of the DNA consensus sequence required for functional activity. J Biol Chem. 1991;266:11632–9.
17. Nioi P, McMahon M, Itoh K, Yamamoto M, Hayes JD. Identification of a novel Nrf2-regulated antioxidant response element (ARE) in the mouse NAD(P)H: quinone

oxidoreductase 1 gene: reassessment of the ARE consensus sequence. Biochem J. 2003;374:337–48.
18. Wasserman WW, Fahl WE. Functional antioxidant responsive elements. Proc Natl Acad Sci USA. 1997;94:5361–6.
19. Nguyen T, Sherratt PJ, Pickett CB. Regulatory mechanisms controlling gene expression mediated by the antioxidant response element. Annu Rev Pharmacol Toxicol. 2003;43:233–60.
20. Liou GY, Storz P. Reactive oxygen species in cancer. Free Radical Res. 2010;44:479–96.
21. Wondrak GT. Redox-directed cancer therapeutics: molecular mechanisms and opportunities. Antioxid Redox Signal. 2009;11:3013–69.
22. Bechtel W, Bauer G. Modulation of intercellular ROS signaling of human tumor cells. Anticancer Res. 2009;29:4559–70.
23. Lau AT, Wang Y, Chiu JF. Reactive oxygen species: current knowledge and applications in cancer research and therapeutic. J Cell Biochem. 2008;104:657–67.
24. Chaiswing L, Oberley TD. Extracellular/microenvironmental redox state. Antioxid Redox Signal. 2010;13:449–65.
25. Maulucci G, Pani G, Fusco S, Papi M, Arcovito G, Galeotti T, Fraziano M, De Spirito M. Compartmentalization of the redox environment in PC-12 neuronal cells. Eur Biophys J. 2010;39:993–9.
26. Gottlieb E, Vousden KH. p53 regulation of metabolic pathways. Cold Spring Harb Perspect Biol. 2010;2:a001040.
27. Han YH, Moon HJ, You BR, Park WH. The effects of MAPK inhibitors on pyrogallol-treated Calu-6 lung cancer cells in relation to cell growth, reactive oxygen species and glutathione. Food Chem Toxicol. 2010;48:271–6.
28. Horak P, Crawford AR, Vadysirisack DD, Nash ZM, DeYoung MP, Sgroi D, Ellisen LW. Negative feedback control of HIF-1 through REDD1-regulated ROS suppresses tumorigenesis. Proc Natl Acad Sci USA. 2010;107:4675–80.
29. Chattopadhyay R, Das S, Maiti AK, Boldogh I, Xie J, Hazra TK, Kohno K, Mitra S, Bhakat KK. Regulatory role of human AP-endonuclease (APE1/Ref-1) in YB-1-mediated activation of the multidrug resistance gene MDR1. Mol Cell Biol. 2008;28:7066–80.
30. DeNicola GM, Karreth FA, Humpton TJ, Gopinathan A, Wei C, Frese K, Mangal D, Yu KH, Yeo CJ, Calhoun ES, Scrimieri F, Winter JM, Hruban RH, Iacobuzio-Donahue C, Kern SE, Blair IA, Tuveson DA. Oncogene-induced Nrf2 transcription promotes ROS detoxification and tumorigenesis. Nature. 2011;475:106–9.
31. Hayes JD, McMahon M. NRF2 and KEAP1 mutations: permanent activation of an adaptive response in cancer. Trends Biochem Sci. 2009;34:176–88.
32. Wakabayashi N, Itoh K, Wakabayashi J, Motohashi H, Noda S, Takahashi S, Imakado S, Kotsuji T, Otsuka F, Roop DR, Harada T, Engel JD, Yamamoto M. Keap1-null mutation leads to postnatal lethality due to constitutive Nrf2 activation. Nat Genet. 2003;35:238–45.
33. Sablina AA, Budanov AV, Ilyinskaya GV, Agapova LS, Kravchenko JE, Chumakov PM. The antioxidant function of the p53 tumor suppressor. Nat Med. 2005;11:1306–13.
34. Levine AJ. Tumor suppressor genes. Bioessays. 1990;12:60–6.
35. Hollstein M, Sidransky D, Vogelstein B, Harris CC. p53 mutations in human cancers. Science. 1991;253:49–53.
36. Ladiges W, Wanagat J, Preston B, Loeb L, Rabinovitch P. A mitochondrial view of aging, reactive oxygen species and metastatic cancer. Aging Cell. 2010;9:462–5.
37. Pani G, Galeotti T, Chiarugi P. Metastasis: cancer cell's escape from oxidative stress. Cancer Metastasis Rev. 2010;29:351–78.
38. Itoh K, Ishii T, Wakabayashi N, Yamamoto M. Regulatory mechanisms of cellular response to oxidative stress. Free Radical Res. 1999;31:319–24.
39. Storz P. Reactive oxygen species in tumor progression. Front Biosci. 2005;10:1881–96.
40. Overmeyer JH, Maltese WA. Death pathways triggered by activated Ras in cancer cells. Front Biosci. 2011;16:1693–713.

41. Lebedeva IV, Su ZZ, Sarkar D, Gopalkrishnan RV, Waxman S, Yacoub A, Dent P, Fisher PB. Induction of reactive oxygen species renders mutant and wild-type K-ras pancreatic carcinoma cells susceptible to Ad.mda-7-induced apoptosis. Oncogene. 2005;24:585–96.
42. Ozben T. Oxidative stress and apoptosis: impact on cancer therapy. J Pharm Sci. 2007;96:2181–96.
43. Maiti AK. Genetic determinants of oxidative stress-mediated sensitization of drug-resistant cancer cells. Int J Cancer. 2012;130:1–9.
44. El-Najjar N, Chatila M, Moukadem H, Vuorela H, Ocker M, Gandesiri M, Schneider-Stock R, Gali-Muhtasib H. Reactive oxygen species mediate thymoquinone-induced apoptosis and activate ERK and JNK signaling. Apoptosis. 2010;15:183–95.
45. Meeran SM, Patel SN, Tollefsbol TO. Sulforaphane causes epigenetic repression of hTERT expression in human breast cancer cell lines. PLoS One. 2010;5:e11457.
46. Su Z, Emdad L, Sauane M, Lebedeva IV, Sarkar D, Gupta P, James CD, Randolph A, Valerie K, Walter MR, Dent P, Fisher PB. Unique aspects of mda-7/IL-24 antitumor bystander activity: establishing a role for secretion of MDA-7/IL-24 protein by normal cells. Oncogene. 2005;24:7552–66.
47. Maiti AK. Gene network analysis of oxidative stress-mediated drug sensitivity in resistant ovarian carcinoma cells. Pharmacogenomics J. 2010;10:94–104.
48. Deeb D, Gao X, Jiang H, Janic B, Arbab AS, Rojanasakul Y, Dulchavsky SA, Gautam SC. Oleanane triterpenoid CDDO-Me inhibits growth and induces apoptosis in prostate cancer cells through a ROS-dependent mechanism. Biochem Pharmacol. 2010;79:350–60.
49. Li MX, Shan JL, Wang D, He Y, Zhou Q, Xia L, Zeng LL, Li ZP, Wang G, Yang ZZ. Human AP endonuclease 1 translocalizes to mitochondria after photodynamic therapy and protects cells from apoptosis. Cancer Sci. 2012;103:882–8.
50. Kim SH, Dass CR. p53-targeted cancer pharmacotherapy: move towards small molecule compounds. J Pharm Pharmacol. 2011;63:603–10.
51. Filomeni G, Piccirillo S, Rotilio G, Ciriolo MR. p38(MAPK) and ERK1/2 dictate cell death/survival response to different pro-oxidant stimuli via p53 and Nrf2 in neuroblastoma cells SH-SY5Y. Biochem Pharmacol. 2012;83:1349–57.
52. Maiti A. Reactive oxygen species reduction is a key underlying mechanism of drug resistance in cancer chemotherapy. Chemotherapy 2012;1:104-8.
53. Bechtel W, Bauer G. Catalase protects tumor cells from apoptosis induction by intercellular ROS signaling. Anticancer Res. 2009;29:4541–57.
54. Masgras I, Carrera S, de Verdier PJ, Brennan P, Majid A, Makhtar W, Tulchinksy E, Jones GD, Roninson IB, Macip S. Reactive oxygen species and mitochondrial sensitivity to oxidative stress determine induction of cancer cell death by p21. J Biol Chem. 2012;287:9845–54.
55. Goh J, Enns L, Fatemie S, Hopkins H, Morton J, Pettan-Brewer C, Ladiges W. Mitochondrial targeted catalase suppresses invasive breast cancer in mice. BMC Cancer. 2011;11:191–203.
56. Hayes JD, McMahon M, Chowdhry S, Dinkova-Kostova AT. Cancer chemoprevention mechanisms mediated through the Keap1-Nrf2 pathway. Antioxid Redox Signal. 2010;13:1713–48.
57. Konstantinopoulos PA, Spentzos D, Fountzilas E, Francoeur N, Sanisetty S, Grammatikos AP, Hecht JL, Cannistra SA. Keap1 mutations and Nrf2 pathway activation in epithelial ovarian cancer. Cancer Res. 2011;71:5081–9.
58. Syed Alwi SS, Cavell BE, Donlevy A, Packham G. Differential induction of apoptosis in human breast cancer cell lines by phenethyl isothiocyanate, a glutathione depleting agent. Cell Stress Chaperones 2012;17:529–38.
59. Fang J, Nakamura H, Iyer AK. Tumor-targeted induction of oxystress for cancer therapy. J Drug Target. 2007;15:475–86.
60. Miele E, Spinelli GP, Di Fabrizio E, Ferretti E, Tomao S, Gulino A. Nanoparticle-based delivery of small interfering RNA: challenges for cancer therapy. Int J Nanomed. 2012;7:3637–57.

61. Dreaden EC, Austin LA, Mackey MA, El-Sayed MA. Size matters: gold nanoparticles in targeted cancer drug delivery. Ther Deliv. 2012;3:457–78.
62. Ming X, Sato K, Juliano RL. Unconventional internalization mechanisms underlying functional delivery of antisense oligonucleotides via cationic lipoplexes and polyplexes. J Control Release. 2011;153:83–92.
63. Chen Y, Bathula SR, Li J, Huang L. Multi-functional nanoparticles delivering siRNA and doxorubicin overcome drug resistance in cancer. J Biol Chem. 2010;285:22639–50.
64. Shubayev VI, Pisanic TR 2nd, Jin S. Magnetic nanoparticles for theragnostics. Adv Drug Deliv Rev. 2009;61:467–77.
65. Khaitan D, Dwarakanath BS. Endogenous and induced oxidative stress in multi-cellular tumour spheroids: implications for improving tumour therapy. Indian J Biochem Biophys. 2009;46:16–24.

# Cancer Stem Cells in Resistance to Cytotoxic Drugs: Implications in Chemotherapy

**Man-Tzu Wang, Hongmei Jiang, Debasish Boral and Daotai Nie**

**Abstract** The efficacy of cytotoxic chemotherapy is limited by drug resistance presented by some cancer cells. Cancer stem cells (CSCs) are a sub-population of tumor cells that can initiate tumor formation. If chemotherapy kills bulk of cells within a tumor but not CSCs, the surviving CSCs can initiate the formation of recurrent tumors. This article discusses the inherent resistance of CSCs toward cytotoxic chemotherapy and some possible mechanisms involved. Approaches to target CSCs to improve the efficacy of chemotherapy will also be discussed.

**Keywords** Cancer stem cells · Drug resistance · Multidrug resistance · ATP binding cassette transporters · Quiescence · Cell survival · Nanog

## Abbreviations

| | |
|---|---|
| ABC | ATP binding cassette |
| ALDH | Aldehyde dehydrogenase |
| BCRP | Breast cancer resistance protein |
| CSCs | Cancer stem cells |
| EGFR | Epidermal growth factor receptor |
| ESC | Embryonic stem cells |
| LRCs | Label retaining cells |
| MDR | Multidrug resistance |
| SP | Side population |

M.-T. Wang · H. Jiang · D. Boral · D. Nie (✉)
Department of Medical Microbiology, Immunology, and Cell Biology,
Southern Illinois University School of Medicine and Simmons Cancer Institute,
PO Box 19626, Springfield, IL 62794, USA
e-mail: dnie@siumed.edu

B. Bonavida (ed.), *Molecular Mechanisms of Tumor Cell Resistance to Chemotherapy*, Resistance to Targeted Anti-Cancer Therapeutics 1, DOI: 10.1007/978-1-4614-7070-0_8,

## Introduction

Cytotoxic chemotherapy is often used in the management of advanced cancer, especially when the disease becomes metastatic. But the efficacy of chemotherapy is often limited by drug resistant cancer cells [1–4]. The major target of cytotoxic chemotherapy is the rapidly dividing cells found within a tumor mass. This has the inherent consequence of collateral damages to other rapidly dividing cells found in the bone marrow, the intestinal and respiratory epithelial lining and cells of the hair bulb. Due to the limited therapeutic windows for most chemotherapeutic agents and steep toxicity curves, the drug-resistant tumor cells cannot be killed simply by dose escalations.

Several possible mechanisms and molecular alterations associated with tumors have been implicated in resistance to chemotherapy, such as hypoxia associated with poor vascularization in tumors [5], activation of pro-surviving signals such as NF-$\kappa$B [6, 7], overexpression of p-glycoprotein [8–10], defects in apoptosis [11–16], and presence of drug-resistant cancer stem cells (CSCs). These mechanisms are not necessarily mutually exclusive; in fact, they often overlap. In this review, we will evaluate the role of CSCs in resistance to cytotoxic drugs and its implications in chemotherapy.

## The Conceptual Framework of Cancer Stem Cells

Tumors are quite heterogeneous. In a given tumor, there is considerable heterogeneity among tumor cells in terms of proliferation, differentiation, and their ability to form tumors when transplanted [4]. It is increasingly appreciated that cancer cells with stem cell-like properties are responsible for the initiation, maintenance and growth of the tumor [17]. Cancer stem cells (CSCs) are defined as a small subpopulation of cells within a tumor that possesses the capability to renew itself and gives rise to a tumor that can recapitulate all the heterogeneous cell lineages of tumor cells within the original tumors [18]. CSCs are also termed as "tumor initiating cells" based on their capacity to initiate tumors when transplanted into immuno-compromised mice. The number of CSCs varies from 5 to 30 % of the total malignant cell population [6]. Studies suggest that, although representing only a small percentage of cells within tumors, the CSCs isolated can reconstitute a new tumor with all the cell types represented in the tumor of origin when transplanted [19].

## Identification and Characterization of Cancer Stem Cells

A study of acute myelogeneous leukemia provided the first experimental evidence on the existence of cancer stem cells. A subpopulation of leukemic cells, which expresses the CD34 surface maker but lacks the CD38 marker, was able to recapitulate leukemia in NOD/SCID mice [20]. The cells exhibited a cell surface immune-phenotype similar to the normal hematopoietic stem cells. In addition to leukemia, CSCs have been isolated from different types of cancers, with different but sometimes overlapping profiles of cell surface markers (Table 1). For example, in breast cancer, as few as 200 $CD44^+/CD24^-$/low cells were able to form tumors in immune deficient mice, whereas injections of 20,000 cells from the remaining population failed to form tumors [21, 22]. The tumorigenic population gave rise to additional $CD44^+/CD24^-$/low epithelial tumors; these cells were able to form tumors in other immune deficient mice when they were serially transplanted [21, 22].

## Role of CSCs in Resistance to Cytotoxic Chemotherapy

A significant implication of the cancer stem cell theory is the intended target of cancer therapy such as chemo- or radio-therapy. If cancer treatment kills most of the cancer cells in the stage of transit amplifying and differentiation without killing the stem cells, the surviving CSCs will eventually lead to the recurrence of tumors. In fact, there are a number of studies demonstrating the increased resistance of CSCs toward chemotherapy [34]. For example, the CD133-positive cells isolated from a glioblastoma patient sample has exhibited significant resistance to the chemotherapeutic agents including temozolomide, carboplatin, and paclitaxel (Taxol) [35]. The increased resistance of CSCs to cytotoxic chemotherapy [34] raises an interesting question regarding how to kill CSCs in a more effective way.

**Table 1** Various CSCs isolated on the basis of cell surface markers by cell sorting techniques

| Cancer | CSC phenotype | Reference |
|---|---|---|
| AML | $CD34^+CD38^-CD90^-$ | [20] |
| Breast cancer | $ESA^+CD44^+CD24^-$/lowLin$^-$, $ALDH1^+$ | [21, 22] |
| Brain cancer | $CD133^+$ | [23] |
| Colon cancer | $CD133^+$ , $CD44^+$ $EpCam^+CD166^+$ | [24, 25] |
| Osteosarcoma | $CD133^+$ | [26] |
| Pancreatic cancer | $CD44^+CD24^+ESA^+$ | [27] |
| Prostate cancer | $CD44^+\alpha2\beta1^+CD133^+$ | [28] |
| CNS | $CD133^+$ | [29] |
| Head and neck | $CD44^+$ | [30] |
| Melanoma | $ABCB5^+$ | [31] |
| Liver | $CD133^+CD13^+$, $ALDH^+$ | [32, 33] |

The mechanism of action of most chemotherapeutic drugs includes the impairment of cell division or mitosis, DNA damages, and induction of apoptosis and other types of cell death. Although the chemotherapeutic drugs kill most cells in a tumor, the key question is whether CSCs can survive and cause tumor recurrence. Indeed, CSCs have an inherent resistance to chemotherapy thorough various mechanisms, including increased functionalities of ATP-binding cassette efflux transporters, increased stress response and pro-surviving signals, and increased DNA repairs. Finally, but not lastly, the quiescence of CSCs can pose significant resistance to chemotherapy whose cytotoxic chemotherapy is more geared to target the rapidly proliferating tumor cells than to those quiescent tumor cells.

## Role of ABC Efflux Transporters in CSC Resistance to Chemotherapy

Multi-drug resistance (MDR), a clinical phenomenon of decreased intracellular drug retention and changed tumor response, is one of the primary factors limiting effective therapy in cancer [36]. Many *in vitro* and *in vivo* models have been developed to study the development of MDR and to assess the potential clinical application of MDR modulators [7, 37]. The differential induction of adenosine triphosphate (ATP)-binding cassette (ABC) transporters has been associated with MDR in many cancers [38, 39]. These transporters transport solutes across membranes, and their roles in drug resistance have been extensively investigated. The ABC transporter gene products, such as ABCB1 (MDR1, P-gp), ABCG2 and BCRP, expressed in tumor cells allow cancer to resist the chemotherapies. For example, ABCB5, a super-family of ABC, has a significant role in chemo-resistance in human melanoma. When ABCB5 mRNA levels are intentionally reduced using siRNAs, sensitivity of melanoma tumor cells to various drugs, including 5-flourouracil, camptothecin, and mitoxantrone, improves considerably [40].

Both normal stem cells and CSCs commonly express drug pumps such as the ABC transporters, ABCB1 and BCRP. Historically, stem cells or CSCs have been isolated as the "side population" (SP) [37, 41], which express active efflux transporters [37, 41]. For example, CSCs enriched from leukemia SP fractions have an increased ability to pump daunorubicin and mitoxantrone out of the cell [42]. The CSCs have been found to be closely associated with the high expression of ABC transporters including ABCB1 and ABCG2. It has been experimentally shown that ABCG2 and other ABC transporters are responsible for the transporting of fluorescent dye out of normal murine and human cancer cells [43]. Among the common chemotherapeutics, vinblastine and paclitaxel are common substrate for ABCB1. Methotrexate, topotecan, and imatibib mesylate are substrates for BCRP. The enhanced expression and activities of the ABC transporters can enable CSCs to pump out chemotherapeutic agents, when they are the substrate for the ABC transporters, which can lead to an increased resistance to chemotherapy.

## Role of Drug Modifying Enzymes in CSC Resistance to Chemotherapy

Aldehyde dehydrogenases (ALDH) (EC 1.2.1.3) are a group of enzymes that catalyze the oxidation (dehydrogenation) of aldehydes, leading to the detoxification of exogenously and endogenously generated aldehydes [44]. ALDH1 activities are frequently used to isolate breast cancer stem cells, and ALDH1 expression is associated with a poor prognosis [45]. Increased ALDH activities were also noted in leukemic CSCs [46–48]. In liver cancer, ALDH is associated with liver CSCs [33]. The increased ALDH expression/activity may confer CSCs with increased resistance to chemotherapeutics such as cyclophosphamide, as ALDH may oxidize and inactivate aldophosphamide/4-hydroxycyclophosphamide (4-HC), the bioactive metabolic byproduct of cyclophosphamide [49].

## Role of Pro-Survival Signaling in CSC Resistance Toward Chemotherapy

Another mechanism of CSC resistance to chemotherapy is the preferential activation of pro-survival signaling. For example, hepatocellular carcinoma CSCs presented a preferential activation of the Akt signaling pathway in resistance to chemotherapy [50]. CD44, a receptor for hyaluronan (HA), is a major marker for CSCs in a variety of cancers. The binding of CD44 by HA can lead to the association of CD44 with the epidermal growth factor receptor (EGFR) [51, 52]. This association can cause the activation of MAP kinase and other cellular signaling pathways which promote cell survival in responses to chemotherapeutic drugs such as cisplatin, methotrexate, and adriamycin [51]. Further, the HA-CD44 interaction can activate the EGFR-elicited cellular signaling pathways without engagement of the ligand EGF [51, 52], which also leads to the resistance toward targeted anti-EGFR therapy. It is of interest to determine whether other CSC surface markers can initiate or potentiate pro-survival signaling in the resistance to chemotherapy.

## Role of Increased DNA Repairs of CSC: Potential Role in Drug Resistance?

In a seminal study, it was found that CSCs isolated from glioblastomas presented an increased resistance toward radiotherapy via the preferential activation of DNA-damage repair mechanisms [23]. It was found that, while radiation caused equal levels of damage to all cancer cells, CSCs repaired the damage more rapidly than non-CSCs, through the ready activation of the DNA damage and replication checkpoints that include the checkpoint kinases, Chk1 and Chk2 [23]. The increased ability of DNA repairs in CSCs can confer resistance toward chemotherapy, especially those that can cause DNA damages and checkpoint controls. For example, as a participant in the cytarabine-induced S-phase checkpoint activation, Chk1 can cause the resistance of AML cells toward cytarabine [53]. The inhibition of Chk1 sensitized AML cells toward cytarabine [53]. It is unknown, however, whether other mechanisms of DNA damage repairs are involved in the resistance of CSCs toward chemo- or radio-therapy. It is likely that the process is multi-factorial. Some unique features of stem cells or CSCs, such as slow cycling or quiescence, can make it possible for the cells to repair the damages.

## Role of CSC Quiescence in Resistance to Chemotherapy

Normal adult stem cells are usually slow-cycling and quiescent. Under steady-state conditions, adult stem cells divide asymmetrically to maintain tissue homeostasis by keeping their number constant. Asymmetric division refers to the asymmetric partitioning of cell components and/or placement of daughter cells relative to external microenvironments. It is a defining strategy used by stem cells to self-renew and at the same time generate progeny cells for differentiation. Selective segregation of template DNA strands is one feature of stem cells that has been utilized to characterize stem cells known as long term label retaining cells (LRCs) [54]. Normal stem cells cycle less frequently than the more differentiated transit-amplifying cells, and therefore, adult stem cells are often LRCs.

Since most cytotoxic chemotherapeutic agents preferentially target the rapidly proliferative cells, another mechanism of CSC drug resistance is the quiescence. CSCs from acute and chronic myelogenous leukemias are relatively quiescent [55, 56], which may cause their resistance to chemotherapy. It is conceivable that the CSC can enter a state of quiescence under substantial selection pressure from chemotherapy. Therefore, there must exist a period of relative lack of proliferation, during which the CSCs can reduce their metabolic rate and enter into a quiescent state to adapt to the new selection pressure.

## Role of Stem Cell Factors in Chemoresistance

The transcription factors Nanog, Oct3/4, and SOX2 constitute the core transcriptional regulatory circuitry in human embryonic stem cells (ESCs) [57, 58]. Ectopic expression of ESC-associated transcription factors can reprogram fibroblasts into pluripotent stem cells [59]. The Nanog expression at the mRNA level was detected in NT2 teratocarcinoma cells [60, 61], malignant cervical cancer [62], testicular carcinoma in site and germ tumors [63], oral carcinoma [64], Ewing's sarcoma [65], seminoma and breast carcinoma [66]. It was found that

Nanog functionally regulates tumor growth [67]. It has been demonstrated that NANOGP8, a pseudogene of NANOG, promotes cancer stem cell characteristics and enables prostate cancer to resist androgen deprivation therapy [68].

In addition to Nanog, Oct3/4 and SOX2 expressions in cancer have been reported. Oct3/4 expression promoted dedifferentiation of melanoma cells to CSC-like cells, with concurrent marked increase in resistance to chemotherapeutic agents [69]. Interestingly, Oct4-induced dedifferentiation was associated with the increased expression of endogenous Oct4, Nanog and Klf4 [69]. The stem cell factors also collaborate with CD44 to promote resistance of CSCs to cisplatin in head and neck squamous cell carcinoma [70]. In this study, it was found that HA stimulates the CD44v3 (an HA receptor) interaction with Oct4-Sox2-Nanog leading to both a complex formation and the nuclear translocation of three CSC transcription factors, stimulation of miR-302 expression in CSCs, cancer stem cell properties that include the resistance to chemotherapeutic agents such as cisplatin [70]. The studies suggest a critical role of stem cell factors in mediating CSC resistance to cytotoxic chemotherapy.

## Can We Target CSCs to Eliminate Drug Resistance and Tumor Recurrence?

Given its role in tumor initiation, CSCs should be targeted to eliminate tumor recurrence after therapeutic interventions such as chemotherapy. However, this effort is made difficult by the inherent resistance of CSCs to chemotherapy. It is important to elucidate how CSCs acquire resistance to chemotherapy so that a strategy can be formulated to sensitize CSCs to chemotherapy. There are a number of promising approaches reported to sensitize CSCs to cytotoxic chemotherapies. For example, colon CSCs, which were resistant to fluorouracil or oxaliplatin, can be sensitized by an interleukin-4 blocking antibody [71]. It was shown that the autocrine stimulation of interleukin-4 receptors on CSCs may contribute to their stemness including the drug-resistant phenotype [71]. The drug resistance of CSCs is likely multi-factorial. By further understanding of the biology of CSC drug resistance and the factors/pathways involved, we can develop means to eliminate those drug resistant cells and, therefore, tumor recurrence will not occur.

**Acknowledgments** We would like acknowledge the support from United States Department of Defense Prostate Cancer Research Program Pre-doctoral Fellowship Award W81XWH-10-1-0636 (M.T.W, H.J.), Breast Cancer Research Program Idea Award W81XWH-08-1-0540 (D.N.), and National Cancer Institute R01CA13445 (D.N.) and R15CA133776 (D.N.).

**Conflicts of Interest** No potential conflicts of interest were disclosed.

## References

1. Bates SE, Regis JI, Robey RW, Zhan Z, Scala S, Meadows BJ. Chemoresistance in the clinic: overview 1994. Bull Cancer. 1994;81(Suppl 2):55s–61s.
2. Gupta AK, McKenna WG, Weber CN, Feldman MD, Goldsmith JD, Mick R, Machtay M, Rosenthal DI, Bakanauskas VJ, Cerniglia GJ, Bernhard EJ, Weber RS, Muschel RJ. Local recurrence in head and neck cancer: relationship to radiation resistance and signal transduction. Clin Cancer Res. 2002;8:885–92.
3. Issa JP, Gharibyan V, Cortes J, Jelinek J, Morris G, Verstovsek S, Talpaz M, Garcia-Manero G, Kantarjian HM. Phase II study of low-dose decitabine in patients with chronic myelogenous leukemia resistant to imatinib mesylate. J Clin Oncol. 2005;23:3948–56.
4. Mansouri A, Henle KJ, Nagle WA. Tumor drug-resistance: a challenge to therapists and biologists. Am J Med Sci. 1994;307:438–44.
5. Harrison L, Blackwell K. Hypoxia and anemia: factors in decreased sensitivity to radiation therapy and chemotherapy? Oncologist. 2004;9(Suppl 5):31–40.
6. Cusack JC Jr, Liu R, Houston M, Abendroth K, Elliott PJ, Adams J, Baldwin AS Jr. Enhanced chemosensitivity to CPT-11 with proteasome inhibitor PS-341: implications for systemic nuclear factor-kappaB inhibition. Cancer Res. 2001;61:3535–40.
7. Camp ER, Li J, Minnich DJ, Brank A, Moldawer LL, MacKay SL, Hochwald SN. Inducible nuclear factor-kappaB activation contributes to chemotherapy resistance in gastric cancer. J Am Coll Surg. 2004;199:249–58.
8. Johnson WW. P-glycoprotein-mediated efflux as a major factor in the variance of absorption and distribution of drugs: modulation of chemotherapy resistance. Methods Find Exp Clin Pharmacol. 2002;24:501–14.
9. Haber M, Smith J, Bordow SB, Flemming C, Cohn SL, London WB, Marshall GM, Norris MD. Association of high-level MRP1 expression with poor clinical outcome in a large prospective study of primary neuroblastoma. J Clin Oncol. 2006;24:1546–53.
10. Hua J, Mutch DG, Herzog TJ. Stable suppression of MDR-1 gene using siRNA expression vector to reverse drug resistance in a human uterine sarcoma cell line. Gynecol Oncol. 2005;98:31–8.
11. Fink D, Nebel S, Aebi S, Zheng H, Cenni B, Nehmé A, Christen RD, Howell SB. The role of DNA mismatch repair in platinum drug resistance. Cancer Res. 1996;56:4881–6.
12. Ioffe ML, White E, Nelson DA, Dvorzhinski D, DiPaola RS. Epothilone induced cytotoxicity is dependent on p53 status in prostate cells. Prostate. 2004;61:243–7.
13. Murata T, Haisa M, Uetsuka H, Nobuhisa T, Ookawa T, Tabuchi Y, Shirakawa Y, Yamatsuji T, Matsuoka J, Nishiyama M, Tanaka N, Naomoto Y. Molecular mechanism of chemoresistance to cisplatin in ovarian cancer cell lines. Int J Mol Med. 2004;13:865–8.
14. Pommier Y, Sordet O, Antony S, Hayward RL, Kohn KW. Apoptosis defects and chemotherapy resistance: molecular interaction maps and networks. Oncogene. 2004;23:2934–49.
15. Righetti SC, Della Torre G, Pilotti S, Ménard S, Ottone F, Colnaghi MI, Pierotti MA, Lavarino C, Cornarotti M, Oriana S, Böhm S, Bresciani GL, Spatti G, Zunino F. A comparative study of p53 gene mutations, protein accumulation, and response to cisplatin-based chemotherapy in advanced ovarian carcinoma. Cancer Res. 1996;56:689–93.
16. Sartorius UA, Krammer PH. Upregulation of Bcl-2 is involved in the mediation of chemotherapy resistance in human small cell lung cancer cell lines. Int J Cancer. 2002;97:584–92.
17. Wicha MS, Liu S, Dontu G. Cancer stem cells: an old idea–a paradigm shift. Cancer Res. 2006;66:1883–90; discussion 95–6.
18. Baumann M, Krause M, Hill R. Exploring the role of cancer stem cells in radioresistance. Nat Rev Cancer. 2008;8:545–54.

19. Clarke MF, Dick JE, Dirks PB, Eaves CJ, Jamieson CH, Jones DL, Visvader J, Weissman IL, Wahl GM. Cancer stem cells–perspectives on current status and future directions: AACR workshop on cancer stem cells. Cancer Res. 2006;66:9339–44.
20. Lapidot T, Sirard C, Vormoor J, Murdoch B, Hoang T, Caceres-Cortes J, Minden M, Paterson B, Caligiuri MA, Dick JE. A cell initiating human acute myeloid leukaemia after transplantation into SCID mice. Nature. 1994;367:645–8.
21. Elliott A, Adams J, Al-Hajj M. The ABCs of cancer stem cell drug resistance. IDrugs. 2010;13:632–5.
22. Al-Hajj M, Wicha MS, Benito-Hernandez A, Morrison SJ, Clarke MF. Prospective identification of tumorigenic breast cancer cells. Proc Natl Acad Sci U S A. 2003;100:3983–8.
23. Bao S, Wu Q, McLendon RE, Hao Y, Shi Q, Hjelmeland AB, Dewhirst MW, Bigner DD, Rich JN. Glioma stem cells promote radioresistance by preferential activation of the DNA damage response. Nature. 2006;444:756–60.
24. Dalerba P, Clarke MF. Cancer stem cells and tumor metastasis: first steps into uncharted territory. Cell Stem Cell. 2007;1:241–2.
25. O'Brien CA, Pollett A, Gallinger S, Dick JE. A human colon cancer cell capable of initiating tumour growth in immunodeficient mice. Nature. 2007;445:106–10.
26. Tirino V, Desiderio V, d'Aquino R, De Francesco F, Pirozzi G, Graziano A, Galderisi U, Cavaliere C, De Rosa A, Papaccio G, Giordano A. Detection and characterization of CD133 + cancer stem cells in human solid tumours. PLoS ONE. 2008;3:e3469.
27. Lee JT, Herlyn M. Old disease, new culprit: tumor stem cells in cancer. J Cell Physiol. 2007;213:603–9.
28. Maitland NJ, Collins AT. Prostate cancer stem cells: a new target for therapy. J Clin Oncol. 2008;26:2862–70.
29. Dell'Albani P. Stem cell markers in gliomas. Neurochem Res. 2008;33:2407–15.
30. Gammon L, Biddle A, Fazil B, Harper L, Mackenzie IC. Stem cell characteristics of cell sub-populations in cell lines derived from head and neck cancers of Fanconi anemia patients. J Oral Pathol Med. 2011;40:143–52.
31. Quintana E, Shackleton M, Foster HR, Fullen DR, Sabel MS, Johnson TM, Morrison SJ. Phenotypic heterogeneity among tumorigenic melanoma cells from patients that is reversible and not hierarchically organized. Cancer Cell. 2010;18:510–23.
32. Gavert N, Vivanti A, Hazin J, Brabletz T, Ben-Ze'ev A. L1-mediated colon cancer cell metastasis does not require changes in EMT and cancer stem cell markers. Mol Cancer Res. 2011;9:14–24.
33. Ma S, Chan KW, Lee TK, Tang KH, Wo JY, Zheng BJ, Guan XY. Aldehyde dehydrogenase discriminates the CD133 liver cancer stem cell populations. Mol Cancer Res. 2008;6:1146–53.
34. Reya T, Morrison SJ, Clarke MF, Weissman IL. Stem cells, cancer, and cancer stem cells. Nature. 2001;414:105–11.
35. Allenspach EJ, Maillard I, Aster JC, Pear WS. Notch signaling in cancer. Cancer Biol Ther. 2002;1:466–76.
36. Leonessa F, Clarke R. ATP binding cassette transporters and drug resistance in breast cancer. Endocr Relat Cancer. 2003;10:43–73.
37. Donnenberg VS, Donnenberg AD. Multiple drug resistance in cancer revisited: the cancer stem cell hypothesis. J Clin Pharmacol. 2005;45:872–7.
38. Minemura M, Tanimura H, Tabor E. Overexpression of multidrug resistance genes MDR1 and cMOAT in human hepatocellular carcinoma and hepatoblastoma cell lines. Int J Oncol. 1999;15:559–63.
39. Doyle LA, Yang W, Abruzzo LV, Krogmann T, Gao Y, Rishi AK, Ross DD. A multidrug resistance transporter from human MCF-7 breast cancer cells. Proc Natl Acad Sci U S A. 1998;95:15665–70.
40. La Porta CA. Drug resistance in melanoma: new perspectives. Curr Med Chem. 2007;14:387–91.

41. Seigel GM, Campbell LM, Narayan M, Gonzalez-Fernandez F. Cancer stem cell characteristics in retinoblastoma. Mol Vis. 2005;11:729–37.
42. Wulf GG, Wang RY, Kuehnle I, Weidner D, Marini F, Brenner MK, Andreeff M, Goodell MA. A leukemic stem cell with intrinsic drug efflux capacity in acute myeloid leukemia. Blood. 2001;98:1166–73.
43. Chiba S. Notch signaling in stem cell systems. Stem Cells. 2006;24:2437–47.
44. Marchitti SA, Brocker C, Stagos D, Vasiliou V. Non-P450 aldehyde oxidizing enzymes: the aldehyde dehydrogenase superfamily. Expert Opin Drug Metab Toxicol. 2008;4:697–720.
45. Ginestier C, Hur MH, Charafe-Jauffret E, Monville F, Dutcher J, Brown M, Jacquemier J, Viens P, Kleer CG, Liu S, Schott A, Hayes D, Birnbaum D, Wicha MS, Dontu G. ALDH1 is a marker of normal and malignant human mammary stem cells and a predictor of poor clinical outcome. Cell Stem Cell. 2007;1:555–67.
46. Pearce DJ, Taussig D, Simpson C, Allen K, Rohatiner AZ, Lister TA, Bonnet D. Characterization of cells with a high aldehyde dehydrogenase activity from cord blood and acute myeloid leukemia samples. Stem Cells. 2005;23:752–60.
47. Pearce DJ, Taussig D, Zibara K, Smith LL, Ridler CM, Preudhomme C, Young BD, Rohatiner AZ, Lister TA, Bonnet D. AML engraftment in the NOD/SCID assay reflects the outcome of AML: implications for our understanding of the heterogeneity of AML. Blood. 2006;107:1166–73.
48. Taussig DC, Pearce DJ, Simpson C, Rohatiner AZ, Lister TA, Kelly G, Luongo JL, Danet-Desnoyers GA, Bonnet D. Hematopoietic stem cells express multiple myeloid markers: implications for the origin and targeted therapy of acute myeloid leukemia. Blood. 2005;106:4086–92.
49. Vasiliou V, Pappa A, Estey T. Role of human aldehyde dehydrogenases in endobiotic and xenobiotic metabolism. Drug Metab Rev. 2004;36:279–99.
50. Ma S, Lee TK, Zheng BJ, Chan KW, Guan XY. CD133 + HCC cancer stem cells confer chemoresistance by preferential expression of the Akt/PKB survival pathway. Oncogene. 2008;27:1749–58.
51. Wang SJ, Bourguignon LY. Hyaluronan and the interaction between CD44 and epidermal growth factor receptor in oncogenic signaling and chemotherapy resistance in head and neck cancer. Arch Otolaryngol–head Neck Surg. 2006;132:771–8.
52. Wang SJ, Bourguignon LY. Hyaluronan-CD44 promotes phospholipase C-mediated Ca2 + signaling and cisplatin resistance in head and neck cancer. Arch Otolaryngol–head Neck Surg. 2006;132:19–24.
53. Schenk EL, Koh BD, Flatten KS, Peterson KL, Parry D, Hess AD, Smith BD, Karp JE, Karnitz LM, Kaufmann SH. Effects of selective checkpoint kinase 1 inhibition on cytarabine cytotoxicity in acute myelogenous leukemia cells in vitro. Clin Cancer Res. 2012;18:5364–73.
54. Welm BE, Tepera SB, Venezia T, Graubert TA, Rosen JM, Goodell MA. Sca-1(pos) cells in the mouse mammary gland represent an enriched progenitor cell population. Dev Biol. 2002;245:42–56.
55. Holyoake T, Jiang X, Eaves C, Eaves A. Isolation of a highly quiescent subpopulation of primitive leukemic cells in chronic myeloid leukemia. Blood. 1999;94:2056–64.
56. Guan Y, Hogge DE. Proliferative status of primitive hematopoietic progenitors from patients with acute myelogenous leukemia (AML). Leukemia. 2000;14:2135–41.
57. Orkin SH. Chipping away at the embryonic stem cell network. Cell. 2005;122:828–30.
58. Boyer LA, Lee TI, Cole MF, Johnstone SE, Levine SS, Zucker JP, Guenther MG, Kumar RM, Murray HL, Jenner RG, Gifford DK, Melton DA, Jaenisch R, Young RA. Core transcriptional regulatory circuitry in human embryonic stem cells. Cell. 2005;122:947–56.
59. Takahashi K, Tanabe K, Ohnuki M, Narita M, Ichisaka T, Tomoda K, Yamanaka S. Induction of pluripotent stem cells from adult human fibroblasts by defined factors. Cell. 2007;131:861–72.

60. Mitsui K, Tokuzawa Y, Itoh H, Segawa K, Murakami M, Takahashi K, Maruyama M, Maeda M, Yamanaka S. The homeoprotein Nanog is required for maintenance of pluripotency in mouse epiblast and ES cells. Cell. 2003;113:631–42.
61. Clark AT, Rodriguez RT, Bodnar MS, Abeyta MJ, Cedars MI, Turek PJ, Firpo MT, Reijo Pera RA. Human STELLAR, NANOG, and GDF3 genes are expressed in pluripotent cells and map to chromosome 12p13, a hotspot for teratocarcinoma. Stem Cells. 2004;22:169–79.
62. Ye F, Zhou C, Cheng Q, Shen J, Chen H. Stem-cell-abundant proteins Nanog, Nucleostemin and Musashi1 are highly expressed in malignant cervical epithelial cells. BMC Cancer. 2008;8:108–13.
63. Hoei-Hansen CE, Almstrup K, Nielsen JE, Brask Sonne S, Graem N, Skakkebaek NE, Leffers H, Rajpert-De Meyts E. Stem cell pluripotency factor NANOG is expressed in human fetal gonocytes, testicular carcinoma in situ and germ cell tumours. Histopathology. 2005;47:48–56.
64. Chiou SH, Yu CC, Huang CY, Lin SC, Liu CJ, Tsai TH, Chou SH, Chien CS, Ku HH, Lo JF. Positive correlations of Oct-4 and Nanog in oral cancer stem-like cells and high-grade oral squamous cell carcinoma. Clin Cancer Res. 2008;14:4085–95.
65. Suvà ML, Riggi N, Stehle JC, Baumer K, Tercier S, Joseph JM, Suvà D, Clément V, Provero P, Cironi L, Osterheld MC, Guillou L, Stamenkovic I. Identification of cancer stem cells in Ewing's sarcoma. Cancer Res. 2009;69:1776–81.
66. Ezeh UI, Turek PJ, Reijo RA, Clark AT. Human embryonic stem cell genes OCT4, NANOG, STELLAR, and GDF3 are expressed in both seminoma and breast carcinoma. Cancer. 2005;104:2255–65.
67. Jeter CR, Badeaux M, Choy G, Chandra D, Patrawala L, Liu C, Calhoun-Davis T, Zaehres H, Daley GQ, Tang DG. Functional evidence that the self-renewal gene NANOG regulates human tumor development. Stem Cells. 2009;27:993–1005.
68. Jeter CR, Liu B, Liu X, Chen X, Liu C, Calhoun-Davis T, Repass J, Zaehres H, Shen JJ, Tang DG. NANOG promotes cancer stem cell characteristics and prostate cancer resistance to androgen deprivation. Oncogene. 30:3833–45.
69. Kumar SM, Liu S, Lu H, Zhang H, Zhang PJ, Gimotty PA, Guerra M, Guo W, Xu X. Acquired cancer stem cell phenotypes through Oct4-mediated dedifferentiation. Oncogene. 2012;31:4898–911.
70. Bourguignon LY, Wong G, Earle C, Chen L. Hyaluronan-CD44v3 interaction with Oct4-Sox2-Nanog promotes miR-302 expression leading to self-renewal, clonal formation, and cisplatin resistance in cancer stem cells from head and neck squamous cell carcinoma. J Biol Chem. 2012;287:32800–24.
71. Todaro M, Alea MP, Di Stefano AB, Cammareri P, Vermeulen L, Iovino F, Tripodo C, Russo A, Gulotta G, Medema JP, Stassi G. Colon cancer stem cells dictate tumor growth and resist cell death by production of interleukin-4. Cell Stem Cell. 2007;1:389–402.

# Two Birds with a Stone: Molecular Cancer Therapy Targeting Signal Transduction and DNA Repair Pathways

**Elisa Zucca, Emmanuele Crespan, Federica Bertoletti, Miroslava Kissova and Giovanni Maga**

**Abstract** The hallmarks of cancer cells are a higher proliferative activity and an aberrant genotype with respect to normal cells. These features can be exploited for the development of selective chemotherapeutic treatments against cancer. In particular, the connections among signal transduction pathways, cell cycle checkpoints and DNA replication and repair have the potential to provide new venues for the treatment of cancer. Here, we will review how the differences existing between normal and tumour cells, with respect to control of cell proliferation and maintenance of the genetic stability, can be exploited in cancer chemotherapy.

**Keywords** DNA repair · DNA polymerase · Cancer · Tyrosine kinase · Signal transduction · Anticancer drugsanticancer drugs

**Abbreviations**

| | |
|---|---|
| BIR | Break induced replication |
| CK2 | Casein kinase II |
| CML | Chronic myeloid leukemia |
| DDR | DNA damage response |
| DSBs | Double strand breaks |
| GISTs | Gastrointestinal stromal tumours |
| HER-2 | Human epidermal growth factor receptor 2 |
| MMEJ | Microhomology-mediated end joining |
| NRTKs | Non receptor tyrosine kinases |
| NSCLC | Non-small cell lung cancer |
| PDGFR | Platelet-derived growth factor receptor |
| PTEN | Phosphatase and tensin homology protein |
| SFK | Src family kinase |

E. Zucca · E. Crespan · F. Bertoletti · M. Kissova · G. Maga (✉)
Institute of Molecular Genetics, IGM-CNR, via Abbiategrasso 207, 27100 Pavia, Italy
e-mail: maga@igm.cnr.it

B. Bonavida (ed.), *Molecular Mechanisms of Tumor Cell Resistance to Chemotherapy*, Resistance to Targeted Anti-Cancer Therapeutics 1, DOI: 10.1007/978-1-4614-7070-0_9,

| | |
|---|---|
| SH2 | Src homology-2 |
| SH3 | Src homology-3 |
| SSA | Single-strand annealing |
| SSBs | Single strand-breaks |
| STKs | Serine-threonine kinases |
| TKs | Tyrosine kinases |
| VEGFR | Vascular endothelial growth factor receptor |

## Introduction

Tumour cells are characterized by a higher proliferative activity with respect to the surrounding cells. This hallmark of cancer cells has been regarded since the beginning as a feature to be exploited for the development of selective chemotherapeutic treatments against cancer. Indeed, anticancer chemotherapy can be regarded as the science of selective toxicity, since it is aimed at reducing the proliferation of cancer cells with minimal perturbation of the homeostasis of normal cells.

In order to achieve selectivity, however, a first requirement to be fulfilled is the identification of a suitable target. Ideally, such a molecular entity should play an essential role in cancer cells, while being dispensable for the normal life of healthy cells. Once a suitable target is identified, small molecules need to be developed which selectively suppress that particular molecular function, without interfering with other similar proteins eventually present in the cells.

In practice, neither of these two goals has been fully achieved in contemporary anticancer chemotherapy. Classical anticancer drugs, in fact, target proteins which are involved in the proliferation of both normal and diseased cells. Selectivity is achieved on the basis that in adult organisms only a small subset of cells have proliferative indexes similar to cancer ones, thus administering the correct doses of drugs for a limited period of time, may achieve suppression of cancer growth without making too much damage to the normal cells.

In recent years, thanks to the advancement of our understanding of the physiology of cancer cells, it was realized that tumourigenesis is almost invariably driven by aberrations in various signal transduction and DNA repair pathways. As a result, cancer cells are dependent, for their survival, on a narrower complement of molecular functions than normal cells. Fostered by these findings, novel targets have been identified, such as tyrosine kinases and DNA repair enzymes. Selective inhibitors of these new targets have been developed and have already entered the clinics. However, after a few years, also these new approaches did not completely fulfill the expectations of the researchers. In many instances, drug resistance readily developed and, thus, reducing the efficacy of the drugs. In addition, alternative pathways, sometimes tumour-specific, can be activated in cancer cells to surrogate for the functions inhibited by the drugs.

As a result, it is now clear that a shift of paradigm is required to develop a new generation of anticancer drugs. Genome-wide analyses of cancer cells proteomes and transcriptomes have provided the experimental evidence for the existence of truly different molecular phenotypes among different tumour types. Thus, cell proliferation cannot be considered anymore, according to the simplistic view of the past, as the hallmark of all cancers since the molecular pathways leading to uncontrolled proliferation are very diverse.

A more fruitful approach should be based on the understanding of the links connecting signal transduction pathways to cell cycle checkpoints and DNA replication and repair. Proliferative signals, which are transduced by tyrosine kinases, beside inducing DNA replication, also activate those pathways that are required to maintain a level of genomic stability compatible with cell survival, including DNA repair. These pathways are the same that also allow tumour cells to cope with the DNA damage induced by classic anticancer agents such as etoposide or cis-Pt. At the same time, tumours are very often defective in one or more DNA repair pathways, thus depending on the remaining ones for their survival. Thus, a strategy aimed to specifically target both signal transduction and DNA damage tolerance pathways may prove to be effective on a large fraction of tumours. An additional benefit of such a combination targeted chemotherapy could be to reduce the emergence of drug resistance. In fact, the genetic barrier required to develop at least one independent mutation for each target will be higher than in the case of monotherapy regimens. To date, several signal-transducing kinases are being regarded as attractive targets for selective cancer chemotherapy. In addition, the realization that in human cells there are at least 15 different DNA polymerases playing non-overlapping roles in DNA replication and repair, has provided entirely new venues for the development of novel cancer drugs.

## Targeting Signal Transducing Pathways

Phosphorylation is a fundamental mechanism used in transduction pathways to propagate the signal to final effectors. The reaction consists in the transfer of the — phosphate from ATP to amino acidic residues in specific peptide substrates. The enzymes responsible for this reaction are tyrosine kinases (TKs) and serine-threonine kinases (STKs). These enzymes regulate multiple aspects of cellular metabolism, determining differentiation, adhesion, motility, genome stability, cell growth and death.

Receptor Tyrosine Kinases (RTKs) are single transmembrane domain receptors resident in the plasma membrane with high affinity for ligands like growth factors, cytokines or hormones. Unlike RTKs, non receptor tyrosine kinases (NRTKs) lack both extracellular and transmembrane domains and can be found free in the cytosol as well as in the nucleus, or linked to the inner cell membranes by myristoylation or palmitoylation modifications. These enzymes harbour protein–protein interaction domains like Src homology-2 (SH2), Src homology-3 (SH3) and pleckstrin homology domains by which NRTKs interact with substrates or regulatory factors.

STKs include a large number of kinases whose activity can be regulated by numerous chemical signals, including DNA damage, cAMP/cGMP, diacylglycerol and Ca2+/calmodulin. Their activity is crucial in the genome stability and its maintenance. Given their roles within the cell, their overexpression or deregulation are linked to the onset, progression and malignancy in a wide range of cancers. Moreover, the deregulation of one or more specific kinases appears to have a positive effect on particular cancer cells survival. Therefore, the specific inhibition of such kinases leads to cancer cell death without impairing the survival of healthy cells. For these reasons, kinases represent ideal candidates for cancer targeted therapy.

## BCR-Abl as the Prototype of Molecular Targeted Chemotherapy

The classic example of a successful therapy targeted to a protein kinase is the inhibition of the Abl kinase in leukemias. NRTKs Abl1 and Abl2 are members of the Abl family of kinases, involved in actin remodelling, cell adhesion and motility, DNA damage response and microbial pathogen response [1]. Equally to the Src Family Kinase (SFK) members, Abl proteins can exist either in an active (open), or inactive (closed) form. The shift from open to close conformation is regulated through self interaction between the regulatory domains SH2 and SH3 and the C-terminal part of the protein. *ABL* genes are constitutively activated by chromosome translocations in various haematopoietic malignancies. Chronic myeloid leukemia (CML) is characterized in almost all cases by a t(9q34;22q11) translocation that fuses the Bcr (breakpoint cluster region) and ABL1 genomic regions. The Bcr-Abl1 fusion gene product (p210) has a constitutive tyrosine kinase activity that leads to the activation of the downstream pathways of Abl, conferring to the haematopoietic cell a tumoral phenotype [2]. A similar translocation occurs in three to five percent of childhood [3] and 20–30 % of adult acute lymphoblastic leukemia (ALL) cases. Additional ABL1 fusion variants like NUP214-ABL1, EML1-ABL1 and ETV6-ABL1 [4–6] have been reported in some other leukemias, such as acute myeloid leukemias (AMLs) [7]. In all of these cases, the cell transformation activity by ABL fusion proteins is inextricably tied to their tyrosine kinase activity. In May 2001, the FDA approved Imatinib as the first-line treatment for CML. Imatinib is a selective TK inhibitor which competes with ATP in binding to the Bcr-Abl protein kinase. In particular, the drug occupies a part of the ATP-binding pocket of the enzyme and stabilizes the inactive, non-ATP-binding form of Bcr-Abl [8]. Imatinib has shown excellent activity against CML, inducing apoptosis in leukemia cells; its introduction in therapy greatly improved the outcome of CML patients with complete haematological remission in more than 90 % of previously treated patients who are resistant to interferon treatment [9, 10].

## Beyond Imatinib: New Generation TK Inhibitors

After Imatinib, many other small-molecule inhibitors specifically targeting different kinases have been approved for the treatment of different cancers or are in clinical trial (Table 1). Erlotinib and Gefitinib are small-molecule inhibitors targeting the Epidermal Growth Factor Receptor (EGFR) Family that are approved in breast and non–small cell lung cancers therapy. EGFR is the cell-surface receptor of extra-cellular protein ligands of the epidermal growth factor family (EGF-family) members. As for other RTKs, binding of the ligand stimulates the dimerization of EGFR resulting in autophosphorylation and, consequently, in the full activation of the kinase domain [11]. In non–small cell lung cancer (NSCLC), which accounts for approximately 85 % of lung cancer cases, EGFR is overexpressed or hyper activated through somatic gain-of-function mutations in exons encoding the EGFR tyrosine kinase domain (in-frame deletions in exon 19 or L858R substitution) [12, 13].

**Table 1** Tyrosine kinase inhibitors in anticancer therapy

| Inhibitor | Target |
|---|---|
| *Small molecule inhibitors* | |
| Axitinib | Multiple targets |
| Bosutinib | Bcr-Abl/SRC |
| Crizotinib | ALK/MET |
| Dastinib | Multiple targets |
| Erlotinib | EGFR |
| Gefitinib | EGFR |
| Imatinib | Bcr-Abl |
| Lapatinib | ErbB1/ErbB2 |
| Nilotinib | Bcr-Abl |
| Pazopanib | VEGFR2/PDGFR/KIT |
| Pegaptanib | VEGFR |
| Ruxolitinib | JAX |
| Sorafenib | Multiple targets |
| Sunitinib | Multiple targets |
| Vandetanib | RET/VEGFR/EGFR |
| Vemurafenib | BRAF |
| *Monoclonal antibodies* | |
| Bevacizumab | VEGFR |
| Cetuximab | ErbB1 |
| Panitumumab | EGFR |
| Ranibizumab | VEGFR |
| Trastuzumab | Erb2 |

Erlotinib and Gefitinib affect the EGFR kinase activity by acting as ATP competing molecules, binding in a reversible fashion to the ATP-binding site of the receptor. Like EGFR, Human Epidermal growth factor Receptor 2 (HER-2) is a member of the EGFR family. HER-2 is a RTK normally involved in signal transduction pathways leading to cell growth and differentiation, migration and apoptosis. It is considered an orphan receptor because none of the Epidermal Growth Factor (EGF) ligands are able to activate it. On the other hand, HER-2 is the preferential dimerization partner of all other members of the EGFR family. HER-2 amplification or overexpression in breast cancer was found to be correlated with aggressive tumour growth and poor clinical prognosis, rendering HER-2 an ideal candidate for chemotherapy [14]. Lapatinib is a selective, potent, small-molecule inhibitor of HER1 and HER-2 that is approved in combination with Capecitabine for the treatment of patients with advanced or metastatic breast tumour stages with overexpression of HER-2 [15, 16]. The binding mechanism resembles that one of Imatinib to Abl, hence, it binds to the inactive HER-2 form [17].

## The Unmet Promises: Drug Resistance to TK Inhibitors

Despite the administration of these drugs that have greatly ameliorated the efficacy of the therapy, in many cases, the onset of drug resistance leads to treatment failure. Drug resistance can be achieved in different ways that are best described in ABL-related tumours. Indeed, the follow-up of CML patients receiving Imatinib showed that primary resistance (no response to Imatinib after the initial treatment) or secondary resistance (development of resistance after achieving an objective response) emerged in 31 % of the patients. In all cases, Imatinib resistance was characterized by the presence of active Bcr-Abl, rather than the activation by the cell of an alternative signaling pathway independent from Bcr-Abl [18] and, thus, indicating that the BCR-ABL signal transduction pathway is crucial to cancer cells survival. The mechanisms of acquired Imatinib resistance were due to BCR-ABL gene amplification or mutation events. In the first case, due to BCR-ABL gene amplification or overexpression, the intracellular Imatinib concentration is not high enough to inhibit all Bcr-Abl molecules in leukaemic cells [19, 20]. In the second case, BCR-ABL mutations emerged upon the selective treatment pressure. Up to now, nearly 100 ABL mutants have been described, some of these appearing with higher frequency than others: 15 single amino-acid substitutions account for more than 85 % of the reported mutations, and 66 % of reported cases occur specifically at seven sites only (G250, Y253, E255, T315, M351, F359, H396). Furthermore, different amino-acid substitutions can involve the same residue like F317C, F317L, and F317V, all showing reduced Imatinib sensitivity [21–23]. Since these mutations in the *BCR-ABL* fusion gene were observed only after Imatinib administration, it indicates that there must have been a low prevalence of mutant cells before the therapy. This idea is supported by the fact that these mutations do not provide any growth advantage in the absence of Imatinib, but are selected specifically only upon drug exertion pressure [24]. Kinase activity is not abrogated by the mutations, although some mutants demonstrate lower enzymatic activity compared with the wild-type BCR-ABL. On the other hand, the kinase activity is enhanced by other type of mutation [25].

It has been shown that besides displaying inhibitory activity towards ABL, Imatinib is active against the RTKs KIT and PDGFRA [26]. KIT is a cytokine receptor expressed on the surface of hematopoietic stem cells and plays a role in cell survival, proliferation, and differentiation; PDGFRA is a member of the platelet-derived growth factor family implicated in mesenchymal cells proliferation. These RTKs are constitutively activated by gain-of-function mutations in most Gastrointestinal Stromal Tumours (GISTs) [27]. Experiments on human tumour cell lines, dependent on the KIT pathway, showed that Imatinib could block the activity of KIT in the GIST cells, arresting therefore proliferation and causing apoptosis. Currently, Imatinib is in phase II and III clinical trials for GIST treatment; the prognosis of GIST patients has dramatically improved after recruitment of Imatinib into the therapeutic arsenal (80 % clinical response rate). However, the mutation resistance issue arising during the Imatinib therapy significantly influences the clinical response also in the case of KIT. As for the BCR-ABL Imatinib resistance model, Imatinib efficacy is affected by the KIT mutations that directly block the drug binding or pose an energetic hindrance disfavouring the closed conformation of the kinase [28, 29].

Also in the case of Erlotinib and Gefitinib, targeting EGFR, after an initial response characterized by tumour regression and improvement in disease-related symptoms, most patients relapse. In response to Erlotinib or Gefitinib treatment, different resistance mechanisms can occur and one of the most frequent is the amplification of the *MET* proto-oncogene [30]. MET is another RTK (also known as Hepatocyte Growth Factor Receptor) and its amplification has also been observed in gastric and esophageal cancers [31, 32]. MET is involved in a pathway distinct from EGFR, and it is normally expressed by cells of epithelial origin where it promotes cell growth and motility [33]. In NSCLC cell lines whose Gefitinib resistance was obtained by continuous drug administration, a marked focal amplification within chromosome 7q31.1–7q33.3, containing the *MET* proto-oncogene, was observed. It is known that Gefitinib leads to the disruption of the signal cascade EGFR/ERBB3/PI3K/Akt [34, 35]. In Gefitinib resistant cell lines, MET amplification leads to PI3K/Akt signaling through ERBB3 activation, independently from EGFR; hence, MET inhibition restores sensitivity to Gefitinib. Resistance towards Erlotinib or Gefitinib can also occur as a consequence of mutations in the *K-RAS* gene, a GTPase of the Ras family, which is a downstream effector of EGFR, [36]. Similarly, treatment of HER-2 overexpressing cancers is not always effective because resistance to Lapatinib develops in some patients during prolonged exposure to the drug. It is known that resistance to Lapatinib in breast cancer arose through AXL overexpression. AXL is a RTK, closely related to MET, involved in cell proliferation and motility and, as the case of MET for NSCLC, its overexpression is associated with poor prognosis and increased invasiveness of several human cancers [37].

In order to avoid drug resistance, design of drugs with inhibitory properties against the mutant kinase forms is being actively pursued. It is the case of the ABL inhibitor Nilotinib, targeting Bcr-Abl with an increased potency (approximately 20-fold) when compared to Imatinib. Even if both drugs act in the same manner, hence, binding to the kinase domain of the inactive ABL form, Nilotinib binding is energetically more stable [38]. Nilotinib remains active towards all Imatinib resistant mutations lying in the activation domain (A-loop), as well as towards mutations spread across the entire Bcr-Abl kinase domain, including the P-loop. In fact, the clinical efficacy of Nilotinib was demonstrated in patients harbouring most Imatinib-resistant mutations, with the exception of Y253H, E255K/V, F359C/V and T315I BCR-ABL mutations. In particular, overcoming the Imatinib resistance mutation T315I, that confers resistance also towards Nilotinib with IC50 value 800-fold greater than against wild type Bcr-Abl, still constitutes a significant unmet medical need [39].

## Overcoming Drug Resistance: The Dual Inhibitor Concept

A more innovative strategy to overcome drug resistance consists in the development of drugs able to inhibit not only the primary target kinase, but also other kinases that contribute to cancer survival. For instance, Src Family Kinases (SFKs) are downstream effectors of Bcr-Abl; Dasatinib, another second generation agent developed for the treatment of CML, besides inhibiting Bcr-Abl, has been shown to successfully inhibit also the SFKs Src, Lyn, Yes, and Lck [40]. In particular, Lyn and Hck are involved in signal transduction pathways downstream of Bcr-Abl. Thus, the activity of Dasatinib, especially towards LYN, contributes to the overall efficacy of this drug in the treatment of CML. Like Imatinib and Nilotinib, Dasatinib acts as a competitor of the ATP substrate, but differently from other drugs, it binds to ABL both in its active and inactive conformation [41]. Compared with Imatinib, Dasatinib has an approximately 300-fold increased potency in antiproliferative assays. Despite the fact that Dasatinib remains active towards the majority of Imatinib-resistance mutants, the mutations T315I/A and F317I still result in a completely resistant phenotype.

Many other drugs showing a dual activity against Abl and SFKs are already in different clinical trial phases and preliminary research investigations. For instance, Bafetinib and Bosutinib are other dual inhibitors of Abl and SFKs that have been tested in clinical trials for CML treatment. Bafetinib targets Abl and Lyn, showing limited inhibition against other SFKs [42]. Despite the fact that Bafetinib is effective in a heavily pretreated (Imatinib, Nilotinib, Dasatinib) population, it lacks any appreciable efficacy against T315I mutation and in blastic CML phases, or in Ph-positive ALL [43]. Another drug in phase III clinical trials for CML treatment, Bosutinib [44], shows similar limitations [39, 45].

The mutation T670I in KIT is selected in GIST tumours after prolonged I-matinib administration. T670 of KIT was identified as one of the key hydrogen bonds for Imatinib binding [46], similarly to T315I in Bcr-Abl. In the same way, the EGFR T790M resistance mutation occurs after prolonged administration of Elrlotinib or Gefitinib in NSCLCs. Also, this mutation is in an analogous position to T315I in Abl and T670I in Kit [47].

In addition to small-molecule inhibitors, monoclonal antibodies (mAb) targeting RTK are also in use in clinical therapy (Table 1). Their use relies on the principle that the targeted receptors are expressed at higher levels on cancer cells than on healthy cells. Several mAbs are approved for cancer therapy or are in clinical trials. Targets of these mAbs are EGFR, HER2, VEGFR, MET, VEGFR2 and IGF1-R. The anti-cancer activity of a mAb is due to different mechanisms: the binding with the receptor can (1) prevent ligand-receptor interaction (2) promote receptor internalization (3) prevent receptor dimerization and activation and (4) induce apoptosis or immune response toward the target cells [48]. As for small-molecule inhibitors, primary and secondary resistance towards these drugs have emerged. Multiple studies showed that primary resistance can be conferred by activating mutations in KRAS, PIK3CA, BRAF or loss of PTEN expression. These mutations negatively correlate with the response to Cetuximab or Panitumumab, two EGFR-targeting mAbs [49]. Secondary resistance is due to (1) overexpression and aberrant phosphorylation of alternative RTK (2) expression of receptor variants (3) increased expression of the target receptor and (4) activation of alternative pathways [48]. To date, both point mutations in the target receptor or rearrangements in the corresponding genomic regions have been observed after mAb treatment.

## Expanding the Dual-Inhibitor Concept: The Quest for Multi-Targeting TK Inhibitors

The drug development research field is still the problem of the unsatisfactory efficacy of to-date approved inhibitors towards the T315I substitution and analogue mutations in other kinases. Some encouraging data are now emerging in clinical trials with other kinase inhibitors. Sunitinib, approved as second line therapy of GIST and renal cell carcinoma (RCC), is a multi-targeted RTKs inhibitor active toward PDGFR, VEGFR, Kit, Ret, CSF-1R and Flt3 [50]. Contrary to Imatinib, it is also active towards several Imatinib-resistant KIT secondary mutations, including the T670I KIT mutant. In the same way, Sorafenib, a small-molecule inhibitor that targets the RTKs vascular endothelial growth factor receptor (VEGFR) and Platelet-derived growth factor receptor (PDGFR) and the STKs C-Raf and B-Raf [51], is also able to inhibit Kit mutation that provides resistance toward Imatinib and Dasatinib.

If the availability of inhibitors that target multiple kinases could result in a higher degree of transduction signal inhibition, and thus in higher efficacy on cancer treatment, it could also result in enhanced toxicity for healthy cells.

The identification of the hierarchical pattern of inhibition of a given compound against all the kinome is now emerging as an essential step in order to estimate the effect of a drug against a selected cancer. Recently, high-throughput screening approaches have been use by Anastassiadis et al. [52] and Davis et al. [53] with the aim to score the inhibition potency and selectivity of 72 and 128 known kinase inhibitors, respectively, against large panels of kinases (442 and 300, respectively). Traditionally, the discovery of kinase inhibitors starts from a high-throughput screening of small molecules with inhibitory properties targeting a kinase of interest. The selectivity of effective compounds is then evaluated against a panel of representative kinases. The same goal can be obtained by screening libraries of compounds against large panels of protein kinases, thus revealing the degree of selectivity of each compound. This method led to the identification of inhibitors that were unexpected for the kinase of interest, revealing multi-targeted inhibitors active on a number of kinase targets larger than predicted. For example, Sunitinib, known to target PDGFR, VEGFR, Kit, RET, CSF-1R and Flt3 [50], shows affinity for RET harboring gatekeeper mutations [RET (V804L/M)], which is resistant to the approved RET inhibitor Vandetanib. In the same way, PKC-412, a compound designed as a protein kinase C (PKC) inhibitor, was shown to be more active against the EGFR mutant T790M. These studies also revealed that compounds designed to target a specific kinase could show higher potency of inhibition against another, unrelated enzyme. For example, DMBI, designed as a Platelet-derived growth factor receptor (PDGFR) inhibitor, is an highly potent inhibitor of FLT3 and TrkC; SB202474, an inactive analog of the p38 MAP kinase inhibitor SB202190 [54], showed significant activity only against the haploid germ cell–specific nuclear protein kinase Haspin. These data open new opportunities for clinical use of drugs already tested for their pharmacological properties.

In order to improve the efficacy of target therapy and to fight/avoid the onset of resistance, a valid strategy should be the analysis of the signalling network surrounding a target kinase. The determination of a unique factor responsible for drug resistance is not possible in many tumours that show heterogeneous resistance due to partial contributions by multiple proteins. Network models paradigm conjecture that signalling pathways are made with no hierarchy and feedback loops and are redundant. This implicates that inhibiting a specific oncogene can lead to the rescue of the signaling by enrichment (gain of function, amplification, overexpression) of proteins that compose the web of interaction with the target. Astsaturov et al. [55] screened siRNA libraries targeting EGFR network in order to identify synthetic lethality with EGFR inhibitors. This approach consented to identify previously uncharacterized genes that can drive resistance modulating EGFR signaling or that can be considerate as concomitant target for the treatment of EGFR hyper activated cancers. For instance, in the case of the EGFR network, the inhibition of SFKs or Aurora A kinase enhances the effect on cancer viability of EGFR inhibitors. The application of this approach to other validated targets could greatly ameliorate the clinical strategy in other malignancies.

## Why Targeting DNA Repair in Anticancer Chemotherapy?

RTKs and NRTKs are involved in transducing proliferative signals, hence, resulting in the activation of DNA replication and cell division. In turn, the high proliferative phenotype of tumour cells is very often accompanied by alterations in their genetic structure, leading to chromosomal aberrations and aneuploidy.

The genomic instability and the high mutation rate, typical features of human cancer cells, are mainly due to defects in DNA repair. In fact, very often at least one of the six major DNA repair pathways (mismatch repair MMR, base excision repair BER, nucleotide excision repair NER, homologous recombination HR, non-homologous end joining NHEJ and translesion synthesis TLS) is indeed defective in tumours. The mechanisms through which cancer cells respond to damaged DNA have important implications in the development of both the tumoral phenotype and the resistance to chemotherapy [56, 57].

Activation of the various DNA repair pathways during cell cycle progression depends on proper regulation of checkpoints, that are signalling cascades, often involving multiple kinases, leading to the activation of specific transcription factors. One of the main transcriptional effectors activated by DNA damage checkpoints is the protein p53. It is, thus, not surprising that *p53* is one of the most frequently mutated gene in human cancers due to its ability to halt cell growth and to modulate apoptosis after checkpoint activation when DNA damage and/or genotoxic stress occur [58].

Germline mutations of *p53* are associated with a disease called Li-Fraumeni Syndrome characterized by an increased risk of developing various cancers with an early age of onset [59]. Moreover, various clinical phenotypes in different types of cancer are associated with somatic *p53* mutations at specific residues [60]. The involvement of *p53* in DNA repair pathways has also a detrimental effect on chemosensitivity, helping cells to resist to DNA damage caused by therapy. An example is the well documented resistance against anthracyclines and mitomycin due to *p53* mutations in breast cancer as well as in haematological malignancies [61].

Polymorphism regarding *BER* genes such as *OGG1*, *APE1*, *MutYH* and *XRCC1* have been examined for their possible effect on cancer development [62]. As an example, a *DDR* gene strongly associated with cancer development is *MutYH*, whose missense mutations, insertions, deletions and duplications give rise to MutYH-associated polyposis, the most common colorectal cancer and polyposis syndrome. *MutYH* recognizes the mismatch 8-oxo-G:A and, through its action, restores the pair 8-oxo-G:C, that can be acted upon by OGG1, another BER glycosylase. The OGG1 polymorphism Ser326Cys is associated with the risk of lung cancer and increased risk of colorectal cancer; these findings confirm the roles of MutYH and OGG1 as essential players in the maintenance of genome stability against oxidative damage. In addition, the polymorphism Asp148Glu of APE1 seems to be associated with hypersensitivity to ionizing radiation and cancer risk and it can affect the prognosis of ovarian, gastro-oesophageal and pancreatico-biliary cancers [62–65]. Finally, polymorphisms of XRCC1, such as Arg399Gln and Arg194Trp, are related to the risk of skin, upper aerodigestive and lung cancers; moreover, it is important to know that these genetic polymorphisms might be associated with overall survival and response to platinum-based chemotherapy in lung cancer patients [62, 66].

Another DNA repair pathway whose alterations are strongly related to tumour development is the MMR pathway. The Lynch syndrome, a tumour predisposition syndrome characterized by colorectal and endometrial cancer and other extracolonic malignancies, is indeed caused by monoallelic germline mutations in mismatch repair genes, such as *MLH1*, *MSH2*, *MSH6* and *PMS2*, while biallelic mutations lead to a more severe scenario called constitutional mismatch repair deficiency (CMMRD). Childhood onset of leukemia/lymphoma, brain tumours and other rare malignancies are all typical features of the CMMRD disease [67, 68].

Interestingly, distinct phenotypes and clinical manifestations are due to different *MMR* gene deficiencies; for example, an increased risk of colon cancer is particularly associated with *MLH1* mutations, whereas *MSH2* mutations have a higher incidence of extracolonic tumours. Also, as shown for clinical features, the chemotherapy resistance has different outcomes based on *MMR* gene loss; for example, the MSH2-deficient cells, but not MLH1-deficient, are sensitive to psoralen, a chemotherapic agent that induces DNA interstrand crosslinks [69].

These examples of DDR defective-associated cancers justify the strong interest in figuring out all the features related with the regulation of DDR to reach a deeper knowledge of cancer development and chemoresponse.

## Targeting DNA Repair Enzymes in Cancer Therapy: The Concept of Synthetic Lethality

Somatic and hereditary mutations in DNA damage response (DDR) genes are thus associated with an increased cancer risk, but they can also offer new venues for cancer treatment. For example, the mutations of *BRCA1* and *BRCA2* genes, involved in double strand breaks repair via HR, are among the most studied, due to their strong correlation with breast and ovarian cancer development. Indeed, it was discovered that, due to their intrinsic deficiency in HR, these BRCA-deficient tumours are particularly sensitive to inhibitors of another DNA repair enzyme: poly(ADP-ribose) polymerase1 (PARP1).

PARP1 has the ability to bind single strand-breaks (SSBs) and facilitate their repair. Loss of PARP1 activity is supposed to cause formation of DNA-SSBs which are subsequently converted to double strand breaks (DSBs). In BRCA-positive cells, these DSBs are repaired by HR, but in BRCA1-or BRCA2- deficient cells they are presumed to accumulate leading to subsequent cells death [70, 71]. This discovery allowed the development of specific PARP1 inhibitors that, in the context of a BRCA1 or BRCA2 deficient genetic backgrounds, proved to be very effective in suppressing tumour growth in Phase I/II clinical trials. Such a synthetic lethality approach, however, did not entirely meet the expectations and failed in Phase III clinical studies. An explanation for this behaviour could be the activation of an alternative DSB repair pathway, such as NHEJ, that allows the survival of HR-deficient cells. However, Patel et al. [71] turn this hypothesis upside down, demonstrating that disabling NHEJ diminished the genomic instability and lethality of PARP inhibition in HR-deficient cells rather than exacerbating it. In this context, an emerging role could be assumed by the X-family DNA polymerases; in fact, they are involved in many of the alternative pathways of the DSBs repair like the Microhomology-Mediated End Joining (MMEJ), Single-strand annealing (SSA), Break Induced Replication (BIR) and others. Recent results make pol$\lambda$ an ideal candidate for a new target therapy due to its ability to promote strand annealing and subsequent elongation between two DNA strands with limited homology (5–10 nt) [72].

## New Candidates for Synthetic Lethality: Repair DNA Polymerases

The case of PARP1 inhibitors highlights a fundamental problem: namely, achieving a true synthetic lethality by suppressing a particular molecular function in the context of the high level of redundancy existing among the cellular metabolic pathways. A possibility to attain such a difficult goal might be to target enzymes common to different DNA repair pathways. The most obvious candidates are, of course, the DNA polymerases (pols). In fact, since DNA pols are essential to several repair pathways, their inhibition might potentially achieve a high level of tumour cell sensitization to chemotherapeutics.

Specialized DNA pols are required to bypass DNA damage lesions that would otherwise cause replication arrest and cell death. In recent years, a number of specialized DNA pols of the X and Y families have been identified. These are characterized by their ability to bypass different classes of lesions and to maintain a high degree of genetic fidelity by incorporating the nucleotide that would normally pair with the undamaged version of the base. Thus, under normal circumstances, specialized pols can be considered as agents that promote genomic stability. Trans-lesion synthesis (TLS), however, needs to be a highly regulated process because, when copying non-cognate lesions or undamaged DNA, the specialized pols have been shown to exhibit reduced fidelity. In addition to a role in mutagenesis, over-expression or increased activity of specialized pols could also result in enhanced TLS capability, allowing cancer cells to better cope with the high environmental stress that results from increased replication rates and higher level of oxidative damage. Moreover, increased TLS could provide cancer cells with an advantage in coping with the DNA damage resulting from the chemotherapeutic assault [73].

The Y-family members, in human cells, are DNA pol $\iota$, DNA pol $\kappa$, DNA pol $\eta$ and Rev1, each of them has a preference for catalysing DNA synthesis across certain kind of lesions. For example, the loss of DNA pol $\eta$ reduces the efficiency to copy cis-syn cyclobutane dimer, one of the main lesion generated by sun exposure, and gives rise to the variant form of xeroderma pigmentosum (XP-V), a disease characterized by high susceptibility to sunlight-induced skin cancer. In addition, it has been shown that pol $\eta$ has a key role in the cellular response to cisplatin and in the cellular resistance to this antitumour drug [74]. Another example is the role of pol $\iota$ in the induction of lung tumours, in fact pol $\iota$ knockout mice treated with urethane, a pulmonary adenoma inducer, did not develop cancer compared to wild type controls. Thus, it suggested that pol $\iota$ deficiency could lead to reduction of lung tumour [75]. On the other hand, Ziv et al. [76] discovered an important role of another Y-family enzyme, pol $\kappa$, in protecting cells from UV damages. Based on these findings, it is becoming clear that the mis-regulation of the Y-family members promotes genetic disorders and can be associated with a malignant phenotype [77].

Beside TLS, the BER and NHEJ DNA repair pathways seem to play a prominent role in promoting genetic instability in cancer cells. Central to these pathways are the X-family enzymes DNA pol $\beta$, $\lambda$, $\mu$ and TdT. TdT has the peculiar characteristic of elongating a single stranded DNA sequence without the need of a template strand. It normally acts during immunoglobulin and T cell receptor gene rearrangements, thus increasing the diversity of these molecules. A DNA pol with closely related amino acid sequence and functional domain organization to TdT is pol $\mu$, whose roles in V(D)J recombination and NHEJ are well described. The over-expression of TdT and pol $\mu$ has been observed in several acute leukemia cells and in Non-Hodgkin's lymphomas, respectively, suggesting a possible role in tumourigenesis [78, 79].

Pol $\beta$ is the smallest pol and is composed of a single 39 kDa polypeptide containing 335 amino acid residues. It is involved in BER pathways (short and long patch BER) and implicated in meiotic events associated with synapsis and recombination and SSB repair [80]. Mice carrying a target disruption of the pol gene show growth retardation and high perinatal lethality; histological examination of the embryos revealed defective neurogenesis, indicating that pol plays an essential role in neural development [81].

The over-expression of pol $\beta$ has been found at both the mRNA and protein levels in many tumour types, in particular in uterus, ovary, prostate and stomach samples. In addition, its ectopic over-expression in cancer cells can increase mutagenesis and enhance resistance to chemotherapeutic agents, including cisplatin, while cancer cells deficient in pol are hypersensitive to oxaliplatin chemotherapy, indicating that BER impairment could affect the therapy outcome [73, 82]. Cells deficient in pol $\beta$ converted into another kind of critical pro-apoptotic lesion during replication. These critical secondary DNA lesions are likely to be unrepaired DNA double strand breaks, which trigger apoptosis in a replication-dependent way by activating the mitochondrial death pathway, i.e. the decline of Bcl-2 level and activation of caspase-9 and caspase-3 [83].

Pol $\lambda$ is a protein of 67–70 kDa which is expressed at highest level in the testis, ovary and fetal liver and it seems to be implicated in short patch BER repair, NHEJ, TLS over 8-oxo-G and V(D)J recombination [84, 85]. Despite the strong expression of pol $\lambda$ seen in testis, POLL null mice were fertile and the only effect seemed to be a modification of heavy chain junctions during V(D)J recombination [86, 87]. Pol $\lambda$ was found to be over-expressed at a significant extent in a range of different tumour types, albeit less frequently than pol $\beta$ [73]. Recently, a new allelic variant of human pol $\lambda$ has been described. It is the result of the amino acid change Arg 438 to Trp and it seems to have a reduced base substitution fidelity. Thus, the ectopic expression of R438W hPol$\lambda$ variant in mammalian cells increases the mutation frequency, affects the DSB repair NHEJ pathway and generates chromosomal aberrations [88]. Another recent study reports that the pol $\lambda$ protein level can be modulated in tumour cells, and in fact, NSCLC that express less protein amount are in a significantly more advanced stage [89]. Furthermore, it has been observed that more than 90 % of leukemic cells of acute lymphocytic leukemia and approximately 30 % of leukemic cells in chronic myelogenous leukemia crisis exhibit elevated TdT activity, which is associated with poor prognosis and response to chemotherapy and reduced survival time. Since leukemic cells also often over-express pol $\lambda$, which has been shown to possess a strong bona fide TdT-like activity, it is possible that both pol $\lambda$ and TdT have an important role in tumourigenesis and progression of the acute leukemia [90]. In a recent study, pol $\lambda$ seems to be involved in the incorporation of therapeutic nucleoside analogs into DNA during BER and NHEJ, thus it may have an impact on the cellular sensitivity to these compounds following DNA damage [91].

Recently, another DNA pol, member of the A-family, called DNA polymerase $\theta$ might be implicated in cancer development. It seems to be involved in tolerance of bulky adducts or in some DNA repair pathways such as BER, DNA interstrand crosslink and DNA break repair. Its overexpression is found in breast, colon and lung cancers and it is usually related to poor prognosis [92, 93]. In addition, a recent study demonstrated that DNA pol $\theta$ knockdown on a panel of tumour cell lines from different primary sites resulted in radiosensitization, whereas having little or no effects on normal tissue cell lines [94].

## Development of Selective Inhibitors of Specialized DNA Polymerases

Currently, several classes of DNA pol inhibitors have been developed. Most of them are non-nucleosidic compounds of natural origin (polypeptides, fatty acid, triterpenoids, sulfolipids, polar lipids, secondary bile acids, phenalenone-derivates, anacardic acids, harbinatic acid, flavonoid derivates and pamoic acid), but only a few are enough specific and active to be potentially considered as drug candidates. Nucleoside analogs (NAs), on the other hand, mimicking the structure of the natural nucleotides, can interact with the catalytic site of the pols and inhibit the DNA synthesis and/or repair by competing with the natural substrates. In addition, the lack of the 3′ hydroxyl group, typical feature of most NAs, prevents the elongation step, determining an abortive DNA replication and single strand breaks formation. Currently, eight NAs have been already approved by FDA for cancer treatment: mercaptopurine, thioguanine, fludarabine and cladribine (purines), cytarabine and gemcitabine (pyrimidines), fluorouracil and capecitabine (fluoropyrimidines); whereas clofarabine (CAFdA), nelarabine, immucillin H (BCX-1777, forodesine) and 8-chloroadenosine (8-Cl-Ado), all novel purine analogs are in advanced clinical phase. Unfortunately, NAs suffer from a number of drawbacks. First, they need to be subjected to three independent cellular phosphorylation steps to be converted to their active forms. Moreover, they can be easily degraded by other enzymes such as nucleosidases and phosphorylases and, finally, their inhibitory effects can affect also replicative polymerases, thus becoming toxic to normal cells [95, 96]. In this scenario, a novel diketo hexenoic acid (DKHA) analog was described as a selective non-nucleoside inhibitor of the template-independent activity of pol $\lambda$ and TdT. Locatelli et al. [90] proved that this compound can selectively suppress cell proliferation of TdT+, but not Tdt$^{-}$, leukemic cells, holding the potential to be further developed as a novel antitumour agent.

## Tying the Ends Together: Targeting Proliferation and Repair in Cancer Cells

The inhibition of different targets in order to obtain synthetic lethality within cancer cells could greatly ameliorate the efficacy of target therapy. In this way, it is possible to hypothesize the concomitant inhibition of targets that belong to different cellular process. A fascinating idea regards the targeting of factors that link signal transduction and DNA repair. In fact, oncogenic kinases activity is linked to DNA repair [97]. For example, it is known that tumours dependent on fusion kinases, like Bcr-Abl, present an elevated number of DNA double-strand breaks (DSBs). These are caused by high ROS levels generated by the altered cell metabolism and also by chemotherapy [98]. The DSBs repair processes require the activity of Werner Helicase/Exonuclease (WRN) which plays a critical role in optimizing DSB repair mechanisms due to its DNA end-processing activities. Slupianek et al. [99] showed that Bcr-Abl is able to enhance WRN expression via c-Myc–induced transactivation and Bcl-xL–dependent inhibition of caspase-mediated cleavage. Moreover, the Bcr-Abl kinase forms a complex with WRN protein which results in constitutive phosphorylation and activation of WRN itself. As result, WRN promotes the survival of Bcr-Abl positive leukemia cells under oxidative and genotoxic stress. Furthermore, in these cells, WRN promote alternative DSBs repair mechanisms such as Homologous Recombination (HR) and Single Strand Annealing (SSA). Additionally, in Bcr-Abl positive leukemia cells,

WRN caused aberrant Non Homologous End Joining (NHEJ) repair products [100]. Altogether, these effects can promote survival of cancer cells, inducing at the same time the accumulation of genetic aberrations in CML, a mechanism by which cancer cells could acquire resistance mutation to the therapy.

Other studies underline a strict interplay between signal transduction proteins and DNA repair factors. For example, inhibition of Chk1, a STK that is activated in response to DNA damage, was found to be synthetically lethal with Src or ERK inhibitors in myeloma and leukemia cells, respectively [101–103]. Also, EGFR was found to interact with BRCA1 in highly aggressive breast cancers. Concomitant inhibition of EGFR with Lapatinib and PARP1 with ABT-888 led to transient DSBs repair deficiency which resulted in the activation of the intrinsic pathway of apoptosis [104].

Casein Kinase II (CK2) is an ubiquitously expressed STK, whose activity is implicated in cell growth and proliferation. Its overexpression is linked to cancerogenesis and attenuated apoptosis induced by chemotherapeutic drugs [105], posing CK2 as an attractive target for cancer treatment. In particular, CK2 inhibits the tumour suppressor activity of promyelocytic leukemia protein (PML) and phosphatase and tensin homology protein (PTEN) [106]. CK2 phosphorylates

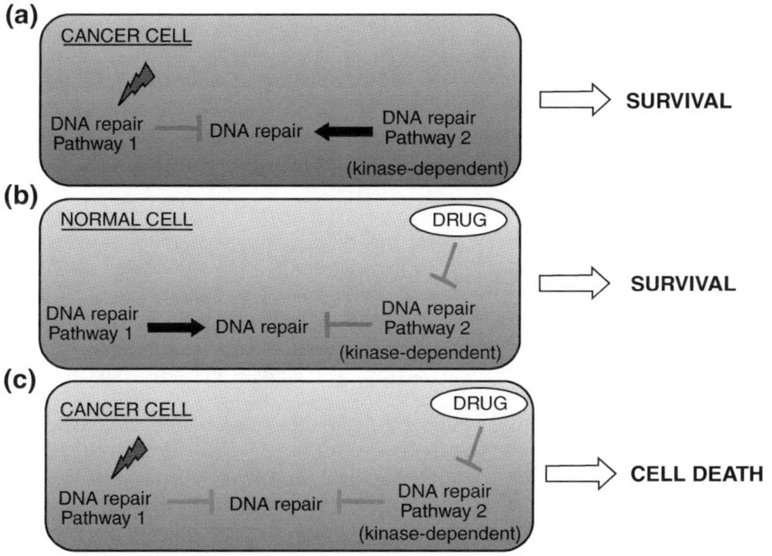


**Fig. 1** *The synthetic lethality approach through targeting DNA repair and signal transduction.* **a** In a cancer cell, a defect in a DNA repair pathway can be overcome through the induction of an alternative pathway, often under the control of signal-transducing kinases. **b** In a normal cell, where all the DNA repair pathways are effective, inhibition of one of them, either by targeting the apical kinase or directly the repair DNA polymerase, will not affect its survival. **c** In a cancer cell, already defective for a given DNA repair mechanism, inhibition of the existing alternative pathways, either by targeting the apical kinase or directly the repair DNA polymerase, will result in apoptosis

XRCC1, a scaffold protein that plays a critical role in DNA base excision repair (BER) by interacting with Ligase III (LigIII). Phosphorylation of XRCC1 in human cell extracts is required for XRCC1-Lig III complex stability and phosphorylation reduction leads to DSBs accumulation [107]. Thus, inhibition of CK2 would lead not only to apoptosis in response to chemotherapeutic drugs, but also in accumulation of DSBs that will enhance the apoptotic effects of the therapy.

## Conclusions

The above-discussed examples highlight the intimate connections between the proliferative signal transduction pathways and the DNA repair. In such a context, inhibition of specialized DNA pols, that act in different repair mechanisms, will reduce the ability of cancer cells to cope with the genotoxic stress imposed by proliferation and/or cancer treatment. Thus, simultaneously targeting signal transduction and DNA repair pathways will drive very efficiently cancer cells to apoptosis (Fig. 1). The investigation of the interplay between signal transduction pathways and DNA repair pathways in cancer cells can lead to the identification of novel therapeutical targets and to the design of cancer treatment strategies that avoid the onset of resistance to the therapy.

**Acknowledgments** Work in authors' laboratory has been partially supported by an Italian Cancer Research Association AIRC grant IG12084 to GM.

**Conflicts of Interest** No potential conflicts of interest were disclosed.

## References

1. Colicelli J. Abl Tyrosine kinases: evolution of function, regulation, and specificity. Sci Signal. 2010;3:re6.
2. Deininger MW, Goldman JM, Melo JV. The molecular biology of chronic Myeloid Leukemia. Blood. 2000;96:3343–56.
3. Carroll WL, Bhojwani D, Min DJ, Raetz E, Relling M, Davies S, Downing JR, Willman CL, Reed JC. Pediatric acute lymphoblastic leukemia. Hematol Am Soc Hematol Educ Program. 2003;2003:102–31.
4. De Keersmaecker K, Graux C, Odero MD, Mentens N, Somers R, Maertens J, Wlodarska I, Vandenberghe P, Hagemeijer A, Marynen P, Cools J. Fusion of eml1 to Abl1 in T-cell acute lymphoblastic leukemia with cryptic T(9;14)(Q34;Q32). Blood. 2005;105:4849–52.
5. Graux C, Cools J, Melotte C, Quentmeier H, Ferrando A, Levine R, Vermeesch JR, Stul M, Dutta B, Boeckx N, Bosly A, Heimann P, Uyttebroeck A, Mentens N, Somers R, MacLeod RA, Drexler HG, Look AT, Gilliland DG, Michaux L, Vandenberghe P, Wlodarska I, Marynen P, Hagemeijer A. Fusion of Nup214 to Abl1 on amplified episomes in T-cell acute lymphoblastic leukemia. Nat Genet. 2004;36:1084–9.

6. Zuna J, Zaliova M, Muzikova K, Meyer C, Lizcova L, Zemanova Z, Brezinova J, Votava F, Marschalek R, Stary J, Trka J. Acute leukemias with Etv6/Abl1 (Tel/Abl) fusion: poor prognosis and prenatal origin. Genes Chromosom Cancer. 2010;49:873–84.
7. Iijima Y, Ito T, Oikawa T, Eguchi M, Eguchi-Ishimae M, Kamada N, Kishi K, Asano S, Sakaki Y, Sato Y. A new Etv6/Tel partner Gene, Arg (Abl-Related Gene or Abl2), identified in an Aml-M3 cell line with a T(1;12)(Q25;P13) translocation. Blood. 2000;95:2126–31.
8. Schindler T, Bornmann W, Pellicena P, Miller WT, Clarkson B, Kuriyan J. Structural mechanism for Sti-571 inhibition of abelson tyrosine kinase. Science. 2000;289:1938–42.
9. Druker BJ, Guilhot F, O'Brien SG, Gathmann I, Kantarjian H, Gattermann N, Deininger MW, Silver RT, Goldman JM, Stone RM, Cervantes F, Hochhaus A, Powell BL, Gabrilove JL, Rousselot P, Reiffers J, Cornelissen JJ, Hughes T, Agis H, Fischer T, Verhoef G, Shepherd J, Saglio G, Gratwohl A, Nielsen JL, Radich JP, Simonsson B, Taylor K, Baccarani M, So C, Letvak L, Larson RA. Five-year follow-up of patients receiving imatinib for chronic myeloid leukemia. N Engl J Med. 2006;355:2408–17.
10. Gambacorti-Passerini C, le Coutre P, Mologni L, Fanelli M, Bertazzoli C, Marchesi E, Di Nicola M, Biondi A, Corneo GM, Belotti D, Pogliani E, Lydon NB. Inhibition of the Abl kinase activity blocks the proliferation of Bcr/Abl+ leukemic cells and induces apoptosis. Blood Cells Mol Dis. 1997;23:380–94.
11. Herbst RS. Review of epidermal growth factor receptor biology. Int J Radiat Oncol Biol Phys. 2004;59:21–6.
12. Rosell R, Moran T, Queralt C, Porta R, Cardenal F, Camps C, Majem M, Lopez-Vivanco G, Isla D, Provencio M, Insa A, Massuti B, Gonzalez-Larriba JL, Paz-Ares L, Bover I, Garcia-Campelo R, Moreno MA, Catot S, Rolfo C, Reguart N, Palmero R, Sanchez JM, Bastus R, Mayo C, Bertran-Alamillo J, Molina MA, Sanchez JJ, Taron M. Screening for epidermal growth factor receptor mutations in lung cancer. N Engl J Med. 2009;361:958–67.
13. Yang CH, Yu CJ, Shih JY, Chang YC, Hu FC, Tsai MC, Chen KY, Lin ZZ, Huang CJ, Shun CT, Huang CL, Bean J, Cheng AL, Pao W, Yang PC. Specific Egfr mutations predict treatment outcome of stage Iiib/Iv patients with chemotherapy-naive non-small-cell lung cancer receiving first-line gefitinib monotherapy. J Clin Oncol. 2008;26:2745–53.
14. Olayioye M. Intracellular signaling pathways of Erbb2/Her-2 and family members. Breast Cancer Res. 2001;3:385–9.
15. Hynes NE, Schlange T. Targeting Adams and Erbbs in lung cancer. Cancer Cell. 2006;10:7–11.
16. Medina PJ, Goodin S. Lapatinib: a dual inhibitor of human epidermal growth factor receptor tyrosine kinases. Clin Ther. 2008;30:1426–47.
17. Wood ER, Truesdale AT, McDonald OB, Yuan D, Hassell A, Dickerson SH, Ellis B, Pennisi C, Horne E, Lackey K, Alligood KJ, Rusnak DW, Gilmer TM, Shewchuk L. A unique structure for epidermal growth factor receptor bound to Gw572016 (Lapatinib): relationships among protein conformation, inhibitor off-rate, and receptor activity in tumour cells. Cancer Res. 2004;64:6652–9.
18. Gambacorti-Passerini CB, Rossi F, Verga M, Ruchatz H, Gunby R, Frapolli R, Zucchetti M, Scapozza L, Bungaro S, Tornaghi L, Rossi F, Pioltelli P, Pogliani E, D'Incalci M, Corneo G. Differences between in vivo and in vitro sensitivity to imatinib of Bcr/Abl plus cells obtained from leukemic patients. Blood Cells Mol Dis. 2002;28:361–72.
19. Mahon FX, Deininger MW, Schultheis B, Chabrol J, Reiffers J, Goldman JM, Melo JV. Selection and characterization of Bcr-Abl positive cell lines with differential sensitivity to the tyrosine kinase inhibitor Sti571: diverse mechanisms of resistance. Blood. 2000;96:1070–9.
20. Weisberg E, Griffin JD. Mechanism of resistance to the Abl tyrosine kinase inhibitor Sti571 in Bcr/Abl-transformed hematopoietic cell lines. Blood. 2000;95:3498–505.
21. Apperley JF, Part I. Mechanisms of resistance to imatinib in chronic myeloid leukaemia. Lancet Oncol. 2007;8:1018–29.
22. Deininger MW, Cortes J, Paquette R, Park B, Hochhaus A, Baccarani M, Stone R, Fischer T, Kantarjian H, Niederwieser D, Gambacorti-Passerini C, So C, Gathmann I, Goldman JM,

Smith D, Druker BJ, Guilhot F. The prognosis for patients with chronic myeloid leukemia who have clonal cytogenetic abnormalities in Philadelphia chromosome-negative cells. Cancer. 2007;110:1509–19.

23. Soverini S, Colarossi S, Gnani A, Rosti G, Castagnetti F, Poerio A, Iacobucci I, Amabile M, Abruzzese E, Orlandi E, Radaelli F, Ciccone F, Tiribelli M, di Lorenzo R, Caracciolo C, Izzo B, Pane F, Saglio G, Baccarani M, Martinelli G. Contribution of Abl kinase domain mutations to imatinib resistance in different subsets of Philadelphia-positive patients: by the gimema working party on chronic myeloid leukemia. Clin Cancer Res. 2006;12:7374–9.
24. Roche-Lestienne C, Soenen-Cornu V, Grardel-Duflos N, Lai JL, Philippe N, Facon T, Fenaux P, Preudhomme C. Several types of mutations of the Abl gene can be found in chronic myeloid leukemia patients resistant to Sti571, and they can pre-exist to the onset of treatment. Blood. 2002;100:1014–8.
25. Griswold IJ, MacPartlin M, Bumm T, Goss VL, O'Hare T, Lee KA, Corbin AS, Stoffregen EP, Smith C, Johnson K, Moseson EM, Wood LJ, Polakiewicz RD, Druker BJ, Deininger MW. Kinase domain mutants of Bcr-Abl exhibit altered transformation potency, kinase activity, and substrate utilization, irrespective of sensitivity to imatinib. Mol Cell Biol. 2006;26:6082–93.
26. Heinrich MC, Griffith DJ, Druker BJ, Wait CL, Ott KA, Zigler AJ. Inhibition of C-kit receptor tyrosine kinase activity by Sti 571, a selective tyrosine kinase inhibitor. Blood. 2000;96:925–32.
27. Duensing A, Medeiros F, McConarty B, Joseph NE, Panigrahy D, Singer S, Fletcher CD, Demetri GD, Fletcher JA. Mechanisms of oncogenic kit signal transduction in primary gastrointestinal stromal tumours (Gists). Oncogene. 2004;23:3999–4006.
28. Heinrich MC, Corless CL, Blanke CD, Demetri GD, Joensuu H, Roberts PJ, Eisenberg BL, von Mehren M, Fletcher CD, Sandau K, McDougall K, Ou WB, Chen CJ, Fletcher JA. Molecular correlates of imatinib resistance in gastrointestinal stromal tumours. J Clin Oncol. 2006;24:4764–74.
29. Mol CD, Dougan DR, Schneider TR, Skene RJ, Kraus ML, Scheibe DN, Snell GP, Zou H, Sang BC, Wilson KP. Structural basis for the autoinhibition and Sti-571 inhibition of C-kit tyrosine kinase. J Biol Chem. 2004;279:31655–63.
30. Engelman JA, Zejnullahu K, Mitsudomi T, Song Y, Hyland C, Park JO, Lindeman N, Gale CM, Zhao X, Christensen J, Kosaka T, Holmes AJ, Rogers AM, Cappuzzo F, Mok T, Lee C, Johnson BE, Cantley LC, Janne PA. Met amplification leads to gefitinib resistance in lung cancer by activating Erbb3 signaling. Science. 2007;316:1039–43.
31. Miller CT, Lin L, Casper AM, Lim J, Thomas DG, Orringer MB, Chang AC, Chambers AF, Giordano TJ, Glover TW, Beer DG. Genomic amplification of met with boundaries within fragile site Fra7g and upregulation of met pathways in esophageal adenocarcinoma. Oncogene. 2006;25:409–18.
32. Smolen GA, Sordella R, Muir B, Mohapatra G, Barmettler A, Archibald H, Kim WJ, Okimoto RA, Bell DW, Sgroi DC, Christensen JG, Settleman J, Haber DA. Amplification of met may identify a subset of cancers with extreme sensitivity to the selective tyrosine kinase inhibitor pha-665752. Proc Natl Acad Sci USA. 2006;103:2316–21.
33. McDermott U, Pusapati RV, Christensen JG, Gray NS, Settleman J. Acquired resistance of non-small cell lung cancer cells to met kinase inhibition is mediated by a switch to epidermal growth factor receptor dependency. Cancer Res. 2010;70:1625–34.
34. Engelman JA, Jänne PA, Mermel C, Pearlberg J, Mukohara T, Fleet C, Cichowski K, Johnson BE, Cantley LC. Erbb-3 mediates phosphoinositide 3-kinase activity in gefitinib-sensitive non-small cell lung cancer cell lines. Proc Natl Acad Sci USA. 2005;102:3788–93.
35. Engelman JA, Mukohara T, Zejnullahu K, Lifshits E, Borrás AM, Gale C-M, Naumov GN, Yeap BY, Jarrell E, Sun J, Tracy S, Zhao X, Heymach JV, Johnson BE, Cantley LC, Jänne PA. Allelic dilution obscures detection of a biologically significant resistance mutation in Egfr-amplified lung cancer. J Clin Invest. 2006;116:2695–706.

36. Pao W, Wang TY, Riely GJ, Miller VA, Pan Q, Ladanyi M, Zakowski MF, Heelan RT, Kris MG, Varmus HE. Kras mutations and primary resistance of lung adenocarcinomas to gefitinib or erlotinib. PLoS Med. 2005;2:e17.
37. Liu L, Greger J, Shi H, Liu Y, Greshock J, Annan R, Halsey W, Sathe GM, Martin A-M, Gilmer TM. Novel mechanism of lapatinib resistance in Her2-positive breast tumour cells: activation of Axl. Cancer Res. 2009;69:6871–8.
38. Weisberg E, Manley PW, Breitenstein W, Bruggen J, Cowan-Jacob SW, Ray A, Huntly B, Fabbro D, Fendrich G, Hall-Meyers E, Kung AL, Mestan J, Daley GQ, Callahan L, Catley L, Cavazza C, Azam M, Neuberg D, Wright RD, Gilliland DG, Griffin JD. Characterization of Amn107, a selective inhibitor of native and mutant Bcr-Abl. Cancer Cell. 2005;7:129–41.
39. Redaelli S, Piazza R, Rostagno R, Magistroni V, Perini P, Marega M, Gambacorti-Passerini C, Boschelli F. Activity of bosutinib, dasatinib, and nilotinib against 18 imatinib-resistant Bcr/Abl mutants. J Clin Oncol. 2009;27:469–71.
40. Olivieri A, Manzione L. Dasatinib: a new step in molecular target therapy. Ann Oncol. 2007;18(6):vi42–6.
41. Tokarski JS, Newitt JA, Chang CY, Cheng JD, Wittekind M, Kiefer SE, Kish K, Lee FY, Borzillerri R, Lombardo LJ, Xie D, Zhang Y, Klei HE. The structure of dasatinib (Bms-354825) bound to activated abl kinase domain elucidates its inhibitory activity against imatinib-resistant abl mutants. Cancer Res. 2006;66:5790–7.
42. Golas JM, Arndt K, Etienne C, Lucas J, Nardin D, Gibbons J, Frost P, Ye F, Boschelli DH, Boschelli F. Ski-606, a 4-anilino-3-quinolinecarbonitrile dual inhibitor of src and abl kinases, is a potent antiproliferative agent against chronic myelogenous leukemia cells in culture and causes regression of K562 xenografts in nude mice. Cancer Res. 2003;63:375–81.
43. Kantarjian H, le Coutre P, Cortes J, Pinilla-Ibarz J, Nagler A, Hochhaus A, Kimura S, Ottmann O. Phase 1 study of inno-406, a dual Abl/Lyn kinase inhibitor, in Philadelphia chromosome-positive leukemias after imatinib resistance or intolerance. Cancer. 2010;116:2665–72.
44. Boschelli F, Arndt K, Gambacorti-Passerini C. Bosutinib: a review of preclinical studies in chronic myelogenous leukaemia. Eur J Cancer. 2010;46:1781–9.
45. Puttini M, Coluccia AML, Boschelli F, Cleris L, Marchesi E, Donella-Deana A, Ahmed S, Redaelli S, Piazza R, Magistroni V, Andreoni F, Scapozza L, Formelli F, Gambacorti-Passerini C. In vitro and in vivo activity of Ski-606, a novel Src-Abl inhibitor, against imatinib-resistant Bcr-Abl(+) neoplastic cells. Cancer Res. 2006;66:11314–22.
46. Guo T, Agaram NP, Wong GC, Hom G, D'Adamo D, Maki RG, Schwartz GK, Veach D, Clarkson BD, Singer S, DeMatteo RP, Besmer P, Antonescu CR. Sorafenib inhibits the imatinib-resistant kitt670i gatekeeper mutation in gastrointestinal stromal tumour. Clin Cancer Res. 2007;13:4874–81.
47. Janne PA. Challenges of detecting Egfr T790m in Gefitinib/Erlotinib-resistant tumours. Lung Cancer. 2008;60(2):S3–9.
48. Sierra JR, Cepero V, Giordano S. Molecular mechanisms of acquired resistance to tyrosine kinase targeted therapy. Mol Cancer. 2010;9:75–85.
49. Bardelli A, Siena S. Molecular mechanisms of resistance to cetuximab and panitumumab in colorectal cancer. J Clin Oncol. 2010;28:1254–61.
50. Prenen H, Cools J, Mentens N, Folens C, Sciot R, Schoffski P, Van Oosterom A, Marynen P, Debiec-Rychter M. Efficacy of the kinase inhibitor Su11248 against gastrointestinal stromal tumour Mutants refractory to imatinib mesylate. Clin Cancer Res. 2006;12:2622–7.
51. Ibrahim N, Yu Y, Walsh WR, Yang JL. Molecular targeted therapies for cancer: sorafenib mono-therapy and its combination with other therapies (Review). Oncol Rep. 2012;27:1303–11.
52. Anastassiadis T, Deacon SW, Devarajan K, Ma H, Peterson JR. Comprehensive assay of kinase catalytic activity reveals features of kinase inhibitor selectivity. Nat Biotechnol. 2011;29:1039–45.

53. Davis MI, Hunt JP, Herrgard S, Ciceri P, Wodicka LM, Pallares G, Hocker M, Treiber DK, Zarrinkar PP. Comprehensive analysis of kinase inhibitor selectivity. Nat Biotechnol. 2011;29:1046–51.
54. Lee JC, Laydon JT, McDonnell PC, Gallagher TF, Kumar S, Green D, McNulty D, Blumenthal MJ, Heys JR, Landvatter SW, et al. A protein kinase involved in the regulation of inflammatory cytokine biosynthesis. Nature. 1994;372:739–46.
55. Astsaturov I, Ratushny V, Sukhanova A, Einarson MB, Bagnyukova T, Zhou Y, Devarajan K, Silverman JS, Tikhmyanova N, Skobeleva N, Pecherskaya A, Nasto RE, Sharma C, Jablonski SA, Serebriiskii IG, Weiner LM, Golemis EA. Synthetic lethal screen of an Egfr-Centered network to improve targeted therapies. Sci Signal. 2010;3:67.
56. Belzile JP, Choudhury SA, Cournoyer D, Chow TY. Targeting DNA repair proteins: a promising avenue for cancer gene therapy. Curr Gene Ther. 2006;6:111–23.
57. Kennedy RD, D'Andrea AD. DNA repair pathways in clinical practice: lessons from pediatric cancer susceptibility syndromes. J Clin Oncol. 2006;24:3799–808.
58. Naccarati A, Polakova V, Pardini B, Vodickova L, Hemminki K, Kumar R, Vodicka P. Mutations and polymorphisms in Tp53 gene–an overview on the role in colorectal cancer. Mutagenesis. 2012;27:211–8.
59. Frebourg T, Friend SH. Cancer risks from germline P53 mutations. J Clin Invest. 1992;90:1637–41.
60. George B, Datar RH, Wu L, Cai J, Patten N, Beil SJ, Groshen S, Stein J, Skinner D, Jones PA, Cote RJ. P53 gene and protein status: the role of P53 alterations in predicting outcome in patients with bladder cancer. J Clin Oncol. 2007;25:5352–8.
61. Knappskog S, Lonning PE. P53 and its molecular basis to chemoresistance in breast cancer. Expert Opin Ther Targets. 2012;16(1):S23–30.
62. Karahalil B, Bohr V, Wilson D 3rd. Impact of DNA polymorphisms in key DNA base excision repair proteins on cancer risk. Hum Exp Toxicol. 2012;31:981–1005.
63. Attar R, Cacina C, Sozen S, Attar E, Agachan B. DNA repair genes in endometriosis. Genet Mol Res. 2010;9:629–36.
64. Chen PL, Yeh KT, Tsai YY, Koeh H, Liu YL, Lee H, Cheng YW. Xrcc1, but not Ape1 and Hogg1 gene polymorphisms is a risk factor for pterygium. Mol Vis. 2010;16:991–6.
65. Lo YL, Jou YS, Hsiao CF, Chang GC, Tsai YH, Su WC, Chen KY, Chen YM, Huang MS, Hu CY, Chen CJ, Hsiung CA. A polymorphism in the Ape1 gene promoter is associated with lung cancer risk. Cancer Epidemiol Biomark Prev. 2009;18:223–9.
66. Cui Z, Yin Z, Li X, Wu W, Guan P, Zhou B. Association between polymorphisms in Xrcc1 gene and clinical outcomes of patients with lung cancer: a meta-analysis. BMC Cancer. 2012;12:71–83.
67. Nielsen M, Morreau H, Vasen HF, Hes FJ. Mutyh-associated polyposis (Map). Crit Rev Oncol Hematol. 2011;79:1–16.
68. Ripperger T, Beger C, Rahner N, Sykora KW, Bockmeyer CL, Lehmann U, Kreipe HH, Schlegelberger B. Constitutional mismatch repair deficiency and childhood leukemia/lymphoma–report on a novel Biallelic Msh6 mutation. Haematologica. 2010;95:841–4.
69. Martin SA, Lord CJ, Ashworth A. Therapeutic targeting of the DNA mismatch repair pathway. Clin Cancer Res. 2010;16:5107–13.
70. Bouwman P, Jonkers J. The effects of deregulated DNA damage signalling on cancer chemotherapy response and resistance. Nat Rev Cancer. 2012;12:587–98.
71. Patel AG, Sarkaria JN, Kaufmann SH. Nonhomologous end joining drives poly(Adp-Ribose) polymerase (Parp) inhibitor lethality in homologous recombination-deficient cells. Proc Natl Acad Sci USA. 2011;108:3406–11.
72. Crespan E, Czabany T, Maga G, Hubscher U. Microhomology-mediated DNA strand annealing and elongation by human DNA polymerases lambda and beta on normal and repetitive DNA sequences. Nucleic Acids Res. 2012;40:5577–90.
73. Albertella MR, Lau A, O'Connor MJ. The overexpression of specialized DNA polymerases in cancer. DNA Repair (Amst). 2005;4:583–93.

74. Brabec V, Malina J, Margiotta N, Natile G, Kasparkova J. Thermodynamic and mechanistic insights into translasion DNA synthesis catalyzed by Y-family DNA polymerase across a Bulky double-base lesion of an antitumour platinum drug. Chemistry. 2012;18:15439–48.
75. Lee GH, Matsushita H. Genetic linkage between Pol iota deficiency and increased susceptibility to lung tumours in mice. Cancer Sci. 2005;96:256–9.
76. Ziv O, Geacintov N, Nakajima S, Yasui A, Livneh Z. DNA polymerase zeta cooperates with polymerases kappa and iota in translesion DNA synthesis across pyrimidine photodimers in cells from Xpv patients. Proc Natl Acad Sci USA. 2009;106:11552–7.
77. Guo C, Kosarek-Stancel JN, Tang TS, Friedberg EC. Y-family DNA polymerases in mammalian cells. Cell Mol Life Sci. 2009;66:2363–81.
78. Chiu A, Pan L, Li Z, Ely S, Chadburn A, Knowles DM. DNA polymerase mu gene expression in B-cell non-Hodgkin's lymphomas: an analysis utilizing in situ hybridization. Am J Pathol. 2002;161:1349–55.
79. Kodama EN, McCaffrey RP, Yusa K, Mitsuya H. Antileukemic activity and mechanism of action of cordycepin against terminal deoxynucleotidyl transferase-positive (Tdt+) leukemic cells. Biochem Pharmacol. 2000;59:273–81.
80. Hubscher U, Maga G, Spadari S. Eukaryotic DNA polymerases. Annu Rev Biochem. 2002;71:133–63.
81. Sugo N, Aratani Y, Nagashima Y, Kubota Y, Koyama H. Neonatal lethality with abnormal neurogenesis in mice deficient in DNA polymerase beta. EMBO J. 2000;19:1397–404.
82. Yang J, Parsons J, Nicolay NH, Caporali S, Harrington CF, Singh R, Finch D, D'Atri S, Farmer PB, Johnston PG, McKenna WG, Dianov G, Sharma RA. Cells deficient in the base excision repair protein, DNA polymerase beta, Are hypersensitive to Oxaliplatin Chemotherapys. Oncogene. 2010;29:463–8.
83. Ochs K, Lips J, Profittlich S, Kaina B. Deficiency in DNA polymerase beta provokes replication-dependent apoptosis via DNA breakage, Bcl-2 decline and caspase-3/9 activation. Cancer Res. 2002;62:1524–30.
84. Di Santo R, Maga G. Human terminal deoxynucleotidyl transferases as novel targets for anticancer chemotherapy. Curr Med Chem. 2006;13:2353–68.
85. Maga G, Hubscher U. Repair and translesion DNA polymerases as anticancer drug targets. Anticancer Agents Med Chem. 2008;8:431–47.
86. Bertocci B, De Smet A, Flatter E, Dahan A, Bories JC, Landreau C, Weill JC, Reynaud CA. Cutting edge: DNA polymerases Mu and Lambda are dispensable for Ig gene hypermutation. J Immunol. 2002;168:3702–6.
87. Bertocci B, De Smet A, Weill JC, Reynaud CA. Nonoverlapping functions of DNA polymerases Mu, Lambda, and terminal deoxynucleotidyltransferase during immunoglobulin V(D)J recombination in vivo. Immunity. 2006;25:31–41.
88. Terrados G, Capp JP, Canitrot Y, Garcia-Diaz M, Bebenek K, Kirchhoff T, Villanueva A, Boudsocq F, Bergoglio V, Cazaux C, Kunkel TA, Hoffmann JS, Blanco L. Characterization of a natural mutator variant of human DNA polymerase Lambda which promotes chromosomal instability by compromising Nhej. PLoS One. 2009;4:e7290.
89. Ohba T, Kometani T, Shoji F, Yano T, Yoshino I, Taguchi K, Kuraoka I, Oda S, Maehara Y. Expression of an X-family DNA polymerase, Pol Lambda, in the respiratory epithelium of non-small cell lung cancer patients with habitual smoking. Mutat Res. 2009;677:66–71.
90. Locatelli GA, Di Santo R, Crespan E, Costi R, Roux A, Hubscher U, Shevelev I, Blanca G, Villani G, Spadari S, Maga G. Diketo Hexenoic acid derivatives are novel selective non-nucleoside inhibitors of mammalian terminal deoxynucleotidyl transferases, with potent cytotoxic effect against leukemic cells. Mol Pharmacol. 2005;68:538–50.
91. Garcia-Diaz M, Murray MS, Kunkel TA, Chou KM. Interaction between DNA polymerase Lambda and anticancer nucleoside analogs. J Biol Chem. 2010;285:16874–9.
92. Lemee F, Bergoglio V, Fernandez-Vidal A, Machado-Silva A, Pillaire MJ, Bieth A, Gentil C, Baker L, Martin AL, Leduc C, Lam E, Magdeleine E, Filleron T, Oumouhou N, Kaina B, Seki M, Grimal F, Lacroix-Triki M, Thompson A, Roche H, Bourdon JC, Wood RD, Hoffmann JS, Cazaux C. DNA polymerase Theta up-regulation is associated with poor

survival in breast cancer, perturbs DNA replication, and promotes genetic instability. Proc Natl Acad Sci USA. 2010;107:13390–5.
93. Pillaire MJ, Selves J, Gordien K, Gourraud PA, Gentil C, Danjoux M, Do C, Negre V, Bieth A, Guimbaud R, Trouche D, Pasero P, Mechali M, Hoffmann JS, Cazaux C. A 'DNA replication' signature of progression and negative outcome in colorectal cancer. Oncogene. 2010;29:876–87.
94. Higgins GS, Prevo R, Lee YF, Helleday T, Muschel RJ, Taylor S, Yoshimura M, Hickson ID, Bernhard EJ, McKenna WG. A small interfering RNA screen of genes involved in DNA repair identifies tumour-specific radiosensitization by POLQ knockdown. Cancer Res. 2010;70:2984–93.
95. Crespan E, Garbelli A, Amoroso A, Maga G. Exploiting the nucleotide substrate specificity of repair DNA polymerases to develop novel anticancer agents. Molecules. 2011;16:7994–8019.
96. Motea EA, Lee I, Berdis AJ. A non-natural nucleoside with combined therapeutic and diagnostic activities against leukemia. ACS Chem Biol. 2012;7:988–98.
97. Skorski T. Oncogenic tyrosine kinases and the DNA-damage response. Nat Rev Cancer. 2002;2:351–60.
98. Koptyra M, Falinski R, Nowicki MO, Stoklosa T, Majsterek I, Nieborowska-Skorska M, Blasiak J, Skorski T. Bcr/Abl kinase induces self-mutagenesis via reactive oxygen species to encode imatinib resistance. Blood. 2006;108:319–27.
99. Slupianek A, Poplawski T, Jozwiakowski SK, Cramer K, Pytel D, Stoczynska E, Nowicki MO, Blasiak J, Skorski T. Bcr/Abl stimulates Wrn to promote survival and genomic instability. Cancer Res. 2011;71:842–51.
100. Nowicki MO, Falinski R, Koptyra M, Slupianek A, Stoklosa T, Gloc E, Nieborowska-Skorska M, Blasiak J, Skorski T. Bcr/Abl oncogenic kinase promotes unfaithful repair of the reactive oxygen species-dependent DNA double-strand breaks. Blood. 2004;104:3746–53.
101. Dai Y, Chen S, Shah R, Pei XY, Wang L, Almenara JA, Kramer LB, Dent P, Grant S. Disruption of Src function potentiates Chk1-inhibitor-induced apoptosis in human multiple myeloma cells in vitro and in vivo. Blood. 2011;117:1947–57.
102. Dai Y, Yu C, Singh V, Tang L, Wang Z, McInistry R, Dent P, Grant S. Pharmacological inhibitors of the mitogen-activated protein kinase (Mapk) kinase/mapk cascade interact synergistically with Ucn-01 to induce mitochondrial dysfunction and apoptosis in human leukemia cells. Cancer Res. 2001;61:5106–15.
103. Shaheen M, Allen C, Nickoloff JA, Hromas R. Synthetic lethality: exploiting the addiction of cancer to DNA repair. Blood. 2011;117:6074–82.
104. Nowsheen S, Cooper T, Stanley JA, Yang ES. Synthetic lethal interactions between Egfr and parp inhibition in human triple negative breast cancer cells. PLoS One. 2012;7:e46614.
105. Meggio F, Pinna LA. One-thousand-and-one substrates of protein kinase Ck2? FASEB J. 2003;17:349–68.
106. Hanif IM, Shazib MA, Ahmad KA, Pervaiz S. Casein kinase Ii: an attractive target for anti-cancer drug design. Int J Biochem Cell Biol. 2010;42:1602–5.
107. Parsons JL, Dianova II, Finch D, Tait PS, Ström CE, Helleday T, Dianov GL. Xrcc1 phosphorylation by Ck2 is required for its stability and efficient DNA repair. DNA Repair. 2010;9:835–41.

# Collateral Sensitivity in Drug-Resistant Tumor Cells

**Mohamed Saeed, Henry Johannes Greten and Thomas Efferth**

**Abstract** Collateral sensitivity is a term for the hypersensitivity of otherwise drug-resistant cells. The selective killing of tumor cells by drugs exerting collateral sensitivity might be used as a novel treatment strategy. In this chapter, we give an overview on drug resistance phenotypes with known collateral sensitivities; furthermore, their molecular and cellular mechanisms were discussed to explain mediation of these hypersensitivities.

**Keywords** ABC transporter · Cancer · Chemotherapy · Collateral sensitivity · Cross resistance · Multidrug resistance · P-glycoprotein

**Abbreviations**

| | |
|---|---|
| ABC transporter | ATP-binding cassette transporter |
| P-gp | P-glycoprotein |

## Cross-Resistance and P-glycoprotein

Chemotherapy still belongs to the main options to treat cancer. A major problem with drug treatment is that tumors frequently develop resistance. Concentrations high enough to kill tumor cell subpopulations frequently cannot be applied because of severe side effects of most classical anti-cancer drugs, e.g. myelosuppression, gastrointestinal mucositis, alopecia, sterility, etc. Clinically two major types of resistance are observed: (1) primary or inherent drug resistance, whereby tumors

M. Saeed · T. Efferth (✉)
Department of Pharmaceutical Biology, Institute of Pharmacy and Biochemistry, University of Mainz, Staudinger Weg 5, 55128 Mainz, Germany
e-mail: efferth@uni-mainz.de

H. J. Greten
Heidelberg School of Traditional Chinese Medicine, Germany and Biomedical Sciences Institute Abel Salazar, University of Porto, Heidelberg, Portugal

B. Bonavida (ed.), *Molecular Mechanisms of Tumor Cell Resistance to Chemotherapy*, Resistance to Targeted Anti-Cancer Therapeutics 1, DOI: 10.1007/978-1-4614-7070-0_10, 

do not respond to chemotherapy from the beginning of therapy and (2) secondary or acquired resistance, whereby tumors initially respond to drugs but, during the course of chemotherapy, drug-resistant subpopulations overgrow the entire tumor and ultimately lead to refractoriness and treatment failure.

During the past three decades, the concept of multidrug resistance has been developed. The basic idea of combination treatment is to kill tumor subpopulations resistant to one drug by another drug in the combination. While this concept undoubtedly is a mainstay of cancer therapy leading to improving survival times of patients during the past decades by constantly optimizing combination regimens, nevertheless, cancers can develop resistance to many drugs at the same time. While cross-resistance of tumors to many different anticancer drugs was a well-known phenomenon already in the early days of cancer chemotherapy [1], a specific profile of cross-resistance was described as multidrug resistance. This profile comprises cross-resistance between anthracyclines, *Vinca* alkaloids, taxanes, epipodophyllotoxins, but not antimetabolites, alkylating agents or platinum compounds. The reason for this classical type of multidrug resistance is a membrane-located efflux transporter which extrudes drugs out of tumor cells. This drug transporter has been termed P-glycoprotein and is encoded by the *MDR1* (*ABCB1*) gene and belongs to the ATP-binding cassette (ABC) transporter family.

A typical, but unusual feature of the P-glycoprotein is that it translocates a wide variety of chemically diverse compounds. While many speculations on the mechanism of action of P-glycoprotein have been made, the only common property of P-glycoprotein substrates is their relative hydrophobic, amphiphilic nature [2, 3]. While initially one binding domain and subsequently two binding sites have been proposed for the P-glycoprotein, more recent investigations suggested multiple different binding sites [4, 5]. An alternative model hypothesized that the P-glycoprotein extrudes diverse drugs by an induced-fit mechanism [6].

Recently, homology models for the P-glycoprotein based on the crystal structure of the bacterial ABC transporter from *Staphylococcus aureus* Sav1866 have been described [7]. Three main membrane-related binding regions were outlined in the P-glycoprotein. The binding region 1 is located at the interface between the membrane and the cytosol and two other binding regions are located in the transmembrane parts of the protein. The regions contain multiple binding pockets. Hence, it is possible that drugs, depending on their structural properties, may bind to either more hydrophobic or more hydrophilic pockets, or even to more than one pocket simultaneously. Additionally, a large binding pocket resides in the protein cavity, which may represent an "escaping" site, where the compounds that bind to any of these regions are released from the protein. Site-directed mutagenesis experiments fit to these putative binding sites of the homology models [8], but a final proof can only be delivered by drug-protein crystal structures. Considering these complex and unresolved situations, the substrate specificity of P-glycoprotein may be much broader than estimated thus far. The binding of established compounds to the P-glycoprotein as a first step in the drug efflux process is illustrated in Figs. 1, 2, 3, 4. By means of a molecular docking approach, we have shown that compounds of diverse drug classes

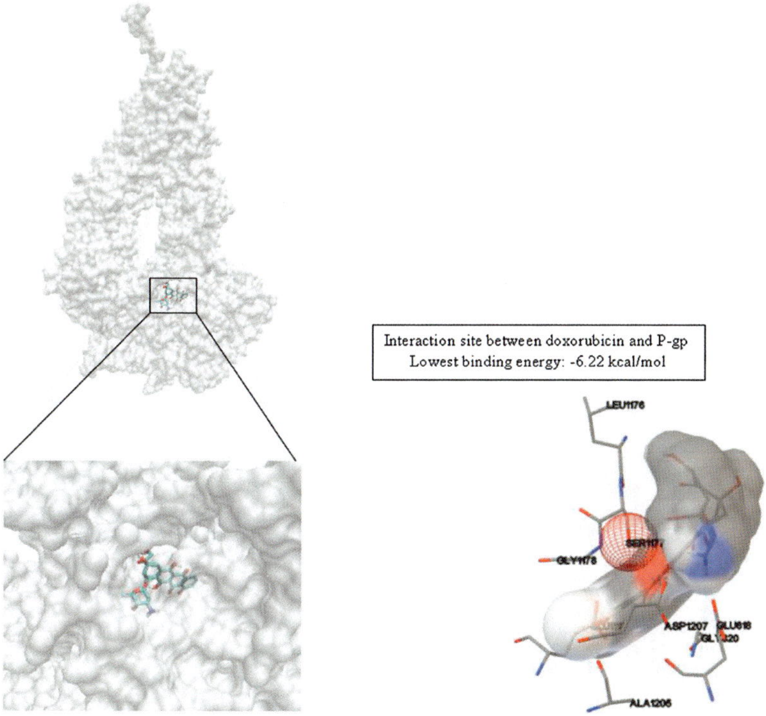


**Fig. 1** Graphical representation of the binding of doxorubicin to the human P-glycoprotein (P-gp). Human P-gp was homology-modeled using MODELLER software [111] with the mouse P-gp crystal structure (PDB-ID 3G5U at www.pdb.org) as template. The docking was performed using AutoDock version 4.2 (http://autodock.scripps.edu/). The best docked position of doxorubicin was illustrated in different styles. *Top* The P-gp surface was calculated and drawn as a *black shadow* (ghost mode) for better contrast with the drug. *Bottom left* For a clearer view of the cavity where the drug was docked, the *box* shows a close-up view of the drug in the cavity. *Bottom right*: Amino acids in close proximity to the docked drug molecule, which constitute the binding cavity

and mode of action bind to different binding sites on the human P-glycoprotein (doxorubicin, Fig. 1; vinblastine, Fig. 2; paclitaxel, Fig. 3; etoposide, Fig. 4).

The human genome contains 49 ABC transporter genes encoding for proteins with diverse functions [9, 10]. After characterization of the classical multidrug resistance phenotypes and the P-glycoprotein, atypical forms have been described with other cross-resistance profiles [11, 12], Atypical multidrug resistance can be caused by other ABC transporters, e.g. multidrug-resistance related proteins (MRPs, ABCCs) or breast cancer resistance protein (BCRP, ABCG2) or proteins not related to the ABC transporters such as DNA topoisomerases, enzymes related to the glutathione redox cycle or DNA repair enzymes [13, 14].

After the description of the calcium channel blocker verapamil as an inhibitor of the P-glycoprotein a plethora of synthetic drugs from diverse pharmacological classes [15] and natural products [16] have been identified. The concept was to

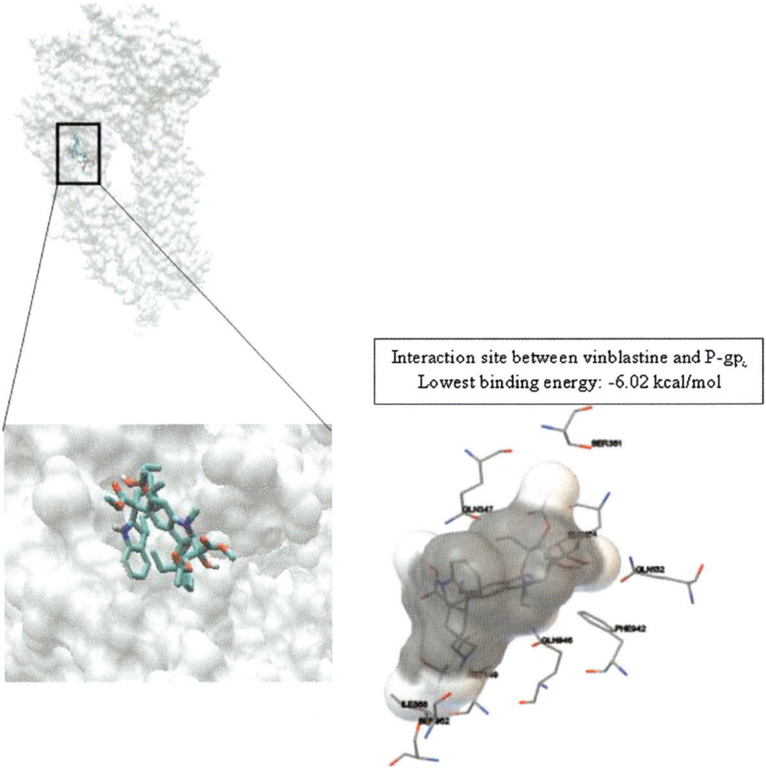


**Fig. 2** Graphical representation of the binding of vinblastine to the human P-glycoprotein. For details see Fig. 1 legend

inhibit the efflux function of the P-glycoprotein and, thereby, to increase anticancer drug concentrations by the combination regimen of cytostatic drugs and P-glycoprotein inhibitors.

Unfortunately, clinical trials with P-glycoprotein inhibitors were not successful to reverse multidrug resistance [17], although the relevance of the P-glycoprotein for treatment failure and worse survival time prognosis of cancer patients has been unambiguously shown [18–20]. Among the reasons for the unfavorable outcome of these clinical trials were: (1) P-glycoprotein inhibitors have been initially developed for the treatment of other diseases than cancer. For instance, verapamil blocks calcium channels in the heart and is an established drug to treat heart rhythm disturbances. These main drug activities may appear as non-tolerable side effects in cancer therapy, (2) The P-glycoprotein is not only expressed in tumors, but also in normal tissues to detoxify potentially harmful xenobiotic compounds taken up with food. High P-glycoprotein expression can be found in the gastrointestinal tract, liver, kidney, blood brain barrier, blood placenta barrier etc. Systemic treatment with P-glycoprotein inhibitors may, therefore, not only affect

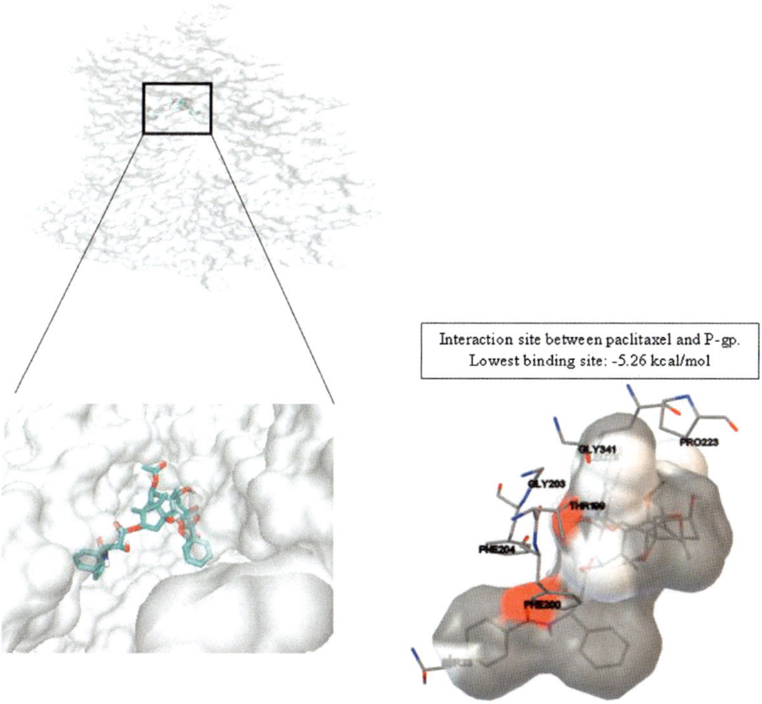


**Fig. 3** Graphical representation of the binding of paclitaxel to the human P-glycoprotein. For details see Fig. 1 legend

P-glycoprotein's functions in multidrug-resistant tumors, but also in healthy tissues and (3) some P-glycoprotein inhibitors are strong inducers of cytochrome P450 monooxygenase (CYP) enzymes in the liver. Up-regulated CYP enzymes metabolize anticancer drugs more rapidly and efficiently leading rather to treatment failure than to reversal of multidrug resistance.

The fact that it was not possible as of yet to establish novel treatment protocols to combat multidrug resistance by the addition of P-glycoprotein inhibitors to standard treatment -protocols resulted with much disappointment in the scientific community.

## Hypersensitivity in Resistant Cells

In the hype to search for P-glycoprotein inhibitors, it was sometimes overseen that multidrug resistance comprises not all classes of anticancer drugs and that P-glycoprotein-expressing tumor cells retain sensitivity e.g. to antimetabolites, alkylating agents or platinum compounds. This finding opened the opportunity to bypass multidrug resistance by treatment with non-cross-resistant drugs.

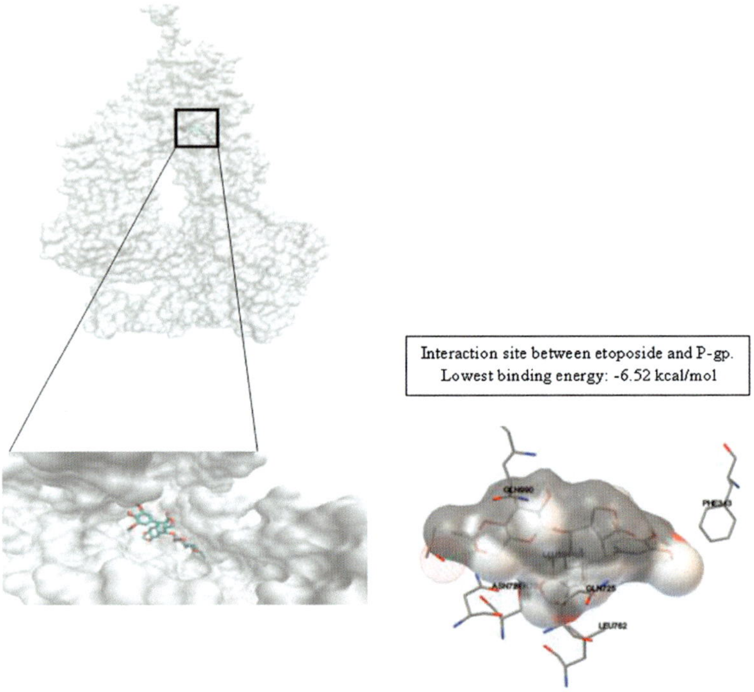


**Fig. 4** Graphical representation of the binding of etoposide to the human P-glycoprotein. For details see Fig. 1 legend

Even more, some of these non-cross-resistant drugs are hypersensitive in multidrug-resistant cells, i.e. they kill multidrug-resistant cells at lower concentrations than their drug-sensitive counterparts. This phenomenon has been termed collateral sensitivity. Hypersensitivity or collateral sensitivity is not a unique feature of multidrug-resistant tumor cells as it has first been described in drug-resistant *Escherichia coli* [21]. Collateral sensitivity was also observed in an otherwise pleiotropic cross-resistant mutant of *Saccharomyces cerevisiae* [22]. These data indicate that collateral sensitivity is a general biological and pharmacological phenomenon, by which organisms react to toxic insults. Collateral sensitive drugs have also been recognized in tumor cell lines, long before the P-glycoprotein was discovered [1]. Bech-Hansen et al. [23] were the first to describe collateral sensitivity in colchicine-selected P-glycoprotein expressing, multidrug-resistant Chinese hamster ovary cells.

The selective killing of P-glycoprotein-expressing tumor cells by taking advantage of collateral sensitivity has been compared to the concept of synthetic lethality [24, 25]. Synthetic lethality is a novel approach that specifically targets cancer cells with specific gene mutations that are not found in normal cells. If two parallel pathways both contribute to an essential cellular process and one pathway is inhibited by a specific mutation, the second pathway may be inhibited by small molecules leading to the death of cancer cells. Normal cells will be spared from the cytotoxic effects of the small molecule, since they escape using the intact first pathway without gene mutation [26, 27]. One example for the concept of synthetic lethality is the inhibition of PARP-1 to specifically kill cancer with mutations in the *BRCA1* or *BRCA2* genes. The idea in the context of multidrug resistance is to find a drug that specifically kills P-glycoprotein-over-expressing multidrug-resistant tumor cells, but not other cells.

Although drugs exerting collateral sensitivity in multidrug-resistant cells do not differentiate between normal and cancerous tissue (because the P-glycoprotein is also expressed in normal organs), it is an intriguing idea to take advantage of collateral sensitivity as "Achilles' heel" to specifically target otherwise refractory tumors.

In the past three to four decades, a plethora of compounds have been identified to exert collateral sensitivity in P-glycoprotein expressing otherwise drug-resistant tumors, including local anesthetics, detergents, antimetabolites, alkylating agents, platinum compounds as well as natural products such as saponins or flavonoids (Table 1). Collateral sensitivity is a phenomenon which is not restricted to the expression of P-glycoprotein, although some studies provided strong evidence by analyzing revertant cell lines or applying siRNA targeting the *MDR1* gene that collateral sensitivity was linked to the presence of the P-glycoprotein (Table 1). It can be speculated that different forms of collateral sensitivity may exist depending on the drug used to select for resistance and the molecular mechanisms of drug resistance in resistant cell lines.

## Mechanisms of Collateral Sensitivity

### *Increased ATP Hydrolysis and ROS Generation by Futile Cycling*

Intriguingly, many P-glycoprotein inhibitors are also substrates of this efflux pump. This means that these substances are exported out of the cell under ATP consumption, diffuse back into the cell, and are exported again and so on. These futile cycles cause increased ATP hydrolysis [28, 29]. The process of multidrug-resistant cells to replenish ATP generates oxidative stress leading both to increased levels of reactive oxygen species (ROS) provided from oxidative phosphorylation and to oxidation of glutathione as an antioxidant stress response. Non-detoxified ROS may lead to induction of apoptosis causing collateral sensitivity of multidrug-resistant cells. These effects have been shown for verapamil as a model compound for P-glycoprotein inhibition [30–32]. Whether this mechanism is also valid for other classes of P-glycoprotein inhibitors should be investigated in the future. Interestingly, verapamil causes collateral sensitivity not only in P-glycoprotein-expressing cells, but also in MRP1-expressing ones, although verapamil is a much weaker substrate for MRP1 than for the P-glycoprotein [33–35]. It has been

**Table 1** Collateral sensitivity in drug-resistant tumors

| Cell line | Tumor type | Collateral sensitivity | Resistance | P-gp expression | Reference |
|---|---|---|---|---|---|
| $CH^RC5$ | Chinese hamster ovary | Local anaesthetics (procaine, xylocaine, propanolal), steroid hormones (1-dehydrotestosterone, corticosterone, 5-beta-pregnan-3,20-dione), triton-x | Colchicine | Yes | Bech-Hansen et al. [23], Ling et al. [53] |
| $CH^RC5$ $CH^RB30$ RPMI 8226 | Human breast cancer | Dexamethasone | Doxorubicin | Yes | Dalton et al. [54] |
| H69AR | Human small cell lung cancer | 1-dehydro-testosterone, lidocaine | Doxorubicin | No | Mirski and Gerlach [55] |
| S180/ADR | Mouse sarcoma | Methotrexate, 5-fluorouracil | Doxorubicin | Yes | Volm et al. [56] |
| CHO-K1 | Chinese hamster ovary | Prednisolone | Actinomycin D | N.A. | Diddens et al. [57] |
| HN-1/DAR-II | Squamous cell carcinoma of the tongue | Cisplatin | Etoposide | N.A. | Lock and Hill [58] |
| HT1080/DR4 | Human fibrosarcoma | Nicardipin | Doxorubicin | No | Cole et al. [59] |
| HL60/MX2 | Human promyelocytic leukemia | Melphalan, mitomycin | Mitoxantrone | No | Harker et al. [60] |
| G3361/CP G3361/ 4HC G3361/ PAM | Human melanoma | Carmustine carmustine carmustine | Cisplatin, hydroxyper-oxycycloph osphamide, melphalan | N.A. | Wang et al. [61] |

(continued)

**Table 1** (continued)

| Cell line | Tumor type | Collateral sensitivity | Resistance | P-gp expression | Reference |
|---|---|---|---|---|---|
| RPMI-8402 CPT-K | Human leukemia | Daunorubicin, doxorubicin, mitoxanthrone, m-AMSA, etoposide, THP-adriamycin, vincritine, vindesine, cyclophosphamide, cisplatin | CPT-11 | N.A. | Oguro et al. [62] |
| RPMI-8402 CPT-T | Human leukemia | Vindesine, cyclophosphamide | CPT-11 | N.A. | Oguro et al. [62] |
| H69AR | Human small cell lung cancer | Buthionine sulfoximine | Doxorubicin | No | Cole et al. [63] |
| DU 145/EMR | Human prostate carcinoma | Cytochalasin B | Estramustine | No | Speicher et al. [64] |
| 9L/ACNU | Rat glioma | L-asparaginase | ACNU | No | Saito et al. [65] |
| SH-SY5Y/VCR | Human neurobastic | Progesterone megestrol acetate | Vincristine | Yes | Fleming et al. [66] |
| NCI-H69/DAU4 | Human small cell lung cancer | Cytarabine, cisplatin, carmustine | Daunorubicin | Yes | Jensen et al. [67] |
| GLC4/ADR | Human small cell lung cancer | Doxycycline oligomycin | Doxorubicin | No | de Jong et al. [68] |
| GM3639/L100 | T cell acute lymphoblastic leukemia | Cyclosporine A | Vincristine | Yes | Larsson et al. [69] |
| CHO VRT 5/15/25 | Chinese hamster ovary | Verapamil, nicardipine | Vincristine | Yes | Stow and Warr [70] |
| RPMI 8226/DOX | Human myeloma | Streptozotocin carmustine (BCNu) | Doxorubicin | Yes | Futscher et al. [42] |

(continued)

**Table 1** (continued)

| Cell line | Tumor type | Collateral sensitivity | Resistance | P-gp expression | Reference |
|---|---|---|---|---|---|
| CHO/ACTD | Chinese hamster ovary | Dexniguldipine | Actinomycin D | Yes | Neumann et al. [71] |
| LI210/VCR | Murine leukemia | 11-deoxycorticosterone, dexamethasone | Vincristine | Yes | Barancik et al. [72] |
| H69/DAU4 | Human small cell lung cancer | 5-fluorouracil, cytarabine, carmustine, bleomycin, camptothecin, cisplatin, hydroxyurea | Danuorubicin<br>teniposide<br>vincristine | Yes<br>No<br>No | Jensen et al. [73] |
| NYH/VM | Human small cell lung cancer | 5-fluorouracil, cytarabine, carmustine, bleomycin, hydroxyurea, methotrexate, mitomycin C | Danuorubicin<br>teniposide<br>vincristine | Yes<br>No<br>No | Jensen et al. [73] |
| H69/VP | Human small cell lung cancer | 5-fluorouracil, mrmustine, cytarabine, carmustine, bleomycin, camptothecin, cisplatin, hydroxyurea, iododoxorubicin | Danuorubicin<br>teniposide<br>vincristine | Yes<br>No<br>No | Jensen et al. [73] |
| AA8-Mtp$^{R10}$ | Chinese hamster ovary | Etoposide, vincristine | Metoprine | No | Assaraf and Slotky [74] |
| AA8-Pyr$^{R10}$ | Chinese hamster ovary | Actinomycin D, colchicine, doxorubicin, etoposide, gramicidin D, vinblastine, vincristine | Pyrimethamine | No | Assaraf and Slotky [74] |
| AA8-Tmp$^{R500}$ | Chinese hamster ovary | Actinomycin D, colchicine, doxorubicin, etoposide, vinblastine | Trimethoprim | No | Assaraf and Slotky[74] |

(continued)

**Table 1** (continued)

| Cell line | Tumor type | Collateral sensitivity | Resistance | P-gp expression | Reference |
|---|---|---|---|---|---|
| $CH^RA3$ $CH^R110$ $CH^RC5$ $CH^RB30$ | Chinese hamster ovary | Polyoxyethaylene amphiphiles (triton X-100 nonidet P-40). dibutylphthalate | Colchicine | Yes | Loe and Sharom [75] |
| LYC5 | Human colon carcinoma | Rotenone, anti-mycin, oligomycin, vincristine, actinomycinD, rhalamine-123, melphelan, doxorubicin | N-(5-indanylsulfonyl)-N'-(4-chlorophenyl) urea (ISCU) | Decreased P-gp | Sosinski et al. [43] |
| KB8-5 | Human epidermoid carcinoma | Rotenone, antimycin, oligomycin, diarylsulfonylurea derivatives | Colchicine | Yes | Sosinski et al. [43] |
| TE-671 MR | Human rhabdomyo-sarcoma xenogrraft | Etoposide | Melphalane | N.A. | Friedman et al. [44] |
| P388/MITOX | Murine leukemia | Cisplatin, bleomycin | Mitoxantrone | Yes | Fichtner et al. [76] |
| P388/ADR | Murine leukemia | Cyclophosphamide | Doxorubicin | Yes | Fichtner et al. [76] |
| P388/MTX | Murine leukemia | Cyclophosphamide, daunorubicin, etoposide, cisplatin | Methotrexate | Yes | Fichtner et al. [76] |
| P388/VCR | Murine leukemia | Cyclophosphamide | Vincristine | Yes | Fichtner et al. [76] |
| RPMI 8226/DOX 40 | Human myeloma | Lovastatin | Doxorubicin | Yes | Holmberg et al. [77] |
| RPMI-7932 VCR4.5/4, VBL-4, BMCOL-1 | Human melanoma | N1, NU-diethylnorspermine | Vincristine, vinblastine, colchicine | Yes | Porter et al. [78] |

(continued)

**Table 1** (continued)

| Cell line | Tumor type | Collateral sensitivity | Resistance | P-gp expression | Reference |
|---|---|---|---|---|---|
| L100 | Human T cell acute lymphoblastic leukemia | Gamma-irradiation, cisplatin | Vincristine | Yes | Cho et al. [79] |
| P388 | Murine leukemia | Cisplatin, alkylating agents | Colchicine | Yes | Demidova et al. [80] |
| CH B30 MCF 7/ AdR | Chinese hamster ovary human breast cancer | Opiates (pentazocine, naloxone, pethidine, morphine) opiates (pentazocine, naloxone, pethidine, morphine) | Colchicine doxorubicin | Yes Yes | Callaghan et al. [40] Callaghan et al. [40] |
| 2007/C13* | Human ovarian carcinoma | Doxorubicin, colcemide, vincristine, vinblastine, paclitaxel | Cisplatin | No | Parekh and Simpkins [81] |
| KB-8-5 KB-8-5-11 KB-C1 | Human epidermoid carcinoma | 2-deoxy-D-glucose | Colchicine | Yes | Bentley et al. [82] |
| M5/CDDP M5/ CDDPc | Murine ovarian reticulosarcoma | Etoposide doxorubicin | Cisplatin | N.A. | Belvedere et al. [83] |
| SUSA/VPC2 | Human testicular teratoma | ICRF-159 | Etoposide | N.A. | Davies et al. [45] |
| H69/BCNU | Human small cell lung cancer | Vindesine, vincristine, taxol, taxotere, doxorubicin, mitoxantrone m-AMSA, etoposide, teniposide, mitomycin cytarabine, gemicitabine | Carmustine | N.A. | Jensen et al. [77] |
| NYH/CIS | Human small cell lung cancer | Teniposide, cytarabine | Cisplatin | N.A. | Jensen et al. [84] |

(continued)

**Table 1** (continued)

| Cell line | Tumor type | Collateral sensitivity | Resistance | P-gp expression | Reference |
|---|---|---|---|---|---|
| NYH/CAM | Human small cell lung cancer | Doxorubicin, etoposide, teniposide, hydroxyurea | Camptothecin | N.A. | Jensen et al. [84] |
| NYH/TPT | Human small cell lung cancer | Teniposide | Topotecan | N.A. | Jensen et al. [84] |
| H69/DAU | Human small cell lung cancer | Cisplatin, cytarabine, gemicitabine | Daunorubicin | Yes | Jensen et al. [84] |
| NYH/VM | Human small cell lung cancer | Cytarabine | Teniposide | N.A. | Jensen et al. [84] |
| H69/VP | Human small cell lung cancer | Mitomycin, cisplatin, cytarabine, hydroxyurea | Etoposide | Yes | Jensen et al. [84] |
| MCF-7/OAP | Human breast cancer | E09 | Oxazaphosphorine | N.A. | Rekha and Sladek [85] |
| CHO/OAR2-3 | Chinese hamster ovary | Cisplatin, nitrogen mustard | Okadaic acid | Yes | Tohda et al. [86] |
| SBC-3/OADM 100 | Human small cell lung cancer | Methotrexate, TNP-351 (antifolate) | Doxorubicin | Yes | Matsuo et al. [46] |
| SW-1573/2R160<br>2R120 | Human lung carcinoma | No 5-fluorouracil, AG337 | Doxorubicin<br>doxorubicin | Yes<br>No | Van Triest et al. [87] |
| C6/VP | Rat glioma in vivo | Lomustine, cisplatin, cytarabine | Etoposide | No | Taki et al. [88] |
| IGROV-1/Pt0.5<br>IGROV-1/Pt-1 | Human ovarian carcinoma | Taxol | Cisplatin | N.A. | Perego et al. [89] |
| A2780/m248<br>A2780/m273 | Human ovarian carcinoma | Tumor necrosis factor | Transfected mutated p53 | N.A. | Sleijfer et al. [90] |
| CEM/ara-C | T cell acute lymphoblastic leukemia | Idarubicin 6-thioguanine, vincristine, 6-mercaptoguanosine | AraC | Yes | Martin-Aragon et al. [91] |

(continued)

**Table 1** (continued)

| Cell line | Tumor type | Collateral sensitivity | Resistance | P-gp expression | Reference |
|---|---|---|---|---|---|
| PC-6/Tax 1-1 | Human oat cell carcinoma | Cisplatin, 5-FU, DX-8951f camptothecin | Taxol | Yes | Ishii et al. [92] |
| PC-6/ADM 2-1 | Human oat cell carcinoma | Cisplatin, camptothecin | Doxorubicin | Yes | Ishii et al. [92] |
| PC-6/VCR29-9 | Human oat cell carcinoma | Cisplatin, 5-fluorouracil, DX-8951f, SN-38, camptothecin | Vincristine | Yes | Ishii et al. [92] |
| PC-6/VP1-1 | Human oat cell carcinoma | | Etoposide | Yes | Ishii et al. [92] |
| PC-14 KDDP PC-9/ CDDP | Human non-small cell lung cancer | Aragusterol | Cisplatin | N.A. | Fuhuoka et al. [93] |
| H69/DAU NYH/VM | Human small cell lung cancer | Gemicitabine, cytarabine gemicitabine, cytarabine | Daunorubicin teniposide | Yes No | Bergman et al. [48, 49] |
| EBG-2/R | Human lung squamous cell carcinoma | Vindesine | Cisplatin/ifos-famide/vindesine | No | Kawai et al. [94] |
| SR-2 | Human small cell lung cancer | Azidothymidine | Cisplatin | N.A. | Savaraj et al. [95] |
| PyrR100 | Chinese hamster ovary | Cholate, etoposide, doxorubicin, vincristine | Pyrimethamine | N.A. | Stark et al. [96] |
| CCRF-CEM | Human T cell acute lymphoblastic leukemia | Cyclosporine A chloroquine dexamethasone | Antirheumatic drug (sulfa-salazine) | N.A. | van der Heijden et al. [97] |
| P388/F11782 | Murine leukemia in vivo | Cisplatin, topotecan, colchicine, *Vinca* alkaloids | F11782 (topoisomerase inhibitor) | N.A. | Kruczyuski et al. [98] |

(continued)

**Table 1** (continued)

| Cell line | Tumor type | Collateral sensitivity | Resistance | P-gp expression | Reference |
|---|---|---|---|---|---|
| K562/D1-9 | Human chronic myeloic leukemia | Arsenic trioxide | Daunorubicin | Yes | Seo et al. [99] |
| | Human renal cell carcinoma | 2-chlorodeoxyadenosine, fludarabine | 2-deoxy-tubercidin | N.A. | Pan and Nelson[100] |
| H69/AR | Small cell lung cancer | Verapamil, buthionine, sulfoximine, apigenin | Doxorubicin | No | Laberge et al. [35] |
| $CH^{R}C5$ | Chinese hamster ovary | Verapamil | Colchicine | Yes | Laberge et al. [28] |
| CEM/ADR 9000 | Human T cell acute lymphoblastic leukemia | Artesunic acid homodimer | Doxorubicin | Yes | Horwedel et al. [101] |
| KB-VIN | Human epidermoid carcinoma | 6,6,8-triethyl-desmosku-montin B | Vincristine | Yes | Nakagawa-Goto et al. [102] |
| KB-V1 | Human epidermoid carcinoma | Tiopronin | Vinblastine | Yes | Goldsborough et al. [52] |
| L1210R | Mouse leukemia | Cold stress | Daunorubicin | Yes | Cerezo et al. [103] |
| MCF-7TaxR | Human breast cancer | Platinum | Paclitaxel | N.A. | Ajabnoor et al. [104] |
| KB-V1, KB-C1, KB-A1 | Human epidermoid carcinoma | 2-deoxy-D-glucose | Vinblastine, colchicine, doxorubicin | Yes | Bell et al. [36] |
| Diverse | Diverse | NSC73306 and derivatives | Vinblastine, colchicine, docorubicin | Yes | Hall et al. [24, 105] |
| MCF-7/ADR | Breast cancer | 2-deoxyglucose | Doxorubicin | N.A. | Kaplan et al. [106] |

(continued)

**Table 1** (continued)

| Cell line | Tumor type | Collateral sensitivity | Resistance | P-gp expression | Reference |
|---|---|---|---|---|---|
| KB-VIN | Human epidermoid carcinoma | 6,8,8-triethyl analogues of desmosdumotin B | Vinblastine | Yes | Nakagawa-Goto et al. [107], Nakagawa-Goto et al. [102] |
| KB-V1 | Human epidermoid carcinoma | 20 NSC compounds | Vinblastine | Yes | Türk et al. [108] |
| MES-Sa Dx 5 | Uterine sarcoma | 20 NSC compounds | Doxorubicin | Yes | Türk et al. [108] |
| CHO-VCR/A CHO-VCR/T | Chinese hamster ovary | Verapamil | Vincristine | | Warr et al. [109, 110] |

hypothesized that verapamil increases MRP1-mediated glutathione efflux rather than being effluxed itself by MRP1 [33]. Increased glutathione efflux reduces the detoxification capacity for ROS, hence, leading to increased apoptosis and collateral sensitivity.

## *Inhibition of ATP Production*

The fact that ATP depletion plays a role for preferential killing by verapamil of multidrug-resistant tumor cells led to the hypothesis that ATP depletion by metabolic inhibitors may also cause collateral sensitivity of multidrug-resistant cells. Indeed, the glycolysis inhibitor 2-deoxy-D-glucose, the electron transport chain inhibitors rotenone and antimycin A are hypersensitive in P-glycoprotein-expressing cells [25, 36, 37]. Tunicamycin reduced GLUT1-mediated glucose transport and potentiated the effects of 2-deoxy-D-gluxose in P-glycoprotein-expressing cells [38]. This indicates that tunicamycin decreased intracellular glucose levels, thus, leading to decreased ATP production and enhancement of collateral sensitivity.

## *Effects on Membrane Fluidity*

Collateral sensitivity also occurs in the absence of P-glycoprotein as shown in Table 1. Membrane-active compounds may account for P-glycoprotein-independent effects. Membrane fluidity has been described as a P-glycoprotein-independent mechanism of doxorubicin resistance affecting the uptake rates in sensitive and multidrug-resistant tumor cells [39].

Membrane-active compounds such as detergents (Triton-X) or local anesthetics perturb the biophysical properties of cell membranes [23]. Thereby, they may not only reverse multidrug-resistance by increasing uptake of co-applied anticancer drugs, but also cause collateral sensitivity if applied alone. As membrane perturbation by these compounds is more pronounced in multidrug-resistant cells than in parental, drug-sensitive cells reversal of resistance and collateral sensitivity seem to be specifically linked to membrane-active drugs such as detergents and local anesthetics [40]. Differences of membrane fluidity occur independently of P-glycoprotein in multidrug-resistant cells, since P-glycoprotein-transfected cells did not show such alterations [41]. Hence, biophysical membrane alterations are rather a consequence of the drug selection process during resistance development than a direct functional linkage to the presence of P-glycoprotein.

## *Alterations of Target Proteins for Collateral Sensitive Drugs*

Tumor cell lines resistant to typical drugs involved in multidrug resistance and drug-resistant cells with other phenotypes may show specific alterations in activity and expression of target proteins for other drugs which then cause collateral sensitivity. It is not known whether these changes occur as specific adaptation during selection pressure by the corresponding resistance-including agent or whether these alterations occur more or less by chance as secondary events.

Double selection of tumor cells with doxorubicin and verapamil led to collateral sensitivity to streptozotocin and carmustine due to functional loss of DNA repair enzyme $O^6$-methyl guanine DNA methyltransferase (AGMT) [42]. Diarylsulfonylurea resistant cells show collateral sensitivity to vincristine and actinomycin D [43]. The hypersensitivity to these two drugs can be explained by decreased expression of P-glycoprotein. Melphalan-resistant xenograft tumors were collateral sensitive to etoposide [44], which could be explained by an increase in protein expression and activity of DNA-topoisomerase II, which is the target protein of etoposide. Davies et al. [45] observed that collateral sensitivity to the DNA topoisomerase II alpha-inhibitor, ICRF-159, was associated with down-regulation of this protein. The collateral sensitivity to methotrexate and the antifolate TNP-351 in doxorubicin-resistant cells was associated by a faster uptake and intracellular accumulation, possibly pointing to increased activity of an uptake transporter [46]. A multidrug-resistant P-glycoprotein expressing cell line revealed collateral sensitivity to 5-fluorouracil due to increased thymidylate synthase protein and mRNA expression and activity. The marine steroid aragusterol provoked collateral sensitivity in cisplatin-resistant cells [47]. Treatment with the compound caused $G_1$ phase arrest in the cell cycle, decreased R6 phosphorylation and expression of cyclins and cyclin-dependent kinases as well as decreased p53 expression. Collateral sensitivities to gemcitabine and cytarabine in daunorubicin- or teniposide-resistant cells may be explained by specific alterations in metabolic enzymes, e.g. increased activity of the gemcitabine and cytarabine-activating deoxycytidine kinase. Furthermore, deoxycytidine deaminase was decreased, which inactivates both drugs [48, 49].

## Conclusion

A synopsis of the literature on collateral sensitivity in drug-resistant tumor cells clearly shows that multiple mechanisms account for this phenotype. As collateral sensitivity has not been investigated in the same detail as the role of P-glycoprotein for multidrug resistance, it is possible that our knowledge on collateral sensitivity is still very incomplete and that we know only the tip of the iceberg as yet. More systematic analyses are required to unravel the full biological and pharmacological relevance.

The molecular mechanisms of some compounds have been significantly investigated (e.g. verapamil) but the modes of action of collateral sensitivity of many other drugs are still sparely known. This is true for both drug resistance-related and -unrelated with ABC transporters. Collateral sensitivity of P-glycoprotein- or MRP-1-expressing cells has been analyzed, but less is known about collateral sensitivity in resistant cells expressing other ABC transporters. For instance, compounds causing collateral sensitivity in P-glycoprotein-expressing cells do not show hypersensitivity in MRP4-, MRP5-, or BCRP-expressing cells [50]. It is also largely unknown whether collateral sensitivity plays a role in the clinical context. In vitro tests of patient samples of chronic lymphocytic leukemia have been performed by Bosanquet and Bell [51]. Comparisons of untreated and treated patients revealed that chlorambucil treatment induced sensitivity to steroids. It still has to be clarified, whether collateral sensitivity is apparent in refractory tumors and whether it can be used to improve survival times of cancer patients. From a pharmacological point of view, the identification of drugs with high degrees of collateral sensitivity is desirable. Most drugs described to exert collateral sensitivity have not been identified in a targeted search for such compounds, but frequently appeared as side products in test panels to characterize cross-resistance profiles. As a consequence, most collateral sensitive drugs only exert modest degrees of collateral sensitivity. As exemplarily shown by Goldsborough et al. [52]. more potent drugs causing hypersensitivity are needed, if collateral sensitivity should be further developed as a novel treatment strategy of otherwise drug-resistant tumors.

**Conflicts of Interest** No potential conflicts of interest were disclosed.

## References

1. Hutchinson DJ. Cross resistance and collateral sensitivity studies in cancer chemotherapy. Adv Cancer Res. 1963;7:235–50.
2. Gottesman MM, Pastan I. Biochemistry of multidrug resistance mediated by the multidrug transporter. Annu Rev Biochem. 1993;62:385–427.
3. Rautio J, Humphreys JE, Webster LO. In vitro P-glycoprotein inhibition assays for assessment of clinical drug interaction potential of new drug candidates: a recommendation for probe substrates. Drug Metab Dispos. 2006;34:786–92.
4. Ayesh S, Shao YM, Stein WD. Co-operative, competitive and non-competitive interactions between modulators of Pglycoprotein. Biochim Biophys Acta. 1996;1316:8–18.
5. Borgnia MJ, Eytan GD, Assaraf YG. Competition of hydrophobic peptides, cytotoxic drugs, and chemosensitizers on a common P-glycoprotein pharmacophore as revealed by its ATPase activity. J Biol Chem. 1996;271:3163–71.
6. Safa AR. Identification and characterization of the binding sites of P-glycoprotein for multidrug resistance-related drugs and modulators. Curr Med Chem Anticancer Agents. 2004;4:1–17.
7. Globisch C, Pajeva IK, Wiese M. Identification of putative binding sites of P-glycoprotein based on its homology model. Chem Med Chem. 2008;3:280–95.
8. Shapiro AB, Ling V. Extraction of Hoechst 33342 from the cytoplasmic leaflet of the plasma membrane by P-glycoprotein. Eur J Biochem. 1997;250:122–9.
9. Efferth T. The human ATP-binding cassette transporter genes: from the bench to the bedside. Curr Mol Med. 2001;1:45–65.
10. Gillet JP, Efferth T, Remacle J. Chemotherapy-induced resistance by ATP-binding cassette transporter genes. Biochim Biophys Acta. 2007;1775:237–62.

11. Danks MK, Yalowich JC, Beck WT. Atypical multiple drug resistance in a human leukemic cell line selected for resistance to teniposide (VM-26). Cancer Res. 1987;47:1297–301.
12. Haber M, Norris MD, Kavallaris M, Bell DR, Davey RA, White L, Stewart BW. Atypical multidrug resistance in a therapy-induced drug-resistant human leukemia cell line (LALW-2): resistance to Vinca alkaloids independent of P-glycoprotein. Cancer Res. 1989;49:5281–7.
13. Volm M, Efferth T. Overcoming resistance in tumors. Dtsch Med Wochenschr. 1994;119:475–9.
14. Efferth T, Grassmann R. Impact of viral oncogenesis on responses to anti-cancer drugs and irradiation. Crit Rev Oncog. 2000;11:165–87.
15. Ford JM, Hait WN. Pharmacology of drugs that alter multidrug resistance in cancer. Pharmacol Rev. 1990;42:155–99.
16. Eichhorn T, Efferth T. P-glycoprotein and its inhibition in tumors by phytochemicals derived from Chinese herbs. J Ethnopharmacol. 2012;141:557–70.
17. Tiwari AK, Sodani K, Dai CL, Ashby CR Jr, Chen ZS. Revisiting the ABCs of multidrug resistance in cancer chemotherapy. Curr Pharm Biotechnol. 2011;12:570–94.
18. Efferth T, Osieka R. Clinical relevance of the MDR-1 gene and its gene-product, P-glycoprotein, for cancer chemotherapy: a meta-analysis. Tumordiagn Ther. 1993; 14:238–43.
19. Tamaki A, Ierano C, Szakacs G, Robey RW, Bates SE. The controversial role of ABC transporters in clinical oncology. Essays Biochem. 2011;50:209–32.
20. Amiri-Kordestani L, Basseville A, Kurdziel K, Fojo AT, Bates SE. Targeting MDR in breast and lung cancer: discriminating its potential importance from the failure of drug resistance reversal studies. Drug Resist Updat. 2012;15:50–61.
21. Sybalski W, Bryson V. Genetic studies on microbial cross resistance to toxic agents: I. Cross resistance of Escherichia coli to fifteen antibiotics. J Bacteriol. 1952;64:489–99.
22. Rank GH, Robertson AJ, Phillips KL. Modification and inheritance of pleiotropic cross resistance and collateral sensitivity in *Saccharomyces cerevisiae*. Genetics. 1975;80:783–93.
23. Bech-Hansen NT, Till JE, Ling V. Pleiotropic phenotype of colchicine-resistant CHO cells: cross-resistance and collateral sensitivity. J Cell Physiol. 1976;88:23–31.
24. Hall MD, Salam NK, Hellawell JL, Fales HM, Kensler CB, Ludwig JA, Szakács G, Hibbs DE, Gottesman MM. Synthesis, activity, and pharmacophore development for isatin-beta-thiosemicarbazones with selective activity toward multidrug-resistant cells. J Med Chem. 2009;52:3191–204.
25. Pluchino KM, Hall MD, Goldsborough AS, Callaghan R, Gottesman MM. Collateral sensitivity as a strategy against cancer multidrug resistance. Drug Resist Updat. 2012;15:98–105.
26. Kaelin WG. The concept of synthetic lethality in the context of anticancer therapy. Nat Rev Cancer. 2005;5:689–98.
27. Chan DA, Giaccia AJ. Harnessing synthetic lethal interactions in anticancer drug discovery. Nat Rev Drug Discov. 2011;10:351–64.
28. Laberge RM, Ambadipudi R, Georges E. P-glycoprotein (ABCB1) modulates collateral sensitivity of a multidrug resistant cell line to verapamil. Arch Biochem Biophys. 2009;491:53–60.
29. Gottesman MM, Ambudkar SV, Xia D. Structure of a multidrug transporter. Nat Biotechnol. 2009;27:546–7.
30. Broxterman HJ, Pinedo HM, Kuiper CM, Kaptein LC, Schuurhuis GJ, Lankelma J. Induction by verapamil of a rapid increase in ATP consumption in multidrug-resistant tumor cells. FASEB J. 1988;2:2278–82.
31. Broxterman HJ, Pinedo HM, Kuiper CM, Schuurhuis GJ, Lankelma J. Glycolysis in P-glycoprotein-overexpressing human tumor cell lines. Effects of resistance-modifying agents. FEBS Lett. 1989;247:405–10.
32. Karwatsky J, Lincoln MC, Georges E. A mechanism for P-glycoprotein-mediated apoptosis as revealed by verapamil hypersensitivity. Biochemistry. 2003;42:12163–73.

33. Trompier D, Chang XB, Barattin R, du Moulinet D'Hardemare A, Di Pietro A, Baubichon-Cortay H. Verapamil and its derivative trigger apoptosis through glutathione extrusion by multidrug resistance protein MRP1. Cancer Res. 2004;64:4950–4956.
34. Rothnie A, Conseil G, Lau AY, Deeley RG, Cole SP. Mechanistic differences between GSH transport by multidrug resistance protein 1 (MRP1/ABCC1) and GSH modulation of MRP1-mediated transport. Mol Pharmacol. 2008;74:1630–40.
35. Laberge RM, Karwatsky J, Lincoln MC, Leimanis ML, Georges E. Modulation of GSH levels in ABCC1 expressing tumor cells triggers apoptosis through oxidative stress. Biochem Pharmacol. 2007;73:1727–37.
36. Bell SE, Quinn DM, Kellett GL, Warr JR. 2-Deoxy-D-glucose preferentially kills multidrug-resistant human KB carcinoma cell lines by apoptosis. Br J Cancer. 1998;78:1464–70.
37. Kaplan O, Jaroszewski JW, Clarke R, Fairchild CR, Schoenlein P, Goldenberg S, Gottesman MM, Cohen JS. The multidrug resistance phenotype: 31P nuclear magnetic resonance characterization and 2-deoxyglucose toxicity. Cancer Res. 1991;51:1638–44.
38. Bentley J, Quinn DM, Pitman RS, Warr JR, Kellett GL. The human KB multidrug-resistant cell line KB-C1 is hypersensitive to inhibitors of glycosylation. Cancer Lett. 1997;115:221–7.
39. Ramu A, Glaubiger D, Magrath IT, Joshi A. Plasma membrane lipid structural order in doxorubicin-sensitive and -resistant P388 cells. Cancer Res. 1983;43:5533–7.
40. Callaghan R, Riordan JR. Collateral sensitivity of multidrug resistant cells to narcotic analgesics is due to effects on the plasma membrane. Biochim Biophys Acta. 1995;1236:155–62.
41. Alemán C, Annereau JP, Liang XJ, Cardarelli CO, Taylor B, Yin JJ, Aszalos A, Gottesman MM. P-glycoprotein, expressed in multidrug resistant cells, is not responsible for alterations in membrane fluidity or membrane potential. Cancer Res. 2003;63:3084–91.
42. Futscher BW, Campbell K, Dalton WS. Collateral sensitivity to nitrosoureas in multidrug-resistant cells selected with verapamil. Cancer Res. 1992;52:5013–7.
43. Sosinski J, Thakar JH, Germain GS, Dias P, Harwood FC, Kuttesch JF, Houghton PJ. Cross-resistance to antitumor diarylsulfonylureas and collateral sensitivity to mitochondrial toxins in a human cell line selected for resistance to the antitumor agent N-(5-indanylsulfonyl)-N'-(4-chlorophenyl)urea. Mol Pharmacol. 1994;45:962–70.
44. Friedman HS, Dolan ME, Kaufmann SH, Colvin OM, Griffith OW, Moschel RC, Schold SC, Bigner DD, Ali-Osman F. Elevated DNA polymerase alpha, DNA polymerase beta, and DNA topoisomerase II in a melphalan-resistant rhabdomyosarcoma xenograft that is cross-resistant to nitrosoureas and topotecan. Cancer Res. 1994;54:3487–93.
45. Davies SL, Bergh J, Harris AL, Hickson ID. Response to ICRF-159 in cell lines resistant to cleavable complex-forming topoisomerase II inhibitors. Br J Cancer. 1997;75:816–21.
46. Matsuo K, Kiura K, Ueoka H, Tabata M, Shibayama T, Matsumura T, Takigawa N, Hiraki S, Harada M. Growth inhibitory effects of antifolates against an adriamycin-resistant human small cell lung cancer cell line. Acta Med Okayama. 1997;51:121–127. Erratum in: Acta Med Okayama. 1997;51:237.
47. Fukuoka K, Yamagishi T, Ichihara T, Nakaike S, Iguchi K, Yamada Y, Fukumoto H, Yoneda T, Samata K, Ikeya H, Nanaumi K, Hirayama N, Narita N, Saijo N, Nishio K. Mechanism of action of aragusterol a (YTA0040), a potent anti-tumor marine steroid targeting the G(1) phase of the cell cycle. Int J Cancer. 2000;88:810–9.
48. Bergman AM, Munch-Petersen B, Jensen PB, Sehested M, Veerman G, Voorn DA, Smid K, Pinedo HM, Peters GJ. Collateral sensitivity to gemcitabine (2',2'-difluorodeoxycytidine) and cytosine arabinoside of daunorubicin- and VM-26-resistant variants of human small cell lung cancer cell lines. Biochem Pharmacol. 2001;61:1401–8.
49. Bergman AM, Pinedo HM, Talianidis I, Veerman G, Loves WJ, van der Wilt CL, Peters GJ. Increased sensitivity to gemcitabine of P-glycoprotein and multidrug resistance-associated protein-overexpressing human cancer cell lines. Br J Cancer. 2003;88:1963–70.

50. Wu CP, Shukla S, Calcagno AM, Hall MD, Gottesman MM, Ambudkar SV. Evidence for dual mode of action of a thiosemicarbazone, NSC73306: a potent substrate of the multidrug resistance linked ABCG2 transporter. Mol Cancer Ther. 2007;6:3287–96.
51. Bosanquet AG, Bell PB. Enhanced ex vivo drug sensitivity testing of chronic lymphocytic leukaemia using refined DiSC assay methodology. Leuk Res. 1996;20:143–53.
52. Goldsborough AS, Handley MD, Dulcey AE, Pluchino KM, Kannan P, Brimacombe KR, Hall MD, Griffiths G, Gottesman MM. Collateral sensitivity of multidrug-resistant cells to the orphan drug tiopronin. J Med Chem. 2011;54:4987–97.
53. Ling V, Kartner N, Sudo T, Siminovitch L, Riordan JR. Multidrug-resistance phenotype in Chinese hamster ovary cells. Cancer Treat Rep. 1983;67:869–74.
54. Dalton WS, Durie BG, Alberts DS, Gerlach JH, Cress AE. Characterization of a new drug-resistant human myeloma cell line that expresses P-glycoprotein. Cancer Res. 1986;46:5125–30.
55. Mirski SE, Gerlach JH, Cole SP. Multidrug resistance in a human small cell lung cancer cell line selected in adriamycin. Cancer Res. 1987;47:2594–8.
56. Volm M, Efferth T, Günther A, Lathan B. Detection of murine S180 cells expressing a multidrug resistance phenotype using different in vitro test systems and a monoclonal antibody. Arzneimittelforschung. 1987;37:862–7.
57. Diddens H, Gekeler V, Neumann M, Niethammer D. Characterization of actinomycin-D-resistant CHO cell lines exhibiting a multidrug-resistance phenotype and amplified DNA sequences. Int J Cancer. 1987;40:635–42.
58. Lock RB, Hill BT. Differential patterns of anti-tumour drug responses and mechanisms of resistance in a series of independently-derived VP-16-resistant human tumour cell lines. Int J Cancer. 1988;42:373–81.
59. Cole SP, Downes HF, Slovak ML. Effect of calcium antagonists on the chemosensitivity of two multidrug-resistant human tumour cell lines which do not overexpress P-glycoprotein. Br J Cancer. 1989;59:42–6.
60. Harker WG, Slade DL, Dalton WS, Meltzer PS, Trent JM. Multidrug resistance in mitoxantrone-selected HL-60 leukemia cells in the absence of P-glycoprotein overexpression. Cancer Res. 1989;49:4542–9.
61. Wang YY, Teicher BA, Shea TC, Holden SA, Rosbe KW. al-Achi A, Henner WD. Cross-resistance and glutathione-S-transferase-pi levels among four human melanoma cell lines selected for alkylating agent resistance. Cancer Res. 1989;49:6185–92.
62. Oguro M, Seki Y, Okada K, Andoh T. Collateral drug sensitivity induced in CPT-11 (a novel derivative of camptothecin)-resistant cell lines. Biomed Pharmacother. 1990;44:209–16.
63. Cole SP, Downes HF, Mirski SE, Clements DJ. Alterations in glutathione and glutathione-related enzymes in a multidrug-resistant small cell lung cancer cell line. Mol Pharmacol. 1990;37:192–7.
64. Speicher LA, Sheridan VR, Godwin AK, Tew KD. Resistance to the antimitotic drug estramustine is distinct from the multidrug resistant phenotype. Br J Cancer. 1991;64:267–73.
65. Saito Y, Nakada Y, Hotta T, Mikami T, Kurisu K, Yamada K, Kiya K, Kawamoto K, Uozumi T. Cross-resistance patterns in ACNU-resistant glioma sublines in culture. J Neurosurg. 1991;75:277–83.
66. Fleming GF, Amato JM, Agresti M, Safa AR. Megestrol acetate reverses multidrug resistance and interacts with P-glycoprotein. Cancer Chemother Pharmacol. 1992;29:445–9.
67. Jensen PB, Roed H, Sehested M, Demant EJ, Vindeløv L, Christensen IJ, Hansen HH. Doxorubicin sensitivity pattern in a panel of small-cell lung-cancer cell lines: correlation to etoposide and vincristine sensitivity and inverse correlation to carmustine sensitivity. Cancer Chemother Pharmacol. 1992;31:46–52.
68. de Jong S, Holtrop M, de Vries H, de Vries EG, Mulder NH. Increased sensitivity of an adriamycin-resistant human small cell lung carcinoma cell line to mitochondrial inhibitors. Biochem Biophys Res Commun. 1992;182:877–85.

69. Larssson R, Fridborg H, Csoka K, Bergh J, Nygren P. Cytotoxic action of cyclosporins on human tumor cell lines is not dependent on immunosuppressive activity. Anticancer Res. 1992;12:1581–5.
70. Stow MW, Warr JR. Reduced influx is a factor in accounting for reduced vincristine accumulation in certain verapamil-hypersensitive multidrug-resistant CHO cell lines. FEBS Lett. 1993;320:87–91.
71. Neumann M, Wilisch A, Diddens H, Probst H, Gekeler V. MDR hamster cells exhibiting multiple altered gene expression: effects of dexniguldipine-HCl (B859–35), cyclosporin A and buthionine sulfoximine. Anticancer Res. 1992;12:2297–302.
72. Barancík M, Docolomanský P, Slezák J, Breier A. Overcoming of vincristine resistance in L1210/VCR cells by several corticosteroids. Collateral sensitivity of resistant cells. Neoplasma. 1993;40:21–5.
73. Jensen PB, Christensen IJ, Sehested M, Hansen HH, Vindeløv L. Differential cytotoxicity of 19 anticancer agents in wild type and etoposide resistant small cell lung cancer cell lines. Br J Cancer. 1993;67:311–20.
74. Assaraf YG, Slotky JI. Characterization of a lipophilic antifolate resistance provoked by treatment of mammalian cells with the antiparasitic agent pyrimethamine. J Biol Chem. 1993;268:4556–66.
75. Loe DW, Sharom FJ. Interaction of multidrug-resistant Chinese hamster ovary cells with amphiphiles. Br J Cancer. 1993;68:342–51.
76. Fichtner I, Stein U, Hoffmann J, Winterfeld G, Pfeil D, Hentschel M. Characterization of four drug-resistant P388 sublines: resistance/sensitivity in vivo, resistance-and proliferation-markers, immunogenicity. Anticancer Res. 1994;14:1995–2003.
77. Holmberg M, Sandberg C, Nygren P, Larsson R. Effects of lovastatin on a human myeloma cell line: increased sensitivity of a multidrug-resistant subline that expresses the 170 kDa P-glycoprotein. Anticancer Drugs. 1994;5:598–600.
78. Porter CW, Ganis B, Rustum Y, Wrzosek C, Kramer DL, Bergeron RJ. Collateral sensitivity of human melanoma multidrug-resistant variants to the polyamine analogue, N1, N11-diethylnorspermine. Cancer Res. 1994;54:5917–24.
79. Cho J, Lee Y, Lutzky J, Redpath L, Slater L. Collateral sensitivity to radiation and cis-platinum in a multidrug-resistant human leukemia cell line. Cancer Chemother Pharmacol. 1995;37:168–72.
80. Demidova NS, Ilyinskaya GV, Shiryaeva OA, Chernova OB, Goncharova SA, Kopnin BP. Decreased sensitivity of multidrug-resistant tumor cells to cisplatin is correlated with sorcin gene co-amplification. Neoplasma. 1995;42:195–201.
81. Parekh H, Simpkins H. Cross-resistance and collateral sensitivity to natural product drugs in cisplatin-sensitive and -resistant rat lymphoma and human ovarian carcinoma cells. Cancer Chemother Pharmacol. 1996;37:457–62.
82. Bentley J, Bell SE, Quinn DM, Kellett GL, Warr JR. 2-deoxy-D-glucose toxicity and transport in human multidrug-resistant KB carcinoma cell lines. Oncol Res. 1996;8:77–84.
83. Belvedere G, Imperatori L, Damia G, Tagliabue G, Meijer C, de Vries EG, D'Incalci M. In vitro and in vivo characterisation of low-resistant mouse reticulosarcoma (M5076) sublines obtained after pulse and continuous exposure to cisplatin. Eur J Cancer. 1996;32A:2011–8.
84. Jensen PB, Holm B, Sorensen M, Christensen IJ, Sehested M. In vitro cross-resistance and collateral sensitivity in seven resistant small-cell lung cancer cell lines: preclinical identification of suitable drug partners to taxotere, taxol, topotecan and gemcitabin. Br J Cancer. 1997;75:869–77.
85. Rekha GK, Sladek NE. Multienzyme-mediated stable and transient multidrug resistance and collateral sensitivity induced by xenobiotics. Cancer Chemother Pharmacol. 1997;40:215–24.
86. Tohda H, Takao M, Kikuchi A, Yasumoto T, Yasui A. Unstable expression of the multi-drug-resistant phenotype in Chinese hamster ovary cells resistant to okadaic acid. Biochem Biophys Res Commun. 1997;232:398–402.

87. van Triest B, Pinedo HM, Telleman F, van der Wilt CL, Jansen G, Peters GJ. Cross-resistance to antifolates in multidrug resistant cell lines with P-glycoprotein or multidrug resistance protein expression. Biochem Pharmacol. 1997;53:1855–66.
88. Taki T, Ohnishi T, Arita N, Hiraga S, Hayakawa T. In vivo etoposide-resistant C6 glioma cell line: significance of altered DNA topoisomerase II activity in multi-drug resistance. J Neurooncol. 1998;36:41–53.
89. Perego P, Romanelli S, Carenini N, Magnani I, Leone R, Bonetti A, Paolicchi A, Zunino F. Ovarian cancer cisplatin-resistant cell lines: multiple changes including collateral sensitivity to Taxol. Ann Oncol. 1998;9:423–30.
90. Sleijfer S, Le TK, de Jong S, Timmer-Bosscha H, Withoff S, Mulder NH. Combined cytotoxic effects of tumor necrosis factor-alpha with various cytotoxic agents in tumor cell lines that are drug resistant due to mutated p53. J Immunother. 1999;22:48–53.
91. Martin-Aragon S, Mukherjee SK, Taylor BJ, Ivy SP, Fu CH, Ardi VC, Avramis VI. Cytosine arabinoside (ara-C) resistance confers cross-resistance or collateral sensitivity to other classes of anti-leukemic drugs. Anticancer Res. 2000;20:139–50.
92. Ishii M, Iwahana M, Mitsui I, Minami M, Imagawa S, Tohgo A, Ejima A. Growth inhibitory effect of a new camptothecin analog, DX-8951f, on various drug-resistant sublines including BCRP-mediated camptothecin derivative-resistant variants derived from the human lung cancer cell line PC-6. Anticancer Drugs. 2000;11:353–62.
93. Fukuoka K, Yamagishi T, Ichihara T, Nakaike S, Iguchi K, Yamada Y, Fukumoto H, Yoneda T, Samata K, Ikeya H, Nanaumi K, Hirayama N, Narita N, Saijo N, Nishio K. Mechanism of action of aragusterol a (YTA0040), a potent anti-tumor marine steroid targeting the G(1) phase of the cell cycle. Int J Cancer. 2000;88:810–9.
94. Kawai H, Kiura K, Tabata M, Yoshino T, Takata I, Hiraki A, Chikamori K, Ueoka H, Tanimoto M, Harada M. Characterization of non-small-cell lung cancer cell lines established before and after chemotherapy. Lung Cancer. 2002;35:305–14.
95. Savaraj N, Wu C, Wangpaichitr M, Kuo MT, Lampidis T, Robles C, Furst AJ, Feun L. Overexpression of mutated MRP4 in cisplatin resistant small cell lung cancer cell line: collateral sensitivity to azidothymidine. Int J Oncol. 2003;23:173–9.
96. Stark M, Rothem L, Jansen G, Scheffer GL, Goldman ID, Assaraf YG. Antifolate resistance associated with loss of MRP1 expression and function in Chinese hamster ovary cells with markedly impaired export of folate and cholate. Mol Pharmacol. 2003;64:220–7.
97. van der Heijden J, de Jong MC, Dijkmans BA, Lems WF, Oerlemans R, Kathmann I, Scheffer GL, Scheper RJ, Assaraf YG, Jansen G. Acquired resistance of human T cells to sulfasalazine: stability of the resistant phenotype and sensitivity to non-related DMARDs. Ann Rheum Dis. 2004;63:131–7.
98. Kruczynski A, Barret JM, Van Hille B, Chansard N, Astruc J, Menon Y, Duchier C, Créancier L, Hill BT. Decreased nucleotide excision repair activity and alterations of topoisomerase IIalpha are associated with the in vivo resistance of a P388 leukemia subline to F11782, a novel catalytic inhibitor of topoisomerases I and II. Clin Cancer Res. 2004;10:3156–68.
99. Seo T, Urasaki Y, Takemura H, Ueda T. Arsenic trioxide circumvents multidrug resistance based on different mechanisms in human leukemia cell lines. Anticancer Res. 2005;25:991–8.
100. Pan BF, Nelson JA. Dihydrodiol dehydrogenase in drug resistance and sensitivity of human carcinomas. Cancer Chemother Pharmacol. 2007;59:697–702.
101. Horwedel C, Tsogoeva SB, Wei S, Efferth T. Cytotoxicity of artesunic acid homo- and heterodimer molecules toward sensitive and multidrug-resistant CCRF-CEM leukemia cells. J Med Chem. 2010;53:4842–8.
102. Nakagawa-Goto K, Chang PC, Lai CY, Hung HY, Chen TH, Wu PC, Zhu H, Sedykh A, Bastow KF, Lee KH. Antitumor agents. 280. Multidrug resistance-selective desmosdumotin B analogues. J Med Chem. 2010;53:6699–705.
103. Cerezo D, Lencina M, Ruiz-Alcaraz AJ, Ferragut JA, Saceda M, Sanchez M, Cánovas M, García-Peñarrubia P, Martín-Orozco E. Acquisition of MDR phenotype by leukemic cells is

associated with increased caspase-3 activity and a collateral sensitivity to cold stress. J Cell Biochem. 2012;113:1416–25.
104. Ajabnoor GM, Crook T, Coley HM. Paclitaxel resistance is associated with switch from apoptotic to autophagic cell death in MCF-7 breast cancer cells. Cell Death Dis. 2012;3:e260.
105. Hall MD, Brimacombe KR, Varonka MS, Pluchino KM, Monda JK, Li J, Walsh MJ, Boxer MB, Warren TH, Fales HM, Gottesman MM. Synthesis and structure-activity evaluation of isatin-$\beta$-thiosemicarbazones with improved selective activity toward multidrug-resistant cells expressing P-glycoprotein. J Med Chem. 2011;54:5878–89.
106. Kaplan O, Navon G, Lyon RC, Faustino PJ, Straka EJ, Cohen JS. Effects of 2-deoxyglucose on drug-sensitive and drug-resistant human breast cancer cells: toxicity and magnetic resonance spectroscopy studies of metabolism. Cancer Res. 19901;50:544–551.
107. Nakagawa-Goto K, Bastow KF, Chen TH, Morris-Natschke SL, Lee KH. Antitumor agents 260. New desmosdumotin B analogues with improved in vitro anticancer activity. J Med Chem. 2008;51:3297–303.
108. Türk D, Hall MD, Chu BF, Ludwig JA, Fales HM, Gottesman MM, Szakács G. Identification of compounds selectively killing multidrug-resistant cancer cells. Cancer Res. 2009;69:8293–301.
109. Warr JR, Brewer F, Anderson M, Fergusson J. Verapamil hypersensitivity of vincristine resistant Chinese hamster ovary cell lines. Cell Biol Int Rep. 1986;10:389–99.
110. Warr JR, Anderson M, Fergusson J. Properties of verapamil-hypersensitive multidrug-resistant Chinese hamster ovary cells. Cancer Res. 1988;48:4477–83.
111. Eswar N, Marti-Renom MA, Webb B, Madhusudhan MS, Eramian D, Shen M, Pieper U, Sali A. Comparative protein structure modeling with MODELLER. In: Current Protocols in Bioinformatics, Supplement 15. Wiley, New York. 2006; p. 5.6.1–5.6.30.

# Human Cancer Resistance to Trail-Apoptotic Pathway-Targeted Therapies

**Anita C. Bellail and Chunhai Hao**

**Abstract** Tumor necrosis factor-related apoptosis inducing ligand (TRAIL) mediates innate and adaptive immunity against the tumorigenesis and tumor progression. TRAIL binds its two death receptors, DR4 and DR5, which activate intracellular pathway of apoptosis for self destruction of tumor cells. To target this apoptotic pathway, recombinant human TRAIL and monoclonal antibodies to DR4 and DR5 have been generated as TRAIL agonists for clinical cancer therapies. A number of TRAIL agonists have passed drug safety evaluation in phase I trials; however, the data from phase II trials thus far are disappointing: TRAIL agonists either in monotherapy or combination have failed to show clinical antitumor activity. In this chapter, we will provide a historic review of the advances and the challenges in the development of TRAIL agonists for clinical treatment of human cancers.

**Keywords** Apoptosis · Cancer · Caspase · Death receptor · Therapies · TNF · TRAIL

**Abbreviations**

| | |
|---|---|
| Apaf1 | Apoptotic protease activating factor 1 |
| Bid | Bcl-2 inhibitory BH3-domain-containing protein |
| tBid | truncated Bid |
| cFLIP | cellular FADD-like interleukin-1$\beta$-converting enzyme-like inhibitory protein |
| CYLD | cylindromatosis |
| CUL3 | Cullin 3 |
| CDDO | Cyano-3,12-dioxooleana-1,9-dien-28-oic acid |

A. C. Bellail
Department of Pathology and Laboratory Medicine, Emory University School of Medicine, Atlanta, GA, USA

C. Hao (✉)
Department of Pathology and Laboratory Medicine, Emory University School of Medicine, 1365-C Clifton Road NE, Atlanta, GA 30322, USA
e-mail: chao@emory.edu

B. Bonavida (ed.), *Molecular Mechanisms of Tumor Cell Resistance to Chemotherapy*, Resistance to Targeted Anti-Cancer Therapeutics 1, DOI: 10.1007/978-1-4614-7070-0_11,

| | |
|---|---|
| CDDO-Me | CDDO-methyl ester |
| CDDO-IM | CDDO-imidazolide |
| CCNU | 1-(2-chloroethyl)-3-cyclohexyl-1-nitrosourea |
| DcR | Decoy receptor |
| DED | death effector domain |
| DD | death domain |
| DISC | death-inducing signaling complex |
| DR | death receptor |
| DIABLO | direct inhibitor of apoptosis binding protein with low pI |
| DFF45 | DNA fragmentation factor 45 |
| DUB | deubiquitinating |
| ERK 1/2 | extracellular signal-regulated kinase 1/2 |
| FADD | Fas-associated death domain |
| FasL | Fas ligand |
| FLICE | FADD-like interleukin-1$\beta$-converting enzyme |
| GPI | glycosyl phosphatidylinositol |
| IAP | inhibitors of apoptosis proteins |
| cIAP | cellular inhibitors of apoptosis proteins |
| IKK$\gamma$ | inhibitor of $\kappa$B (I$\kappa$B) kinase $\gamma$ |
| mAb | monoclonal antibody |
| mTOR | mammalian target of rapamycin |
| NHL | non-Hodgkin lymphoma |
| NSCLC | non-small cell lung carcinoma |
| NF-$\kappa$B | nuclear factor-$\kappa$B |
| OTU | N-terminal ovarian tumor domain |
| PEA-15 | phosphopritein enriched in astrocytes-Mr 15,000 |
| PED | phosphoprotein enriched in diabetes |
| PLAC | preligand assembly complex |
| PI3K | phosphatidylinositide-3-kinase |
| rhTRAIL | recombinant human TRAIL |
| RIP | receptor interacting protein |
| Smac | second mitochondria-derived activator of caspase |
| TNF | tumor necrosis factor |
| TNFSF | TNF ligand superfamily member |
| TNFR | TNF receptor |
| TRAIL | tumor necrosis factor-related apoptosis inducing ligand |
| TRAILR | TRAIL receptor |
| TRADD | TNFR1-associated death domain |
| TRAF2 | TNFR-associated factor 2 |
| TNFAIP3 | TNF$\alpha$-induced protein 3 |
| TNFSF15 | TNF ligand superfamily member 15 |
| UB | ubiquitin |
| XIAP | X-linked inhibitor of apoptosis |

## Introduction

Apoptosis is a genetically programmed cell death that eliminates unwanted cells and maintains tissue homeostasis under physiological and pathological conditions. More than 230,000 articles have been published on apoptosis in the last four decades since Kerr, Wyllie and Currie first used the term of apoptosis for programmed cell death in 1972 [1]. More than 70 % of these publications occurred in the last decade after Brenner, Horvitz and Sulston were awarded with the 2002 Nobel Prize in Medicine for their work on programmed cell death. These efforts have contributed to our understanding of programmed cell death in the aspects of biology, biochemistry, genetics, physiology, pathology and pharmacology, ultimately leading to the genesis of a new class of apoptotic pathway-targeted therapeutic agents. To date, thousands of patients have received apoptotic pathway-targeted therapies in treating various human diseases.

Human cancers are genetic diseases in which genomic alterations result in the deregulation of cell growth and death pathways and thus lead to the relentless cancer cell growth at expense of cancer cell death [2]. The ultimate goal of apoptotic pathway-targeted therapies is to restore the endogenous programmed cell death and drive cancer cells into self destruction [3, 4]. To date, two classic pathways of apoptosis have been well characterized for therapeutic targeting: the death receptors-mediated extrinsic pathway [5] and the mitochondrial intrinsic pathway [6]. Therapeutic agents targeting these two cell death pathways have been developed and entered clinical trials for cancer treatments [7–9].

Death receptors belong to the tumor necrosis factor receptor (TNFR) superfamily that can trigger apoptosis upon binding of ligands of the tumor necrosis factor (TNF) family. TRAIL is a TNF family ligand that binds death receptor-4 (DR4) and DR5 on the cell surface and activates intracellular pathway of programmed cell death [7, 10]. Recombinant human TRAIL (rhTRAIL) and DR4 and DR5 agonistic antibodies have been developed as TRAIL agonists for clinical trials [11, 12]. Despite the excitement from phase I trials, however, the data from phase II trials have clearly established the resistance of human cancers to the treatment of TRAIL agonists [13]. In this chapter, we will provide a historic review of the development of death receptors-targeted cancer therapies and discuss the challenges that we currently face in cancer resistance to the targeted therapies.

## Historic Review of TNF Family in Cancer therapies

In 1975, TNFα, the prototype of TNF family ligands was isolated from the serum of mice treated with endotoxin and named as such because it caused tumor necrosis [14]. Nineteen TNF family ligands have been identified as type II transmembrane proteins [15] and seventeen of them have C-terminal TNF homology domains that bind the cysteine repeat domains of receptors of the TNF receptor (TNFR)

**Table 1** The TNF Family Ligands and Death Receptors and Toxicities

| Ligands | Death Receptors | Therapeutic Agent | Toxicity |
|---|---|---|---|
| TNFα | TNFR1 | rh TNFα | Hypotension and hepatotoxicity |
| FasL | Fas | Fas antibody | Hepatotoxicity |
| TNFSF15 | DR3 | | |
| TRAIL | DR4, DR5 | tagged rhTRAIL | Toxicity to hepatocytes and neural cells |

superfamily. The TNFR family receptors are expressed by various cells in different organ systems; in contrast, however, almost all of the TNF family ligands are expressed by cells of the immune system, including B and T lymphocytes, natural killer and dendritic cells, and monocytes [16]. Ligand-receptor interactions initiate multiple signaling pathways promoting cell survival, death, differentiation, and inflammation in the receptor-expressing cells, depending on the activation state of the cells and the expression levels of the TNFR superfamily receptors.

Twenty-eight receptors of the TNFR family have been reported [7] and classified into three groups: death receptors, decoy receptors and TNFR-associated factor binding receptors [17]. The death receptors have cysteine rich extracellular domains that recognize their death ligands and homologous cytoplasmic domains that activate intracellular apoptosis pathway [18]. The following four sets of death ligands and death receptors have been reported (Table 1): TNFα [14, 19, 20] and TNFR1 (CD120a, DR1, p55, p60) [21]; Fas ligand (FasL, CD95L) [22] and Fas (APO-1, CD95, DR2) [23]; TNF ligand superfamily member 15 (TNFSF15, TL1A) [24–26] and DR3 (Apo3, LARD, TR3, TRAMP, WSL-1) [27–31]; TRAIL [32, 33] and TRAIL receptor-1 (TRAILR1, APO-2, DR4) [34, 35] or TRAILR2 (DR5, KILLER, TRICK2) [35–38]. Since their discovery, preclinical and clinical efforts have been made for development of these pathways-targeted therapeutic agents for clinical treatments of cancers. The first major roadblock in translation of these therapeutic agents into successful clinical therapies was the hepatotoxicity that these agents caused in mice and human.

## *TNFR1 and Fas Targeted Cancer Therapies and Toxicities*

In 1984, *TNFα* gene was cloned from human [19] and recombinant human TNFα (rhTNFα) was soon generated and studies showed its antitumor activity but also toxicity to normal cells in mice [39]. Despite the toxicity of rhTNFα observed in mice, phase I and II trials of rhTNFα were launched in the 1990s and dose-limiting side effects of hypotension and hepatotoxicity were observed in patients [40–44]. The systemic use of rhTNFα as a therapeutic agent has therefore been abandoned since. In 1989, an agonistic antibody against Fas was shown to have the ability to activate Fas-mediated apoptosis [45]. Systemic injection of the Fas agonistic antibody in mice, however, caused profound hepatotoxicity [46] so that clinical trials of

Fas agonistic antibody have never been considered. Currently, clinical use of TNFα and FasL is limited to local delivery to avoid the systemic side effects [47, 48].

## *Tagged rhTRAIL-Induced Hepatotoxicity*

TRAIL was identified in 1995 by Immunex [32] and 1996 by Genentech [33] and rhTRAIL was soon generated as a TRAIL agonist for cancer therapies [49, 50]. Studies of these recombinant proteins, however, revealed the toxicities of tagged forms of rhTRAIL to normal cells: a leucine zipper-fused rhTRAIL killed isolated human astrocytes in culture [49]; a polyhistidine-tagged rhTRAIL was toxic to isolated human hepatocytes [51]; and a Flag-tagged rhTRAIL induced apoptosis of neural cells in human brain slice culture [52]. These in vitro studies raised the same safety concern as those expressed with rhTNFα and Fas agonistic antibody and triggered a call for "steering anti-cancer drugs away from the TRAIL" [53]. In contrast, however, studies of non-tagged native sequence soluble forms of rhTRAIL showed that non-tagged rhTRAIL was not toxic to normal cells in non-human primates [50] and isolated human astrocytes in culture [54]. A mouse monoclonal antibody (mAb) against DR5 (TRA-8) was generated as a TRAIL agonist for having antitumor activity without causing liver injury in mice [55]. Systemic injection of non-tagged rhTRAIL inhibited the growth of cancer xenografts but caused no injury to human hepatocytes in chimeric mice reconstituted with human livers [56]. It appears that non-tagged rhTRAIL contains sufficient bound zinc ions and forms the homotrimers whereas polyhistidine-tagged rhTRAIL has poor zinc ion coordination and forms disulfide-liked dimmers that cause hepatotoxicity [57]. Moreover, normal human cells express the cell surface DR4 and/or DR5 not sufficiently enough to engage intracellular death pathway [56, 58].

## *Phase I Trials of the Safety of TRAIL Agonists*

Preclinical findings that non-tagged rhTRAIL and its agonistic antibody are non-toxic to normal cells have provided renewed impetus for the launch of clinical trials of these TRAIL-apoptotic pathway-targeted therapeutic agents. In 2005, phase I trials of full human mAb against DR4 (HGS-ETR1, Human Genome Sciences) and DR5 (HGS-ETR2, Human Genome Sciences) were reported at the 96th Annual Meeting of the American Association for Cancer Research (AACR), revealing that these DR4 and DR5 agonistic antibodies were safe and well tolerated in patients. At the 42nd Annual Meeting of the American Society of Clinical Oncology (ASCO) in 2006, a phase I trial of rhTRAIL (Apo2L/TRAIL, Genentech/Amgen) showed that monotherapy with this no-tagged rhTRAIL was well tolerated in patients with no dose-limiting toxicity. In addition, this phase I trial showed the monotherapy of rhTRAIL was associated with partial response and stable disease in patients with advanced cancer. These preclinical and clinical studies have finally put TRAIL toxicity issue behind and, as a result, a number of DR4/DR5 agonistic antibodies have been generated and quickly entered clinical trials as TRAIL agonist [11].

## TRAIL-Induced Apoptosis in Cancer Immunity

Since its discovery in the middle 1990s [32, 33], TRAIL has been well studied for its physiological functions. The studies of mice treated with TRAIL-neutralizing antibody revealed for the first time that TRAIL is required for natural killer cells-mediated innate immunity against cancer progression and metastasis in mice [59, 60]. Studies of TRAIL-deficient mice further showed that TRAIL plays a critical role in immunosurveillance against tumorigenesis [61]. In addition, TRAIL contributes to T lymphocyte-mediated adaptive and dendritic cell-mediated innate immunity against cancers [62, 63]. Therefore, these studies have established TRAIL as a natural cancer killer of immune system in immunosurveillance and it can drive cancer cells into self destruction through activation of the endogenous pathway of programmed cell death [16].

### *TRAIL-Induced Extrinsic Pathway of Apoptosis*

TRAIL is a type II transmembrane protein with a short intracellular amino-terminal and a long extracellular carboxy-terminal [32, 33]. The extracellular carboxy-terminal containing the receptor-binding domain [64] can be cleaved to yield a biologically active soluble protein of the amino acids 114–281 of the full TRAIL protein [65]. TRAIL has two death receptors, DR4 [34, 35] and DR5 [36–38]; both are type I transmembrane proteins with two extracellular cysteine-rich domains that can bind TRAIL and a cytoplasmic death domain (DD) that can engage intracellular machinery of apoptosis. DR4 and DR5 exist either as homotrimers or heterotrimers linked through an interaction between the pre-ligand assembly domains in the extracellular termini [66]. Upon TRAIL ligation, DR4 and DR5 recruit the intracellular adaptor, Fas-associated death domain (FADD) [36, 67]. FADD contains a carboxy-terminal DD and an amino-terminal death effector domain (DED) [68], through which, FADD recruits DED-containing apoptosis-initiating protease caspase-8 [69–71] and caspase-10 [71–74] to the receptors for the formation of a death-inducing signaling complex (DISC), similar to the one reported in Fas apoptotic pathway [75, 76]. In the DISC (Fig. 1), caspase-8 and caspase-10 are dimerized and cleaved, become enzymatically active, and cleave apoptosis effector caspase-3 and caspase-7 in the execution of programmed cell death [77].

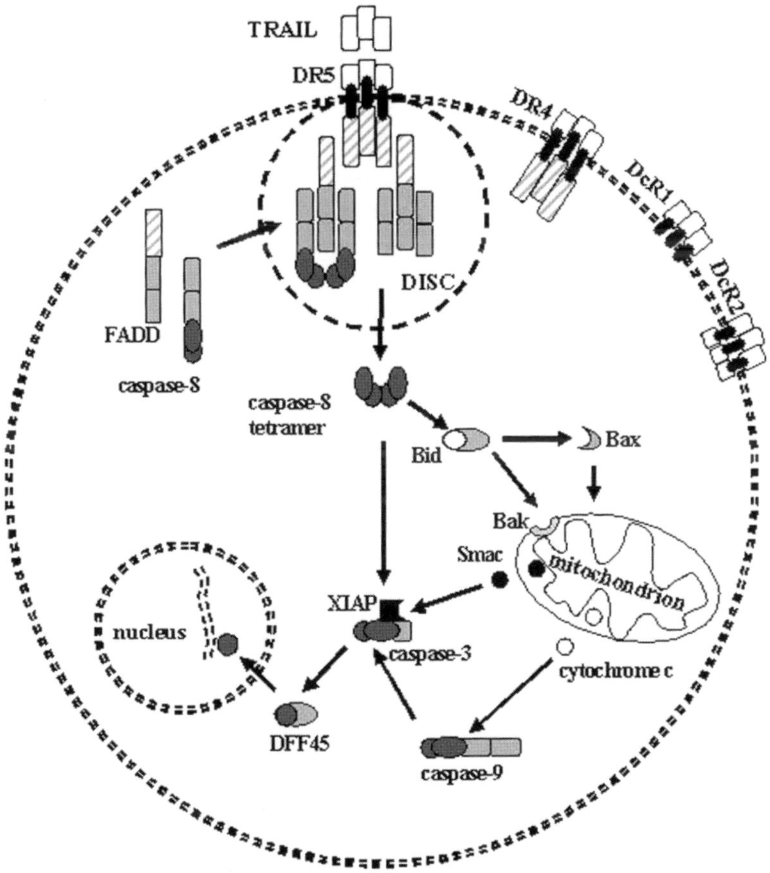


**Fig. 1** TRAIL-induced apoptosis through the extrinsic and intrinsic pathways

## *TRAIL-Induced Intrinsic Pathway of Apoptosis*

The intrinsic mitochondrial pathway of apoptosis is required for TRAIL-induced apoptosis (Fig. 1). Once cleaved and enzymatically activated in the DISC, caspase-8 active enzyme was released from the DISC and cleaves Bcl-2 inhibitory BH3-domain-containing protein (Bid) in cytoplasm [78]. The truncated Bid (tBid) in turn interacts with Bax and Bak and induces the protein oligomerization in mitochondrial membrane, leading to the change of the membrane potential and subsequent release from the mitochondria of cytochrome *c* [79] and second mitochondria-derived activator of caspase (Smac) or direct inhibitor of apoptosis binding protein with low pI (DIABLO) [80, 81]. In the cytosol, cytochrome *c* binds to Apoptotic protease activating factor 1 (Apaf1) and recruits dATP and caspase-9 for the formation of apoptosome (Fig. 1). Herein caspase-9 is cleaved through autoproteolysis and becomes active and cleaves downstream effector caspase-3 [82]. The caspase-3 cleavage, however, is inhibited in cancers due to the expression of X-linked inhibitor of apoptosis (XIAP) and interaction of XIAP and caspase-3 [83]. Therefore, TRAIL-induced extrinsic pathway can be rescued by the release of mitochondrial Smac that binds XIAP, removes it from caspase-3 and thus releases its inhibition of caspase-3 cleavage [84].

## *Caspase-8 in TRAIL-Induced Apoptosis*

Caspase-8 (also termed FADD-like interleukin-1$\beta$-converting enzyme, FLICE) is an apoptosis-initiating caspase that is synthesized as a zymogen and exists in two isoforms (p55, p53); each consists of two DED domains and a protease domain made of two subunits, p18 and p12 [85]. Through its DEDs, caspase-8 zymogens are recruited to the DISC where they become cleaved through two-step autoproteolytic processes: the first intra-molecular cleavage generates p12 subunit [86, 87] and the second cleavage results in the release of the p18 subunit from the DED and p10 from the p12 subunit [88, 89]. The caspase-8 p18 and p10 subunits become enzymatically active and cleave downstream caspase-3 and caspase-7 [77]. Caspase-3 and caspase-7 are effector caspases and each of them consists of a protease domain with a large and small subunit. Once cleaved and activated, caspase-3 cleaves downstream DNA fragmentation factor 45 (DFF45) in the execution of programmed cell death [90].

It remains debated about how a caspase-8 zymogen becomes cleaved and activated when there is no protease above it. An induced proximity model initially suggested that caspase-8 zymogens were brought into close proximity in the DISC for the autoproteolytic cleavage [91, 92]. Further in vitro biochemical studies proposed the dimerization model that caspase-8 zymogens first form dimers through an interaction of alanine 397 residues of their protease domains and become proteolytically active and cleaved to form the enzymatically active p18 and p10 tetramer [93–95]. However, this model of caspase-8 activation remains to be established in vivo in normal or cancerous cells. In addition, the question remains why there are two apoptosis initiators, caspase-8 and caspase-10 in the DISC.

## *TRAIL's Antitumor Activity*

The cytotoxicity of rhTRAIL was first reported in cancer cell lines originated from various organs and its antitumor activity was shown in the cell lines-derived xenografts in mice [49, 50, 96]. The expression of DR4 and/or DR5 on the surface of cancer cell lines and in some cancer tissues were reported [54, 97–100] and the cytotoxicity of TRAIL agonists were shown due to the induction of apoptosis in cancer cell lines [54, 55, 101–103]. TRAIL-induced apoptosis was further shown to involve the DISC assembly through DR4 and/or DR5-mediated recruitment of

FADD, caspase-8 and caspase-10 [71] in the lipid rafts [104]. In the DISC, caspase-8 and caspase-10 are cleaved through autoproteolysis and once released from the DISC, caspase-8 and caspase-10 enzymatically active subunits cleave Bid and thus induce mitochondrial release of cytochrome *c* and Smac, leading to the cleavage of XIAP and caspase-3 in the execution of TRAIL-induced apoptosis [99].

The preclinical studies have developed therapeutic modalities for TRAIL distribution in tumors. Systemic administration by intraperitoneal and intravenous injection of rhTRAIL and DR4/DR5 agonistic antibody was shown to inhibit the growth of tumor xenografts in mice [55, 96]. A local convection-enhanced delivery effectively distributed rhTRAIL throughout xenografts in rat brains [105] and intratumor injection of TRAIL-secreting neural stem cells inhibited the growth of brain tumor xenografts in mice [106]. Furthermore, intratumor injection of an adenovirus vector specifically replicating in tumor cells and expressing TRAIL significantly inhibited the growth of tumor xenografts in mice [107, 108]. These in vitro and in vivo preclinical studies suggest that TRAIL-apoptotic pathway exists in cancer cells and thus can be targeted for clinical cancer therapies.

## Trail Resistance in Human Cancers

The understanding of TRAIL physiology in tumor immunosurveillance boosts enthusiasms in the development of TRAIL agonists for cancer therapies [16]. On the other hand, however, the notion that TRAIL mediates innate and adaptive immunity against cancer predicts that cancers occur in patients by evasion of TRAIL-mediated immunosurveillance and are resistant to the treatment of TRAIL agonists. This seems to be true since studies of human cancer cell lines from the earlier stage showed that only a fraction of the cell lines undergo apoptosis under TRAIL treatment [54] and most if not all in vivo studies utilized TRAIL sensitive cancer cell lines in generation of xenografts [50, 56, 109]. While it remains under intensive investigation about how cancers invade the immunosurveillance, studies of cancer human cancer cell lines and derived xenografts have revealed the multiple checkpoints of TRAIL resistance from the upstream death receptors down to the cleavage of effector caspases.

### *Genomic Defects in TRAIL Apoptotic Pathway*

Genetic mutations, epigenetic silencing and chromosomal alterations have been identified in human cancers, resulting in the loss or silencing of apoptotic genes of TRAIL apoptotic pathway. Somatic inactive mutations of *Bax* gene were first identified [110], leading to TRAIL resistance in colon cancer cells [111]. Somatic inactive mutations of *DR4* and *DR5* were reported in non-Hodgkin lymphoma (NHL) and lung, head, neck, gastric and breast cancers [112–116]. The mutations occur in the DD of DR4/DR5 and block DR4/DR5-mediated intracellular signal transduction [117, 118]. *DR4* and *DR5* genes map to human chromosome 8p12-23 and cytogenetic analysis revealed loss of chromosome 8p12-23 in TRAIL-resistant glioblastoma cell lines [119]. Caspase-8 is silenced in neuroblastoma by DNA methylation, leading to the cancer resistance to the treatments of TRAIL agonists [120, 121]. Somatic inactive mutations of *caspase-8* were reported in colorectal and gastric carcinomas [122, 123] and loss of the *caspase-8* loci on chromosome 2q33-34 was identified in glioblastoma cell lines [119]. Inactive mutations and genomic loss of TRAIL apoptotic genes in human cancers lead to the cancer resistance to the treatments of TRAIL agonists.

## Inhibition by Decoy Receptors

In addition to DR4 and DR5, TRAIL has two membrane-associated decoy receptors, DcR1 (TRAIL-R3) [124–126] and DcR2 (TRAIL-R4) [127–129]. DcR1 is a glycosyl phosphatidylinositol (GPI)-anchored membrane protein with cysteine-rich extracellular domains but no cytoplasmic DD. DcR2 is a type I transmembrane protein with extracellular domains and a cytoplasmic DD. Through their extracellular domains, DcR1 and DcR2 interact with rhTRAIL and inhibit TRAIL-induced apoptosis either by competing with death receptors for TRAIL binding [130, 131] or by interrupting the homotrimeric formation of death receptors [132]. In addition, the interaction of TRAIL and DcR2 activates the nuclear factor-$\kappa$B (NF-$\kappa$B) pathway for cancer cell survival [129]. These studies have suggested DcR1/DcR2 act as TRAIL inhibitors and therefore raised concerns about the usefulness of rhTRAIL as compared to DR4/DR5 agonistic antibodies. This concern, however, is mainly based on the studies in transfectants and analysis of human cancer cell lines and tissues has detected neither DcR1 nor DcR1 and thus rule them out in TRAIL resistance [133].

## DISC Modifications and Inhibition of Caspase-8 Cleavage

The DISC can be modified by intracellular adaptors that either inhibit caspase-8 or activate cell survival pathways (Fig. 2). Receptor interacting protein (RIP) and TNFR1-associated death domain (TRADD), first identified as DD adaptor proteins in TNFR1 and Fas pathway [134–136], were found in the DR4/DR5-mediated DISC in transfectants [38], but analysis of human cancer cell lines detected RIP, but not TRADD in the DISC and showed that RIP recruits an inhibitor of $\kappa$B (I$\kappa$B) kinase $\gamma$ (IKK$\gamma$), through which IKK$\alpha$ and IKK$\beta$ are subsequently recruited to the DISC for the activation of IKK$\alpha/\beta$ kinases, the phosphorylation of I$\kappa$B, and the activation of nuclear factor-$\kappa$B (NF-$\kappa$B) [133, 137–139].

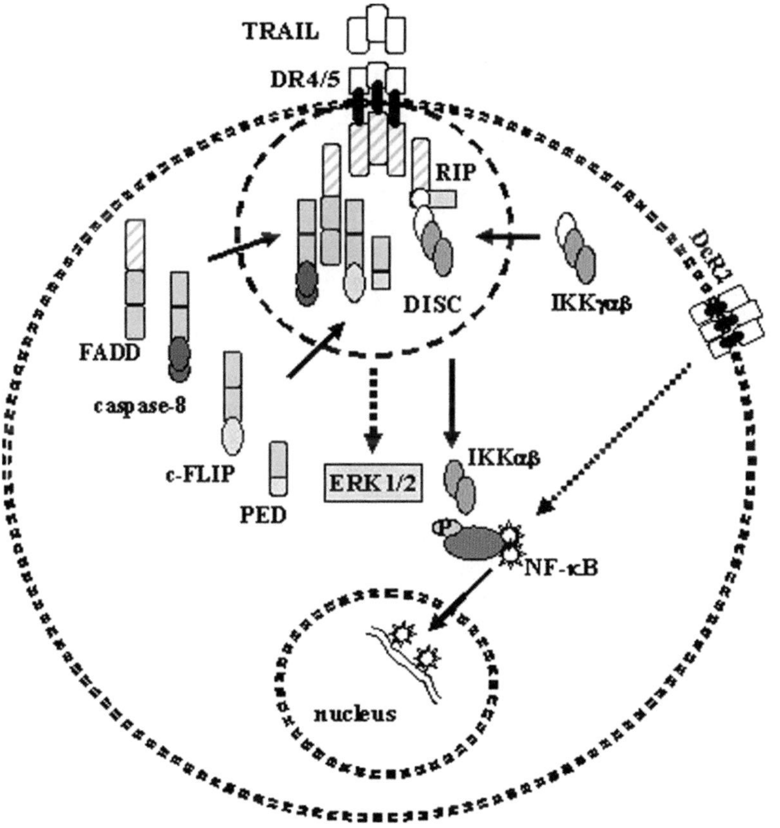


**Fig. 2** The DISC modification in TRAIL-induced non-apoptotic signals

FADD is an adaptor protein that interacts with DR4/DR5 through its DD and caspase-8 through its DED. Through the DED interaction, FADD can also recruit two DED adaptors to the DISC: one is cellular FADD-like interleukin-1$\beta$-converting enzyme [FLICE]-inhibitory protein (c-FLIP) [140] and another is phosphoprotein enriched in diabetes or phosphoprotein enriched in astrocytes-15 kDa (PED/PEA-15) [141]. The c-FLIP exists in two isoforms: a short form (c-FLIP$_S$) consisting of two DEDs, and a long form (c-FLIP$_L$) containing two DEDs and a caspase-like domain that lacks catalytic activity [140, 142]. PED/PEA-15 possess a DED but not a caspase-like domain [141, 143]. Through the DED, c-FLIP and PED/PEA-15 interact with caspase-8 and inhibit caspase-8 cleavage in the DISC [71, 144]. In addition, c-FLIP and PED/PEA-15 may interact with RIP and TNFR-associated factor 2 (TRAF2) and thus link the DISC to NF-$\kappa$B and extracellular signal-regulated kinase 1/2 (ERK1/2) pathway [145–147].

## NF-κB Pathway

Treatment of human cancer cell lines with rhTRAIL can activate NF-κB and ERK1/2 signaling pathways through RIP, c-FLIP and PED/PEA-15; however, the question then is whether NF-κB and ERK1/2 activity in cancer cells lead to the cell resistance to TRAIL. Some have reported that NF-κB activity inhibits TRAIL-induced apoptosis in lymphoid cells [148] and promotes the growth of ductal carcinoma cells [149], thus raising concern about the therapeutic use of rhTRAIL and its agonists [150]. These effects might be cell type-dependent as no such effects were observed in other cancer cells such as non-small cell lung carcinoma (NSCLC) [104]. NF-κB activity has been shown to upregulate anti-apoptotic genes *c-FLIP* and *Mcl-1* and *cIAP2* [151–153] as well as apoptotic genes *TRAIL* and *DR5* [154, 155]. TRAIL-induced NF-κB activity requires IκB phosphorylation and NF-κB nuclear translocation; however, targeting these steps affects neither c-FLIP expression nor TRAIL resistance [156, 157]. In contrast, knockdown of RIP, c-FLIP and PED releases the inhibition of caspase-8 cleavage and converts TRAIL- resistant cells to the sensitive phenotype [104, 133, 158, 159]. These studies suggest that the DISC is the molecular switch that controls caspase-8 cleavage and thereby the intracellular cell death and survival machinery.

## Inhibition of Mitochondrial Pathway and Effector Caspases

TRAIL-induced apoptosis requires the activation of the intrinsic mitochondrial pathway that is regulated by the Bcl-2 family proteins. The Bcl-2 family can be divided into three subfamilies: Bcl-2, Bax and BH3-only proteins [160]. Bid is a BH3-only protein and once cleaved by caspase-8, tBid interacts with Bax and Bak of the Bax subfamily, resulting in the change of mitochondrial membrane potential whereas Bcl-2 and Bcl-$X_L$ of the Bcl-2 subfamily interact with Bax and Bak to maintain the mitochondrial membrane potential [161, 162]. Overexpression of either Bcl-2 or Bcl-$X_L$ blocks TRAIL-induced apoptosis [163–165] and inactivation of Bax leads to the cell resistance to TRAIL [111]. However, it is largely unknown how these Bcl-2 family proteins are expressed in human cancers and consequently how they interact with each other and control the mitochondrial pathway.

The family of inhibitors of apoptosis proteins (IAP) consists of XIAP, cellular IAP1 (cIAP1), cIAP2, survivin, and livin. IAPs interact with caspase-3 and caspase-7 to prevent their cleavage and enzymatic activation [166]. XIAP, cIAP1, and cIAP2 have been shown to be highly expressed in various human cancer cell lines and tissues in correlation with poor clinical prognosis [83, 167–169]. Inhibition of the expression of XIAP and survivin enhances TRAIL-induced apoptosis in cancer cells [170, 171]. Targeting of XIAP by Smac peptide [172], Smac mimic [173] and XIAP antisense RNA [174] has been shown to enhance TRAIL-induced apoptosis in already sensitive cancer cell lines; thus, the ability of these XIAP-targeted agents to overcome TRAIL resistance in resistant cell lines remains to be seen. In addition, it is unclear whether Bcl-2 and/or IAP family proteins constitute the checkpoints that block TRAIL apoptotic pathway in any given human cancers.

## *Ubiquitin Regulation of TRAIL Pathway*

The caspase-8 dimerization and cleavage are the critical upstream events in TNF family-induced apoptosis [93, 94, 175, 176] and post-translational modification of proteins by ubiquitin regulates these biochemical processes [177]. Ubiquitin (UB) is covalently attached to lysine residues of the substrate proteins through the catalytic reactions mediated by UB-activating (E1), conjugating (E2) and ligase (E3) and removed by deubiquitinating (DUB) enzymes [178]. UB has seven lysine (K) residues; each of them and its N-terminal methionine (M1) can be linked to the C-terminal glycine residue of another UB to form polyUB chains [179], which regulate TNF$\alpha$-induced signal [177]. Upon TNF$\alpha$ binding, TNF receptor 1 (TNFR1) recruits receptor-interacting protein 1 (RIP1), cellular inhibitor of apoptosis protein 1 and 2 (cIAP1 and cIAP2) and TNFR-associated factor 2 (TRAF2) for the assembly of TNFR1-associated complex I [173]. TRAF2, cIAP1 and cIAP2 E3 ligases that activate nuclear factor-$\kappa$B (NF-$\kappa$B) through attaching polyUB chains to RIP1 [180] and binding of the polyUB chain to I$\kappa$B kinase $\gamma$ (IKK$\gamma$) [181]. TRAF2 and RIP1 then detach from TNFR1 and recruit FADD and caspase-8 for the assembly of the cytoplasm complex II [182], where cylindromatosis (CYLD), a DUB removes the polyUB chains from RIP1 and promotes caspase-8 cleavage for TNF$\alpha$-induced apoptosis [183].

Unlike TNFR1, however, DR4 and DR5 recruit FADD and caspase-8 in the assembly of a plasma membrane bound DISC where caspase-8 becomes dimerized and cleaved, initiating apoptosis [69, 70]. Cullin 3 (CUL3), an E3 ligase adds K48 and K63-linked polyUB chains to caspase-8 and facilitates its dimerization and cleavage in the DISC, where A20 DUB removes the polyUB chains from caspase-8 [184]. A20 (TNF$\alpha$-induced protein 3; TNFAIP3) is well known for its anti-inflammatory activities [185] through its N-terminal ovarian tumor domain (OTU) that acts as a DUB and removes K63-linked polyUB chains from RIP1, TRAF6 and RIP2, thus restricting TNFR1, Toll-like receptor and nucleotide-binding oligomerization domain-induced NF-$\kappa$B signal [186–188]. In contrast to the role of A20 in the TNFR1 pathway, the A20 C-terminal Zinc finger (Znf) domain of an E3 ligase [189] can mediate RIP1 polyubiquitination through K63-linked polyUB chains in TRAIL-induced DISC [190]. In TRAIL-resistant glioblastoma cell lines, A20 and DR5 form a plasma membrane-bound preligand assembly complex (PLAC) under physiologic conditions and TRAIL treatment leads to the recruitment of caspase-8 to the PLAC for the assembly of the DISC, in which the polyUB chains attached to RIP1 bind to the caspase-8 protease domain and inhibit caspse-8 cleavage and the initiation of TRAIL-induced apoptosis. It remains to be seen whether this revised TRAIL model can be applied in other cancers and whether

A20 E3 ligase inhibitors, which are currently under development, can overcome TRAIL resistance for combination cancer therapy with TRAIL agonists.

## TRAIL-Based Combination Therapies

The genetic defects in TRAIL apoptotic genes have been detected only in a small fraction of human cancers [119]. The vast majority of human cancers appears to have intact TRAIL apoptotic pathway and thus can be therapeutically targeted by TRAIL agonists. Unfortunately, the pathway is often inhibited in human cancers at multiple points in the apoptotic pathway through numerous mechanisms [11, 191]. Therapeutic agents targeting these resistance points have been developed for combination therapies with TRAIL agonists. Moreover, cancer-addicted oncogenes can inhibit cancer cell death pathway and thus drive cancer growth; thus, oncogene-targeted therapeutic agents may have potential for synergistic antitumor activity with TRAIL agonists. Finally, chemotherapy and radiation therapy remain the standard clinical treatment of human cancers and have shown synergistic antitumor activity with TRAIL agonists.

### *Targeting of the DISC Modification and Mitochondrial Pathway*

A number of therapeutic agents that target the TRAIL inhibitors, c-FLIP, Bcl-2 and XIAP have been developed as single agents or in combination with TRAIL agonists for cancer therapies. 2-Cyano-3,12-dioxooleana-1,9-dien-28-oic acid (CDDO) is a synthetic oleanane triterpenoid [192] and its derivatives, CDDO-methyl ester (CDDO-Me) and CDDO-imidazolide (CDDO-Im) have been shown to be able to inhibit c-FLIP expression and promote TRAIL-induced apoptosis in cancer cell lines [193, 194]. In a phase I trial of CDDO, sponsored through the collaboration between the National Cancer Institute and Reata Pharmaceuticals, thromboembolic side effect was revealed and considered as dose-limiting toxicity [195]. Antisense oligodeoxynucleotides against Bcl-2 (Genasense; Genta Inc., Berkeley Heights, JN) [196], XIAP (AEG35156; Aegera Therapeutics, Montreal, Quebec, Canada) [197] and survivin (LY2181308; Eli Lilly and Company, IN) [198] have passed phase I trials of drug safety and pharmacokinetic analysis. In addition, a number of Smac mimetics with a high affinity for XIAP including synthetic Smac N-terminal peptide [172], small molecule Smac mimic [173], Smac peptidomimetic (LBW242, Novartis, Cambridge, MA) [199] and SM-164 [200] have been generated for clinical treatment of cancers. It remains to be seen whether the combination of these agents with TRAIL agonists may lead to effective cancer therapies.

## *Targeting Oncogenes-Driven Signaling Pathways*

Mutations and/or deletions of tumor suppressors such as *TP53* and *PTEN* are commonly seen in human cancers [201], but it is unclear whether the genomic status of each of the genes correlates to TRAIL sensitivity. Studies of glioblastoma cell lines failed to show this correlation [54, 99], but other studies suggest that DR5 is p53 target gene and p53 rescue compound can induce the expression of DR5 [37] and thus enhance TRAIL killing of cancer cells [202]. Oncogene-driven pathways such as mammalian target of rapamycin (mTOR) have been reported to contribute to TRAIL resistance through induction of c-FLIP$_S$ expression [203]. Other mechanisms in TRAIL resistance include DNA methyltransferases through the up-regulation of PED/PEA-15 [204], heat shock protein 90$\alpha$ facilitating c-FLIP$_S$ recruitment to the DISC [205], microRNA-21 inhibiting TRAIL killing of tumor cells [206] and the ubiquitin–proteasome pathway blocking c-FLIP$_S$ degradation [207].

Therapeutic agents targeting oncogene-driven signaling pathways have been developed for clinical treatments of cancers. These therapeutic agents have also been evaluated in preclinical studies for the synergistic antitumor activity in combination with TRAIL agonists, including proteasome inhibitor bortezomib [208, 209], epidermal growth factor receptor inhibitors [210], histone deacetylase inhibitor [211, 212], DNA methylation inhibitor [204], cyclooxygenase 2 inhibitors [213, 214], mTOR inhibitor [203], phosphatidylinositide-3-kinase (PI3K) inhibitor [215], kinase inhibitor sorafenib [153], and CD20 antibody Rituximab [216, 217]. The molecular mechanisms by which these therapeutic agents can overcome the cancer resistance to TRAIL and the clinical value of the combination with TRAIL agonists for cancer therapies remain to be investigated.

## *Combination of TRAIL Agonists with Chemotherapy and Radiation*

Chemotherapeutic agents can interact with DNA and form intra-strand cross-links and thus affect intracellular signal pathways [218]. Unfortunately, human cancers eventually become resistant to chemotherapy. In addition, chemotherapy non-specifically targets both cancer and normal cells and causes a broad range of unwanted side effects at therapeutic doses. On the other hand, studies have shown that chemotherapeutic agents enhance TRAIL-induced apoptosis in cancer cells [98, 219–229]; although the molecular mechanisms in the synergistic effects remain controversial. Some studies showed that the treatment with cisplatin and etoposide up-regulate DR4/DR5 mRNA [219], but others observed that cisplatin has no effect on the expression of DR4 and DR5 protein [222]. In addition, studies showed that camptothecin up-regulates Bax [98]; cisplatin activates JNK/p38 [227]; cisplatin down-regulates c-FLIP$_S$ [98] and 1-(2-chloroethyl)-3-cyclohexyl-

1-nitrosourea (CCNU), temozolomide and topotecan have the synergistic cytotoxicity with TRAIL [229] perhaps through down-regulation of c-$FLIP_S$ [99]. While the mechanisms remain controversial, these studies suggest that chemotherapy can enhance TRAIL apoptotic effects on human cancer cell lines and derived xenografts.

Radiation therapy is the standard adjuvant treatment of human cancers. Studies have shown the synergistic cytotoxic effects of radiation with TRAIL agonists. Combination treatment with rhTRAIL and ionizing radiation can kill human glioblastoma cells through activation of the extrinsic and intrinsic apoptotic pathways [230]. Radiation treatment can up-regulate caspase-8 [231] and DR5 [232] and combination treatment of radiation and TRA-8 antibody increases the survival of mice with intracranial glioblastoma xenografts [232]. Clearly, the combination treatment of TRAIL agonists with radiation requires the presence of TRAIL death machinery in human cancers.

## Clinical Trials of TRAIL Agonists

The unraveling of TRAIL apoptotic pathway in cancer cells has resulted in a rapid development of the pathway-targeted cancer therapies. A number of therapeutic agents and modalities have been developed, of which rhTRAIL and DR4/DR5 agonistic antibodies have been under clinical investigation [11, 12]. The advantage in clinical use of rhTRAIL is that the recombinant ligand can target DR4 and DR5. On the other hand, antibodies have been proven to be effective clinical therapeutics because they selectively target specific antigens and have a much longer half-life than recombinant proteins. Two major categories of DR4 and DR5 mAbs have been developed: humanized mouse and fully human monoclonal antibodies (Table 2). Several of DR4 and DR5 agonistic antibodies and rhTRAIL have been evaluated in phase I-II trials for their safety, pharmacokinetics and therapeutic efficacy [9, 11].

### *Clinical Trials of rhTRAIL*

A soluble form of non-tagged rhTRAIL (Apo2L/TRAIL, dulanermin) is currently under clinical development jointly by Genentech and Amgen for cancer therapy [233]. The data from phase I trial were first reported at the 42nd ASCO Annual Meeting in 2006 and published in 2010 [234]: 41 patients with advanced cancers and non-Hodgkin lymphoma were enrolled; the monotherapy was well-tolerated with partial response in 3 % and stable disease in 53 % of patients. The data of phase Ib trial of 24 patients with advanced non-small-cell lung carcinoma (NSCLC) was then published, showing that the addition of paclitaxel, carboplatin and bevacizumab (PCB) to rhApo2L/TRAIL (dulanermin) improved the response

**Table 2** Clinical Trails of TRAIL Agonists

| Target | Agonist | Molecular type | Sponsoring company | Clinical trial |
|---|---|---|---|---|
| DR4/5 | Apo2L/TRAIL (dulanermin) | rhTRAIL | Genentech/Amgen | Phase I/II |
| DR4 | HGS-ETR1 (mapatumumab) | fully human mAb | HGS | Phase I/II |
| DR5 | HGS-ETR2 (lexatumumab) | fully human mAb | HGS | Phase I/II |
| DR5 | HGS-TR2J (KMTR2) | fully human mAb | HGS and Kirin Pharma | discontinued |
| DR5 | CS-1008 (tigatuzumab) | humanized mouse mAb | Daiichi Sankyo | Phase I |
| DR5 | AMG 655 (conatumumab) | fully human mAb | Amgen | Phase I/II |
| DR5 | PRO95780 (drozitumab) | fully human mAb | Genentech | Phase I/II |
| DR5 | LBY135 | chimeric mouse/human mAb | Novartis | Phase I |

rates such as progression-free survival of patients [235]. It was therefore a surprise when the data from the phase II trial of 213 patients with advanced NSCLC were published in 2011, showing that the addition of PCB to rhApo2L/TRAIL (dulanermin) did not improve the response rates [236]. The small sample size and lack of control in the phase Ib may explain the difference in the outcome between the phase Ib and II trials. The phase II trial, however, has clearly established the resistance of advanced NSCLC to the TRAIL treatment, either in monotherapy or in combination with chemotherapy.

## *Humanized Monoclonal Antibodies against DR5*

In 2001, a mouse mAb against human DR5 (TRA-8) was generated by immunizing mice with a fusion protein of DR5 extracellular domain and IgG [55]. This mouse mAb was then humanized for clinical treatment of cancers by Daiichi Sankyo [237]. The data of phase I trial of the humanized $IgG_1$ mAb (CS-1008, tigatuzumab) was published in 2010: 17 patients were enrolled with 16 of them diagnosed with carcinoma and 1 patient with lymphoma; all these patients tolerated well of the monotherapy of tigatuzumab and 7 of the 17 patients presented with stable disease [238]. At the 2008 ASCO Annual Meeting, Novartis presented the data from a phase I trial of a chimeric (mouse/human) mAb, LYB135, either in monotherapy or combination with capecitabine in treating patients with advanced solid cancers: LBY135 was well tolerated in 56 patients enrolled and showed signs of clinical activity.

At the 2012 ASCO Annual Meeting, the preliminary data from the phase II trial of tigatuzumab (CS-1008) in combination with carboplatin and paclitaxel were reported: 100 patients with naïve metastatic and unrespectable NSCLC were enrolled and the treatments did not improve the efficacy of carboplatin and paclitaxel. A phase II trial of tigatuzumab (CS-1008) in combination with the nanoparticle albumin-bound paclitaxel is currently ongoing in treating patients with metastatic, triple-negative breast cancers.

## *Fully Human Monoclonal Antibodies Against DR4*

At the 93rd AACR Annual Meeting in 2002, Human Genome Sciences in collaboration with Cambridge Antibody Technology reported for the first time the generation of fully human monoclonal antibodies against human DR4 and DR5 through phage display: the DR4 agonistic human mAb HGS-ETR1 (mapatumumab) and the DR5 human mAb HGS-ETR2 (lexatumumab) and HGS-TR2J (KMTR2). The antibodies bind DR4/DR5 directly and trigger apoptotic cell death in human cancer cell lines [239]. The data from the phase I trials of HGS-ETR1 showed that the monotherapy of HGS-ETR was well tolerated in patients [240, 241]. Phase II trials of HGS-ETR1 in treating patients with NSCLC [242] and colorectal cancer [243] showed no clinical activity of a single agent. At the 2010 ASCO Annual Meeting, the data from phase II trial showed that the combination of HGS-ERT1 with paclitaxel and carboplatin did not improve the response rates in the patients with advanced cancer.

## *Fully Human Monoclonal Antibodies Against DR5*

The data of a phase I trial of the first fully human mAb HGS-ETR2 (lexatumumab) in treating patients with advanced solid cancer were published in 2007 and 2010: the 10 mg/kg dose was identified as the maximum tolerated dose; dose-limiting toxicity was reached at 20 mg/kg; and twelve of the patients had stable disease [244, 245]. At the 2007 ASCO Annual Meeting, the data from a phase Ib trial showed that patients tolerated well with HGS-ETR2 in combination with gemcitabine, pemetrexed, doxorubicin or FOLFIRI and tumor shrinkage was observed in the FOLFIRI and the doxorubicin combination. HGS-ETR2 is currently in phase I and II trials as a single agent or in combination with chemotherapeutic agents. HGS-TR2J (KMTR2) is a fully human mAb derived from trans-chromosomal mice expressing a human Ig locus immunized with the DR5 extracellular domain under the co-development of Human Genome Sciences and Kirin Pharma [246]. Recently, however, the companies announced to discontinue clinical development of this DR5 mAb.

At the 2007 ASCO Annual Meeting, Genentech reported the generation of a fully human DR5 mAb, Apomab (PRO95780, drozitumab) through phage display [247] and the preliminary data of phase I trial: this mAb was well tolerated in patients with advanced cancers. At the 2010 ASCO Annual Meeting, the data from two phase II trials were released, showing that PRO95780 (drozitumab) plus rituximab in treating patients with relapsed NHL and in combination of PCB in treating patients with NSCLC did not improve the response rates in the patients.

At the 2007 AACR Annual Meeting, Amgen reported its pre-clinical study of a fully human DR5 agonistic mAb, AMG655 (conatumumab). The preliminary data from phase I trial were then reported at the 2007 ASCO Annual Meeting and the phase I trial data were published: showing that this mAb was well tolerated in Japanese patients [248] and the patients in US [249]. The subsequent data from a phase I/II trial showed that the addition of AMG655 to doxorubicin was safe but did not improved cancer control in patients with soft tissue sarcoma [250].

## *Conclusions and Future Directions*

Since the discovery of TNF$\alpha$ in 1975 [14], a tremendous amount of efforts have been made with the hope that these efforts will lead to an effective clinical treatment of human cancers. 30 years later in 2005, TRAIL agonists of the TNF family finally passed the phase I trials of drug safety test and became available for clinical treatment of cancers. Unfortunately, the data from phase II trials of therapeutic efficacy evaluation have been disappointing: TRAIL agonists, alone or in combination, have failed to show antitumor activities. More data are expected to be released from ongoing clinical trials of TRAIL agonists in treating various human cancers. The hope is that these phase II trials of efficacy test may identify the types of cancers and/or the fraction of the type cancer that can be effectively treated with TRAIL agonists. Clearly, however, the vast majority of human cancers are resistant to the treatment of TRAIL agonists, alone or in combination.

TRAIL mediates the innate and adaptive immunity against cancer. This physiological role of TRAIL suggests that cancers occur in patients by escape from TRAIL-mediated immunity and thus predicts that cancers are resistant to the treatment of TRAIL agonists. In consideration of this possibility, most phase II trials started with a combination of TRAIL agonists with various cancer therapeutic agents, but still failed to show antitumor activity of TRAIL agonists. This is truly a surprise and a disappointment, but it does bring us back to the fundamental question: how cancers evade the immunosurveillance. Understanding how TRAIL-mediated immunity works against cancer will certainly lead to the smart design of TRAIL-based cancer therapies.

The current molecular models of TRAIL modulations in human cancer have been established mainly through studies of cancer cell lines and cell lines-derived xenografts. There is no doubt about the value of these cancer research models in establishment of the fundamental principles. At this stage of the development of TRAIL agonists, it may become more important that we analyze human cancer tissues and tissues-derived stem cell models genetically and functionally. This may allow us to determine whether TRAIL apoptotic pathway exists and can be therapeutically targeted, in particular in each individual patient. If TRAIL resistance occurs, this approach may allow us to identify the roadblock for targeted therapy to remove the resistance. It seems that we know much more about molecules, cell lines, and animal xenografts, but very little about human cancers when it comes to how TRAIL pathway works. It has been a 35-year journey from the discovery of TNFα to the clinical trials of TRAIL agonists. It may take the same amount of time and efforts to develop this new class of cancer therapeutics as effective clinical therapeutics; but this will only occur, if we finally understand the therapeutic targets – patient's cancers.

**Acknowledgments** The work in our laboratory has been supported in part by grants from the Alberta Heritage Foundation for Medical Research, Canadian Institutes of Health Research, National Cancer Institute of Canada, Georgia Cancer Coalition and National Institutes of Health of the United States.

**Conflict of Interest** No potential conflicts of interest were disclosed.

## References

1. Kerr JF, Wyllie AH, Currie AR. Apoptosis: a basic biological phenomenon with wide-ranging implications in tissue kinetics. Br J Cancer. 1972;26:239–57.
2. Vogelstein B, Kinzler KW. Cancer genes and the pathways they control. Nat Med. 2004;10:789–99.
3. Letai AG. Diagnosing and exploiting cancer's addiction to blocks in apoptosis. Nat Rev Cancer. 2008;8:121–32.
4. Reed JC. Drug insight: cancer therapy strategies based on restoration of endogenous cell death mechanisms. Nat Clin Pract Oncol. 2006;3:388–98.
5. Ashkenazi A, Dixit VM. Death receptors: signaling and modulation. Science. 1998;281:1305–8.
6. Green DR, Reed JC. Mitochondria and apoptosis. Science. 1998;281:1309–12.
7. Ashkenazi A. Targeting death and decoy receptors of the tumour-necrosis factor superfamily. Nat Rev Cancer. 2002;2:420–30.
8. Ferrin G, Linares CI, Muntane J. Mitochondrial drug targets in cell death and cancer. Curr Pharm Des. 2011;17:2002–16.
9. Storey S. Targeting apoptosis: selected anticancer strategies. Nat Rev Drug Discov. 2008;7:971–2.
10. Johnstone RW, Frew AJ, Smyth MJ. The TRAIL apoptotic pathway in cancer onset, progression and therapy. Nat Rev Cancer. 2008;8:782–98.
11. Bellail AC, Qi L, Mulligan P, Chhabra V, Hao C. TRAIL agonists on clinical trials for cancer therapy: the promises and the challenges. Rev Recent Clin Trials. 2009;4:34–41.
12. Fox NL, Humphreys R, Luster TA, Klein J, Gallant G. Tumor Necrosis Factor-related apoptosis-inducing ligand (TRAIL) Receptor-1 and Receptor-2 agonists for cancer therapy. Expert Opin Biol Ther. 2010;10:1–18.
13. Bellail AC, Hao C. The roadmap of TRAIL apoptotic pathway-targeted cancer therapies: what is next? Expert Rev Anticancer Ther. 2012;12:547–9.
14. Carswell EA, et al. An endotoxin-induced serum factor that causes necrosis of tumors. Proc Natl Acad Sci U S A. 1975;72:3666–70.

15. Sun M, Fink PJ. A new class of reverse signaling costimulators belongs to the TNF family. J Immunol. 2007;179:4307–12.
16. Smyth MJ, et al. Nature's TRAIL–on a path to cancer immunotherapy. Immunity. 2003;18:1–6.
17. Watts TH. TNF/TNFR family members in costimulation of T cell responses. Annu Rev Immunol. 2005;23:23–68.
18. Lavrik I, Golks A, Krammer PH. Death receptor signaling. J Cell Sci. 2005;118:265–7.
19. Pennica D, et al. Human tumour necrosis factor: precursor structure, expression and homology to lymphotoxin. Nature. 1984;312:724–9.
20. Beutler B, et al. Identity of tumour necrosis factor and the macrophage-secreted factor cachectin. Nature. 1985;316:552–4.
21. Lewis M, et al. Cloning and expression of cDNAs for two distinct murine tumor necrosis factor receptors demonstrate one receptor is species specific. Proc Natl Acad Sci U S A. 1991;88:2830–4.
22. Suda T, Takahashi T, Golstein P, Nagata S. Molecular cloning and expression of the Fas ligand, a novel member of the tumor necrosis factor family. Cell. 1993;75:1169–78.
23. Itoh N, et al. The polypeptide encoded by the cDNA for human cell surface antigen Fas can mediate apoptosis. Cell. 1991;66:233–43.
24. Migone TS, et al. TL1A is a TNF-like ligand for DR3 and TR6/DcR3 and functions as a T cell costimulator. Immunity. 2002;16:479–92.
25. Meylan F, et al. The TNF-family receptor DR3 is essential for diverse T cell-mediated inflammatory diseases. Immunity. 2008;29:79–89.
26. Wang EC. On death receptor 3 and its ligands. Immunology. 2012;137:114–6.
27. Chinnaiyan AM, et al. Signal transduction by DR3, a death domain-containing receptor related to TNFR-1 and CD95. Science. 1996;274:990–2.
28. Marsters SA, et al. Apo-3, a new member of the tumor necrosis factor receptor family, contains a death domain and activates apoptosis and NF-kappa B. Curr Biol. 1996;6:1669–76.
29. Bodmer JL, et al. TRAMP, a novel apoptosis-mediating receptor with sequence homology to tumor necrosis factor receptor 1 and Fas(Apo-1/CD95). Immunity. 1997;6:79–88.
30. Screaton GR, et al. LARD: a new lymphoid-specific death domain containing receptor regulated by alternative pre-mRNA splicing. Proc Natl Acad Sci U S A. 1997;94:4615–9.
31. Tan KB, et al. Characterization of a novel TNF-like ligand and recently described TNF ligand and TNF receptor superfamily genes and their constitutive and inducible expression in hematopoietic and non-hematopoietic cells. Gene. 1997;204:35–46.
32. Wiley SR, et al. Identification and characterization of a new member of the TNF family that induces apoptosis. Immunity. 1995;3:673–82.
33. Pitti RM, et al. Induction of apoptosis by Apo-2 ligand, a new member of the tumor necrosis factor cytokine family. J Biol Chem. 1996;271:12687–90.
34. Pan G, et al. The receptor for the cytotoxic ligand TRAIL. Science. 1997;276:111–3.
35. Schneider P, et al. Characterization of two receptors for TRAIL. FEBS Lett. 1997;416:329–34.
36. Walczak H, et al. TRAIL-R2: a novel apoptosis-mediating receptor for TRAIL. EMBO J. 1997;16:5386–97.
37. Wu GS, et al. KILLER/DR5 is a DNA damage-inducible p53-regulated death receptor gene. Nat Genet. 1997;17:141–3.
38. Chaudhary PM, et al. Death receptor 5, a new member of the TNFR family, and DR4 induce FADD-dependent apoptosis and activate the NF-kappaB pathway. Immunity. 1997;7:821–30.
39. Tracey KJ, et al. Shock and tissue injury induced by recombinant human cachectin. Science. 1986;234:470–4.
40. Creagan ET, Kovach JS, Moertel CG, Frytak S, Kvols LK. A phase I clinical trial of recombinant human tumor necrosis factor. Cancer. 1988;62:2467–71.

41. Creaven PJ, et al. A phase I clinical trial of recombinant human tumor necrosis factor given daily for five days. Cancer Chemother Pharmacol. 1989;23:186–91.
42. Lenk H, Tanneberger S, Muller U, Ebert J, Shiga T. Phase II clinical trial of high-dose recombinant human tumor necrosis factor. Cancer Chemother Pharmacol. 1989;24:391–2.
43. Schiller JH, et al. Biological and clinical effects of intravenous tumor necrosis factor-alpha administered three times weekly. Cancer Res. 1991;51:1651–8.
44. Skillings J, et al. A phase II study of recombinant tumor necrosis factor in renal cell carcinoma: a study of the National Cancer Institute of Canada Clinical Trials Group. J Immunother. (1991); 11:67–70 (1992).
45. Trauth BC, et al. Monoclonal antibody-mediated tumor regression by induction of apoptosis. Science. 1989;245:301–5.
46. Ogasawara J, et al. Lethal effect of the anti-Fas antibody in mice. Nature. 1993;364:806–9.
47. Grunhagen DJ, de Wilt JH, ten Hagen TL, Eggermont AM. Technology insight: Utility of TNF-alpha-based isolated limb perfusion to avoid amputation of irresectable tumors of the extremities. Nat Clin Pract Oncol. 2006;3:94–103.
48. ElOjeimy S, et al. FasL gene therapy: a new therapeutic modality for head and neck cancer. Cancer Gene Ther. 2006;13:739–45.
49. Walczak H, et al. Tumoricidal activity of tumor necrosis factor-related apoptosis-inducing ligand in vivo. Nat Med. 1999;5:157–63.
50. Ashkenazi A, et al. Safety and antitumor activity of recombinant soluble Apo2 ligand. J Clin Invest. 1999;104:155–62.
51. Jo M, et al. Apoptosis induced in normal human hepatocytes by tumor necrosis factor-related apoptosis-inducing ligand. Nat Med. 2000;6:564–7.
52. Nitsch R, et al. Human brain-cell death induced by tumour-necrosis-factor-related apoptosis-inducing ligand (TRAIL). Lancet. 2000;356:827–8.
53. Nagata S. Steering anti-cancer drugs away from the TRAIL. Nat Med. 2000;6:502–3.
54. Hao C, et al. Induction and intracellular regulation of tumor necrosis factor-related apoptosis-inducing ligand (TRAIL) mediated apotosis in human malignant glioma cells. Cancer Res. 2001;61:1162–70.
55. Ichikawa K, et al. Tumoricidal activity of a novel anti-human DR5 monoclonal antibody without hepatocyte cytotoxicity. Nat Med. 2001;7:954–60.
56. Hao C, et al. TRAIL inhibits tumor growth but is nontoxic to human hepatocytes in chimeric mice. Cancer Res. 2004;64:8502–6.
57. Lawrence D, et al. Differential hepatocyte toxicity of recombinant Apo2L/TRAIL versions. Nat Med. 2001;7:383–5.
58. Song JH, Bellail A, Tse MC, Yong VW, Hao C. Human astrocytes are resistant to Fas ligand and tumor necrosis factor-related apoptosis-inducing ligand-induced apoptosis. J Neurosci. 2006;26:3299–308.
59. Takeda K, et al. Involvement of tumor necrosis factor-related apoptosis-inducing ligand in surveillance of tumor metastasis by liver natural killer cells. Nat Med. 2001;7:94–100.
60. Takeda K, et al. Critical role for tumor necrosis factor-related apoptosis-inducing ligand in immune surveillance against tumor development. J Exp Med. 2002;195:161–9.
61. Cretney E, et al. Increased susceptibility to tumor initiation and metastasis in TNF-related apoptosis-inducing ligand-deficient mice. J Immunol. 2002;168:1356–61.
62. Schmaltz C, et al. T cells require TRAIL for optimal graft-versus-tumor activity. Nat Med. 2002;8:1433–7.
63. Taieb J, et al. A novel dendritic cell subset involved in tumor immunosurveillance. Nat Med. 2006;12:214–9.
64. Cha SS, et al. 2.8 A resolution crystal structure of human TRAIL, a cytokine with selective antitumor activity. Immunity. 1999;11:253–61.
65. Mariani SM, Krammer PH. Differential regulation of TRAIL and CD95 ligand in transformed cells of the T and B lymphocyte lineage. Eur J Immunol. 1998;28:973–82.

66. Clancy L, et al. Preligand assembly domain-mediated ligand-independent association between TRAIL receptor 4 (TR4) and TR2 regulates TRAIL-induced apoptosis. Proc Natl Acad Sci U S A. 2005;102:18099–104.
67. Schneider P, et al. TRAIL receptors 1 (DR4) and 2 (DR5) signal FADD-dependent apoptosis and activate NF-kappaB. Immunity. 1997;7:831–6.
68. Chinnaiyan AM, O'Rourke K, Tewari M, Dixit VM. FADD, a novel death domain-containing protein, interacts with the death domain of Fas and initiates apoptosis. Cell. 1995;81:505–12.
69. Bodmer JL, et al. TRAIL receptor-2 signals apoptosis through FADD and caspase-8. Nat Cell Biol. 2000;2:241–3.
70. Kischkel FC, et al. Apo2L/TRAIL-dependent recruitment of endogenous FADD and caspase-8 to death receptors 4 and 5. Immunity. 2000;12:611–20.
71. Xiao C, Yang BF, Asadi N, Beguinot F, Hao C. Tumor necrosis factor-related apoptosis-inducing ligand-induced death-inducing signaling complex and its modulation by c-FLIP and PED/PEA-15 in glioma cells. J Biol Chem. 2002;277:25020–5.
72. Wang J, Chun HJ, Wong W, Spencer DM, Lenardo MJ. Caspase-10 is an initiator caspase in death receptor signaling. Proc Natl Acad Sci U S A. 2001;98:13884–8.
73. Kischkel FC, et al. Death receptor recruitment of endogenous caspase-10 and apoptosis initiation in the absence of caspase-8. J Biol Chem. 2001;276:46639–46.
74. Sprick MR, et al. Caspase-10 is recruited to and activated at the native TRAIL and CD95 death-inducing signalling complexes in a FADD-dependent manner but can not functionally substitute caspase-8. EMBO J. 2002;21:4520–30.
75. Kischkel FC, et al. Cytotoxicity-dependent APO-1 (Fas/CD95)-associated proteins form a death-inducing signaling complex (DISC) with the receptor. EMBO J. 1995;14:5579–88.
76. Muzio M, et al. FLICE, a novel FADD-homologous ICE/CED-3-like protease, is recruited to the CD95 (Fas/APO-1) death–inducing signaling complex. Cell. 1996;85:817–27.
77. Boldin MP, Goncharov TM, Goltsev YV, Wallach D. Involvement of MACH, a novel MORT1/FADD-interacting protease, in Fas/APO-1- and TNF receptor-induced cell death. Cell. 1996;85:803–15.
78. Li H, Zhu H, Xu CJ, Yuan J. Cleavage of BID by caspase 8 mediates the mitochondrial damage in the Fas pathway of apoptosis. Cell. 1998;94:491–501.
79. Luo X, Budihardjo I, Zou H, Slaughter C, Wang X. Bid, a Bcl2 interacting protein, mediates cytochrome c release from mitochondria in response to activation of cell surface death receptors. Cell. 1998;94:481–90.
80. Du C, Fang M, Li Y, Li L, Wang X. Smac, a mitochondrial protein that promotes cytochrome c-dependent caspase activation by eliminating IAP inhibition. Cell. 2000;102:33–42.
81. Verhagen AM, et al. Identification of DIABLO, a mammalian protein that promotes apoptosis by binding to and antagonizing IAP proteins. Cell. 2000;102:43–53.
82. Li P, et al. Cytochrome c and dATP-dependent formation of Apaf-1/caspase-9 complex initiates an apoptotic protease cascade. Cell. 1997;91:479–89.
83. Wagenknecht B, et al. Expression and biological activity of X-linked inhibitor of apoptosis (XIAP) in human malignant glioma. Cell Death Differ. 1999;6:370–6.
84. Deng Y, Lin Y, Wu X. TRAIL-induced apoptosis requires Bax-dependent mitochondrial release of Smac/DIABLO. Genes Dev. 2002;16:33–45.
85. Scaffidi C, Medema JP, Krammer PH, Peter ME. FLICE is predominantly expressed as two functionally active isoforms, caspase-8/a and caspase-8/b. J Biol Chem. 1997;272:26953–8.
86. Walker NP, et al. Crystal structure of the cysteine protease interleukin-1 beta-converting enzyme: a (p20/p10)2 homodimer. Cell. 1994;78:343–52.
87. Wilson KP, et al. Structure and mechanism of interleukin-1 beta converting enzyme. Nature. 1994;370:270–5.
88. Medema JP, et al. FLICE is activated by association with the CD95 death-inducing signaling complex (DISC). EMBO J. 1997;16:2794–804.

89. Yang X, Chang HY, Baltimore D. Autoproteolytic activation of pro-caspases by oligomerization. Mol Cell. 1998;1:319–25.
90. Liu X, Zou H, Slaughter C, Wang X. DFF, a heterodimeric protein that functions downstream of caspase-3 to trigger DNA fragmentation during apoptosis. Cell. 1997;89:175–84.
91. Muzio M, Stockwell BR, Stennicke HR, Salvesen GS, Dixit VM. An induced proximity model for caspase-8 activation. J Biol Chem. 1998;273:2926–30.
92. Salvesen GS, Dixit VM. Caspase activation: the induced-proximity model. Proc Natl Acad Sci U S A. 1999;96:10964–7.
93. Boatright KM, et al. A unified model for apical caspase activation. Mol Cell. 2003;11:529–41.
94. Donepudi M, Mac Sweeney A, Briand C, Grutter MG. Insights into the regulatory mechanism for caspase-8 activation. Mol Cell. 2003;11:543–9.
95. Chang DW, Xing Z, Capacio VL, Peter ME, Yang X. Interdimer processing mechanism of procaspase-8 activation. EMBO J. 2003;22:4132–42.
96. Roth W, et al. Locoregional Apo2L/TRAIL eradicates intracranial human malignant glioma xenografts in athymic mice in the absence of neurotoxicity. Biochem Biophys Res Commun. 1999;265:479–83.
97. Spierings DC, et al. Expression of TRAIL and TRAIL death receptors in stage III non-small cell lung cancer tumors. Clin Cancer Res. 2003;9:3397–405.
98. Song JH, Song DK, Herlyn M, Petruk KC, Hao C. Cisplatin down-regulation of cellular Fas-associated death domain-like interleukin-1beta-converting enzyme-like inhibitory proteins to restore tumor necrosis factor-related apoptosis-inducing ligand-induced apoptosis in human melanoma cells. Clin Cancer Res. 2003;9:4255–66.
99. Song JH, et al. TRAIL triggers apoptosis in malignant glioma cells through extrinsic and intrinsic pathways. Brain Pathol. 2003;13:539–53.
100. Younes M, Georgakis GV, Rahmani M, Beer D, Younes A. Functional expression of TRAIL receptors TRAIL-R1 and TRAIL-R2 in esophageal adenocarcinoma. Eur J Cancer. 2006;42:542–7.
101. Pollack IF, Erff M, Ashkenazi A. Direct stimulation of apoptotic signaling by soluble Apo2 l/tumor necrosis factor-related apoptosis-inducing ligand leads to selective killing of glioma cells. Clin Cancer Res. 2001;7:1362–9.
102. Knight MJ, Riffkin CD, Muscat AM, Ashley DM, Hawkins CJ. Analysis of FasL and TRAIL induced apoptosis pathways in glioma cells. Oncogene. 2001;20:5789–98.
103. Pukac L, et al. HGS-ETR1, a fully human TRAIL-receptor 1 monoclonal antibody, induces cell death in multiple tumour types in vitro and in vivo. Br J Cancer. 2005;92:1430–41.
104. Song JH, et al. Lipid rafts and nonrafts mediate tumor necrosis factor related apoptosis-inducing ligand induced apoptotic and nonapoptotic signals in non small cell lung carcinoma cells. Cancer Res. 2007;67:6946–55.
105. Saito R, et al. Convection-enhanced delivery of tumor necrosis factor-related apoptosis-inducing ligand with systemic administration of temozolomide prolongs survival in an intracranial glioblastoma xenograft model. Cancer Res. 2004;64:6858–62.
106. Ehtesham M, et al. Induction of glioblastoma apoptosis using neural stem cell-mediated delivery of tumor necrosis factor-related apoptosis-inducing ligand. Cancer Res. 2002;62:7170–4.
107. Griffith TS, Broghammer EL. Suppression of tumor growth following intralesional therapy with TRAIL recombinant adenovirus. Mol Ther. 2001;4:257–66.
108. Wohlfahrt ME, Beard BC, Lieber A, Kiem HP. A capsid-modified, conditionally replicating oncolytic adenovirus vector expressing TRAIL Leads to enhanced cancer cell killing in human glioblastoma models. Cancer Res. 2007;67:8783–90.
109. Jin H, et al. Apo2 ligand/tumor necrosis factor-related apoptosis-inducing ligand cooperates with chemotherapy to inhibit orthotopic lung tumor growth and improve survival. Cancer Res. 2004;64:4900–5.

110. Rampino N, et al. Somatic frameshift mutations in the BAX gene in colon cancers of the microsatellite mutator phenotype. Science. 1997;275:967–9.
111. LeBlanc H, et al. Tumor-cell resistance to death receptor–induced apoptosis through mutational inactivation of the proapoptotic Bcl-2 homolog Bax. Nat Med. 2002;8:274–81.
112. Lee SH, et al. Alterations of the DR5/TRAIL receptor 2 gene in non-small cell lung cancers. Cancer Res. 1999;59:5683–6.
113. Fisher MJ, et al. Nucleotide substitution in the ectodomain of trail receptor DR4 is associated with lung cancer and head and neck cancer. Clin Cancer Res. 2001;7:1688–97.
114. Park WS, et al. Inactivating mutations of KILLER/DR5 gene in gastric cancers. Gastroenterology. 2001;121:1219–25.
115. Shin MS, et al. Mutations of tumor necrosis factor-related apoptosis-inducing ligand receptor 1 (TRAIL-R1) and receptor 2 (TRAIL-R2) genes in metastatic breast cancers. Cancer Res. 2001;61:4942–6.
116. Lee SH, et al. Somatic mutations of TRAIL-receptor 1 and TRAIL-receptor 2 genes in non-Hodgkin's lymphoma. Oncogene. 2001;20:399–403.
117. McDonald ER 3rd, Chui PC, Martelli PF, Dicker DT, El-Deiry WS. Death domain mutagenesis of KILLER/DR5 reveals residues critical for apoptotic signaling. J Biol Chem. 2001;276:14939–45.
118. Bin L, et al. Tumor-derived mutations in the TRAIL receptor DR5 inhibit TRAIL signaling through the DR4 receptor by competing for ligand binding. J Biol Chem. 2007;282:28189–94.
119. Li YC, et al. Genomic alterations in human malignant glioma cells associate with the cell resistance to the combination treatment with tumor necrosis factor-related apoptosis-inducing ligand and chemotherapy. Clin Cancer Res. 2006;12:2716–29.
120. Hopkins-Donaldson S, et al. Loss of caspase-8 expression in highly malignant human neuroblastoma cells correlates with resistance to tumor necrosis factor-related apoptosis-inducing ligand-induced apoptosis. Cancer Res. 2000;60:4315–9.
121. Eggert A, et al. Resistance to tumor necrosis factor-related apoptosis-inducing ligand (TRAIL)-induced apoptosis in neuroblastoma cells correlates with a loss of caspase-8 expression. Cancer Res. 2001;61:1314–9.
122. Kim HS, et al. Inactivating mutations of caspase-8 gene in colorectal carcinomas. Gastroenterology. 2003;125:708–15.
123. Soung YH, et al. CASPASE-8 gene is inactivated by somatic mutations in gastric carcinomas. Cancer Res. 2005;65:815–21.
124. Degli-Esposti MA, et al. Cloning and characterization of TRAIL-R3, a novel member of the emerging TRAIL receptor family. J Exp Med. 1997;186:1165–70.
125. Sheridan JP, et al. Control of TRAIL-induced apoptosis by a family of signaling and decoy receptors. Science. 1997;277:818–21.
126. Mongkolsapaya J, et al. Lymphocyte inhibitor of TRAIL (TNF-related apoptosis-inducing ligand): a new receptor protecting lymphocytes from the death ligand TRAIL. J Immunol. 1998;160:3–6.
127. Marsters SA, et al. A novel receptor for Apo2L/TRAIL contains a truncated death domain. Curr Biol. 1997;7:1003–6.
128. Pan G, Ni J, Yu G, Wei YF, Dixit VM. TRUNDD, a new member of the TRAIL receptor family that antagonizes TRAIL signalling. FEBS Lett. 1998;424:41–5.
129. Degli-Esposti MA, et al. The novel receptor TRAIL-R4 induces NF-kappaB and protects against TRAIL-mediated apoptosis, yet retains an incomplete death domain. Immunity. 1997;7:813–20.
130. Sheikh MS, et al. The antiapoptotic decoy receptor TRID/TRAIL-R3 is a p53-regulated DNA damage-inducible gene that is overexpressed in primary tumors of the gastrointestinal tract. Oncogene. 1999;18:4153–9.
131. Liu X, Yue P, Khuri FR, Sun SY. Decoy receptor 2 (DcR2) is a p53 target gene and regulates chemosensitivity. Cancer Res. 2005;65:9169–75.

132. Merino D, et al. Differential inhibition of TRAIL-mediated DR5-DISC formation by decoy receptors 1 and 2. Mol Cell Biol. 2006;26:7046–55.
133. Bellail AC, et al. DR5-mediated DISC controls caspase-8 cleavage and initiation of apoptosis in human glioblastomas. J Cell Mol Med. 2010;14:1303–1317.
134. Stanger BZ, Leder P, Lee TH, Kim E, Seed B. RIP: a novel protein containing a death domain that interacts with Fas/APO-1 (CD95) in yeast and causes cell death. Cell. 1995;81:513–23.
135. Hsu H, Huang J, Shu HB, Baichwal V, Goeddel DV. TNF-dependent recruitment of the protein kinase RIP to the TNF receptor-1 signaling complex. Immunity. 1996;4:387–96.
136. Hsu H, Shu HB, Pan MG, Goeddel DV. TRADD-TRAF2 and TRADD-FADD interactions define two distinct TNF receptor 1 signal transduction pathways. Cell. 1996;84:299–308.
137. Lin Y, et al. The death domain kinase RIP is essential for TRAIL (Apo2L)-induced activation of IkappaB kinase and c-Jun N-terminal kinase. Mol Cell Biol. 2000;20:6638–45.
138. Harper N, Farrow SN, Kaptein A, Cohen GM, MacFarlane M. Modulation of tumor necrosis factor apoptosis-inducing ligand- induced NF-kappa B activation by inhibition of apical caspases. J Biol Chem. 2001;276:34743–52.
139. Varfolomeev E, et al. Molecular determinants of kinase pathway activation by Apo2 ligand/ tumor necrosis factor-related apoptosis-inducing ligand. J Biol Chem. 2005;280:40599–608.
140. Irmler M, et al. Inhibition of death receptor signals by cellular FLIP. Nature. 1997;388:190–5.
141. Condorelli G, et al. PED/PEA-15 gene controls glucose transport and is overexpressed in type 2 diabetes mellitus. EMBO J. 1998;17:3858–66.
142. Scaffidi C, Schmitz I, Krammer PH, Peter ME. The role of c-FLIP in modulation of CD95-induced apoptosis. J Biol Chem. 1999;274:1541–8.
143. Condorelli G, et al. PED/PEA-15: an anti-apoptotic molecule that regulates FAS/TNFR1-induced apoptosis. Oncogene. 1999;18:4409–15.
144. Yang BF, Xiao C, Roa WH, Krammer PH, Hao C. Calcium/Calmodulin-dependent protein kinase II regulation of c-FLIP expression and phosphorylation in modulation of fas-mediated signaling in malignant glioma cells. J Biol Chem. 2003;278:7043–50.
145. Kataoka T, et al. The caspase-8 inhibitor FLIP promotes activation of NF-kappaB and Erk signaling pathways. Curr Biol. 2000;10:640–8.
146. Kataoka T, Tschopp J. N-terminal fragment of c-FLIP(L) processed by caspase 8 specifically interacts with TRAF2 and induces activation of the NF-kappaB signaling pathway. Mol Cell Biol. 2004;24:2627–36.
147. Krueger J, Chou FL, Glading A, Schaefer E, Ginsberg MH. Phosphorylation of phosphoprotein enriched in astrocytes (PEA-15) regulates extracellular signal-regulated kinase-dependent transcription and cell proliferation. Mol Biol Cell. 2005;16:3552–61.
148. Jeremias I, et al. Inhibition of nuclear factor kappaB activation attenuates apoptosis resistance in lymphoid cells. Blood. 1998;91:4624–31.
149. Trauzold A, et al. TRAIL promotes metastasis of human pancreatic ductal adenocarcinoma. Oncogene. 2006;25(56):7434-9.
150. Malhi H, Gores GJ. TRAIL resistance results in cancer progression: a TRAIL to perdition? Oncogene. 2006;25:7333–5.
151. Kreuz S, Siegmund D, Scheurich P, Wajant H. NF-kappaB inducers upregulate cFLIP, a cycloheximide-sensitive inhibitor of death receptor signaling. Mol Cell Biol. 2001;21:3964–73.
152. Micheau O, Lens S, Gaide O, Alevizopoulos K, Tschopp J. NF-kappaB signals induce the expression of c-FLIP. Mol Cell Biol. 2001;21:5299–305.
153. Ricci MS, et al. Reduction of TRAIL-induced Mcl-1 and cIAP2 by c-Myc or sorafenib sensitizes resistant human cancer cells to TRAIL-induced death. Cancer Cell. 2007;12:66–80.
154. Baetu TM, Kwon H, Sharma S, Grandvaux N, Hiscott J. Disruption of NF-kappaB signaling reveals a novel role for NF-kappaB in the regulation of TNF-related apoptosis-inducing ligand expression. J Immunol. 2001;167:3164–73.

155. Shetty S, et al. Transcription factor NF-kappaB differentially regulates death receptor 5 expression involving histone deacetylase 1. Mol Cell Biol. 2005;25:5404–16.
156. Steele LP, Georgopoulos NT, Southgate J, Selby PJ, Trejdosiewicz LK. Differential susceptibility to TRAIL of normal versus malignant human urothelial cells. Cell Death Differ. 2006;13:1564–76.
157. Song JH, et al. Lipid rafts and non-rafts mediate TRAIL-induced apoptotic and non-apoptotic signals in non-small cell lung carcinoma cells. Cancer Res. 2007;67:1–10.
158. Sharp DA, Lawrence DA, Ashkenazi A. Selective knockdown of the long variant of cellular FLICE inhibitory protein augments death receptor-mediated caspase-8 activation and apoptosis. J Biol Chem. 2005;280:19401–9.
159. Wang P, et al. Inhibition of RIP and c-FLIP enhances TRAIL-induced apoptosis in pancreatic cancer cells. Cell Signal. 2007;19:2237–46.
160. Cory S, Adams JM. The Bcl2 family: regulators of the cellular life-or-death switch. Nat Rev Cancer. 2002;2:647–56.
161. Gross A, Jockel J, Wei MC, Korsmeyer SJ. Enforced dimerization of BAX results in its translocation, mitochondrial dysfunction and apoptosis. EMBO J. 1998;17:3878–85.
162. Wei MC, et al. tBID, a membrane-targeted death ligand, oligomerizes BAK to release cytochrome c. Genes Dev. 2000;14:2060–71.
163. Hinz S, et al. Bcl-XL protects pancreatic adenocarcinoma cells against CD95- and TRAIL-receptor-mediated apoptosis. Oncogene. 2000;19:5477–86.
164. Munshi A, et al. TRAIL (APO-2L) induces apoptosis in human prostate cancer cells that is inhibitable by Bcl-2. Oncogene. 2001;20:3757–65.
165. Fulda S, Meyer E, Debatin KM. Inhibition of TRAIL-induced apoptosis by Bcl-2 overexpression. Oncogene. 2002;21:2283–94.
166. Salvesen GS, Duckett CS. IAP proteins: blocking the road to death's door. Nat Rev Mol Cell Biol. 2002;3:401–10.
167. Bockbrader KM, Tan M, Sun Y. A small molecule Smac-mimic compound induces apoptosis and sensitizes TRAIL- and etoposide-induced apoptosis in breast cancer cells. Oncogene. 2005;24:7381–8.
168. Mizutani Y, et al. Overexpression of XIAP expression in renal cell carcinoma predicts a worse prognosis. Int J Oncol. 2007;30:919–25.
169. Karikari CA, et al. Targeting the apoptotic machinery in pancreatic cancers using small-molecule antagonists of the X-linked inhibitor of apoptosis protein. Mol Cancer Ther. 2007;6:957–66.
170. Chawla-Sarkar M, et al. Downregulation of Bcl-2, FLIP or IAPs (XIAP and survivin) by siRNAs sensitizes resistant melanoma cells to Apo2L/TRAIL-induced apoptosis. Cell Death Differ. 2004;11:915–23.
171. Cummins JM, et al. X-linked inhibitor of apoptosis protein (XIAP) is a nonredundant modulator of tumor necrosis factor-related apoptosis-inducing ligand (TRAIL)-mediated apoptosis in human cancer cells. Cancer Res. 2004;64:3006–8.
172. Fulda S, Wick W, Weller M, Debatin KM. Smac agonists sensitize for Apo2L/TRAIL- or anticancer drug-induced apoptosis and induce regression of malignant glioma in vivo. Nat Med. 2002;8:808–15.
173. Li L, et al. A small molecule Smac mimic potentiates TRAIL- and TNFalpha-mediated cell death. Science. 2004;305:1471–4.
174. Naumann U, et al. Adenoviral expression of XIAP antisense RNA induces apoptosis in glioma cells and suppresses the growth of xenografts in nude mice. Gene Ther. 2007;14:147–61.
175. Hughes MA, et al. Reconstitution of the death-inducing signaling complex reveals a substrate switch that determines CD95-mediated death or survival. Mol Cell. 2009;35:265–79.
176. Oberst A, et al. Inducible dimerization and inducible cleavage reveal a requirement for both processes in caspase-8 activation. J Biol Chem. 2010;285:16632–42.

177. Skaug B, Jiang X, Chen ZJ. The role of ubiquitin in NF-kappaB regulatory pathways. Annu Rev Biochem. 2009;78:769–96.
178. Bhoj VG, Chen ZJ. Ubiquitylation in innate and adaptive immunity. Nature. 2009;458:430–7.
179. Ikeda F, Crosetto N, Dikic I. What determines the specificity and outcomes of ubiquitin signaling? Cell. 2010;143:677–81.
180. Ea CK, Deng L, Xia ZP, Pineda G, Chen ZJ. Activation of IKK by TNFalpha requires site-specific ubiquitination of RIP1 and polyubiquitin binding by NEMO. Mol Cell. 2006;22:245–57.
181. Rahighi S, et al. Specific recognition of linear ubiquitin chains by NEMO is important for NF-kappaB activation. Cell. 2009;136:1098–109.
182. Micheau O, Tschopp J. Induction of TNF receptor I-mediated apoptosis via two sequential signaling complexes. Cell. 2003;114:181–90.
183. Wang L, Du F, Wang X. TNF-alpha induces two distinct caspase-8 activation pathways. Cell. 2008;133:693–703.
184. Jin Z, et al. Cullin3-based polyubiquitination and p62-dependent aggregation of caspase-8 mediate extrinsic apoptosis signaling. Cell. 2009;137:721–35.
185. Lee EG, et al. Failure to regulate TNF-induced NF-kappaB and cell death responses in A20-deficient mice. Science. 2000;289:2350–4.
186. Shembade N, Ma A, Harhaj EW. Inhibition of NF-kappaB signaling by A20 through disruption of ubiquitin enzyme complexes. Science. 2010;327:1135–9.
187. Boone DL, et al. The ubiquitin-modifying enzyme A20 is required for termination of Toll-like receptor responses. Nat Immunol. 2004;5:1052–60.
188. Hitotsumatsu O, et al. The ubiquitin-editing enzyme A20 restricts nucleotide-binding oligomerization domain containing 2-triggered signals. Immunity. 2008;28:381–90.
189. Wertz IE, et al. De-ubiquitination and ubiquitin ligase domains of A20 downregulate NF-kappaB signalling. Nature. 2004;430:694–9.
190. Bellail AC, Olson JJ, Yang X, Chen ZJ, Hao C. A20 ubiquitin ligase-mediated polyubiquitination of RIP1 inhibits caspase-8 cleavage and TRAIL-induced apoptosis in glioblastoma. Cancer Discov. 2012;2:140–55.
191. Hao C, Song JH, Vilimanovich U, Kneteman NM. Modulation of TRAIL signaling complex. Vitam Horm. 2004;67:81–99.
192. Honda T, et al. Synthetic oleanane and ursane triterpenoids with modified rings A and C: a series of highly active inhibitors of nitric oxide production in mouse macrophages. J Med Chem. 2000;43:4233–46.
193. Zou W, et al. c-Jun NH2-terminal kinase-mediated up-regulation of death receptor 5 contributes to induction of apoptosis by the novel synthetic triterpenoid methyl-2-cyano-3,12-dioxooleana-1, 9-dien-28-oate in human lung cancer cells. Cancer Res. 2004;64:7570–8.
194. Hyer ML, et al. Synthetic triterpenoids cooperate with tumor necrosis factor-related apoptosis-inducing ligand to induce apoptosis of breast cancer cells. Cancer Res. 2005;65:4799–808.
195. Speranza G, et al. Phase I study of the synthetic triterpenoid, 2-cyano-3, 12-dioxoolean-1, 9-dien-28-oic acid (CDDO), in advanced solid tumors. Cancer Chemother Pharmacol. 2012;69:431–8.
196. Jansen B, et al. Chemosensitisation of malignant melanoma by BCL2 antisense therapy. Lancet. 2000;356:1728–33.
197. Dean E, et al. Phase I trial of AEG35156 administered as a 7-day and 3-day continuous intravenous infusion in patients with advanced refractory cancer. J Clin Oncol. 2009;27:1660–6.
198. Tanioka M, et al. Phase I study of LY2181308, an antisense oligonucleotide against survivin, in patients with advanced solid tumors. Cancer Chemother Pharmacol. 2011;68:505–11.

199. Chauhan D, et al. Targeting mitochondrial factor Smac/DIABLO as therapy for multiple myeloma (MM). Blood. 2007;109:1220–7.
200. Lu J, et al. Therapeutic potential and molecular mechanism of a novel, potent, nonpeptide, Smac mimetic SM-164 in combination with TRAIL for cancer treatment. Mol Cancer Ther. 2011;10:902–14.
201. Ishii N, et al. Frequent co-alterations of TP53, p16/CDKN2A, p14ARF, PTEN tumor suppressor genes in human glioma cell lines. Brain Pathol. 1999;9:469–79.
202. Weinmann L, et al. A novel p53 rescue compound induces p53-dependent growth arrest and sensitises glioma cells to Apo2L/TRAIL-induced apoptosis. Cell Death Differ. 2008;15:718–29.
203. Panner A, James CD, Berger MS, Pieper RO. mTOR controls FLIPS translation and TRAIL sensitivity in glioblastoma multiforme cells. Mol Cell Biol. 2005;25:8809–23.
204. Eramo A, et al. Inhibition of DNA methylation sensitizes glioblastoma for tumor necrosis factor-related apoptosis-inducing ligand-mediated destruction. Cancer Res. 2005;65:11469–77.
205. Panner A, Murray JC, Berger MS, Pieper RO. Heat shock protein 90alpha recruits FLIPS to the death-inducing signaling complex and contributes to TRAIL resistance in human glioma. Cancer Res. 2007;67:9482–9.
206. Corsten MF, et al. MicroRNA-21 knockdown disrupts glioma growth in vivo and displays synergistic cytotoxicity with neural precursor cell delivered S-TRAIL in human gliomas. Cancer Res. 2007;67:8994–9000.
207. Koschny R, et al. Bortezomib sensitizes primary human astrocytoma cells of WHO grades I to IV for tumor necrosis factor-related apoptosis-inducing ligand-induced apoptosis. Clin Cancer Res. 2007;13:3403–12.
208. Mitsiades CS, et al. TRAIL/Apo2L ligand selectively induces apoptosis and overcomes drug resistance in multiple myeloma: therapeutic applications. Blood. 2001;98:795–804.
209. Kandasamy K, Kraft AS. Proteasome inhibitor PS-341 (VELCADE) induces stabilization of the TRAIL receptor DR5 mRNA through the 3′-untranslated region. Mol Cancer Ther. 2008;7:1091–100.
210. Gibson EM, Henson ES, Haney N, Villanueva J, Gibson SB. Epidermal growth factor protects epithelial-derived cells from tumor necrosis factor-related apoptosis-inducing ligand-induced apoptosis by inhibiting cytochrome c release. Cancer Res. 2002;62:488–96.
211. Guo F, et al. Cotreatment with histone deacetylase inhibitor LAQ824 enhances Apo-2L/ tumor necrosis factor-related apoptosis inducing ligand-induced death inducing signaling complex activity and apoptosis of human acute leukemia cells. Cancer Res. 2004;64:2580–9.
212. Frew AJ, et al. Combination therapy of established cancer using a histone deacetylase inhibitor and a TRAIL receptor agonist. Proc Natl Acad Sci U S A. 2008;105:11317–22.
213. Liu X, Yue P, Zhou Z, Khuri FR, Sun SY. Death receptor regulation and celecoxib-induced apoptosis in human lung cancer cells. J Natl Cancer Inst. 2004;96:1769–80.
214. Martin S, et al. Cyclooxygenase-2 inhibition sensitizes human colon carcinoma cells to TRAIL-induced apoptosis through clustering of DR5 and concentrating death-inducing signaling complex components into ceramide-enriched caveolae. Cancer Res. 2005;65:11447–58.
215. Poh TW, Huang S, Hirpara JL, Pervaiz S. LY303511 amplifies TRAIL-induced apoptosis in tumor cells by enhancing DR5 oligomerization, DISC assembly, and mitochondrial permeabilization. Cell Death Differ. 2007;14:1813–25.
216. Daniel D, et al. Cooperation of the proapoptotic receptor agonist rhApo2L/TRAIL with the CD20 antibody rituximab against non-Hodgkin lymphoma xenografts. Blood. 2007;110:4037–46.
217. Maddipatla S, Hernandez-Ilizaliturri FJ, Knight J, Czuczman MS. Augmented antitumor activity against B-cell lymphoma by a combination of monoclonal antibodies targeting TRAIL-R1 and CD20. Clin Cancer Res. 2007;13:4556–64.

218. Siddik ZH. Cisplatin: mode of cytotoxic action and molecular basis of resistance. Oncogene. 2003;22:7265–79.
219. Lacour S, et al. Anticancer agents sensitize tumor cells to tumor necrosis factor-related apoptosis-inducing ligand-mediated caspase-8 activation and apoptosis. Cancer Res. 2001;61:1645–51.
220. Keane MM, Ettenberg SA, Nau MM, Russell EK, Lipkowitz S. Chemotherapy augments TRAIL-induced apoptosis in breast cell lines. Cancer Res. 1999;59:734–41.
221. Nagane M, et al. Increased death receptor 5 expression by chemotherapeutic agents in human gliomas causes synergistic cytotoxicity with tumor necrosis factor-related apoptosis-inducing ligand in vitro and in vivo. Cancer Res. 2000;60:847–53.
222. Ferreira CG, Span SW, Peters GJ, Kruyt FA, Giaccone G. Chemotherapy triggers apoptosis in a caspase-8-dependent and mitochondria-controlled manner in the non-small cell lung cancer cell line NCI-H460. Cancer Res. 2000;60:7133–41.
223. Gibson SB, Oyer R, Spalding AC, Anderson SM, Johnson GL. Increased expression of death receptors 4 and 5 synergizes the apoptosis response to combined treatment with etoposide and TRAIL. Mol Cell Biol. 2000;20:205–12.
224. Nimmanapalli R, et al. Pretreatment with paclitaxel enhances apo-2 ligand/tumor necrosis factor-related apoptosis-inducing ligand-induced apoptosis of prostate cancer cells by inducing death receptors 4 and 5 protein levels. Cancer Res. 2001;61:759–63.
225. Singh TR, Shankar S, Chen X, Asim M, Srivastava RK. Synergistic interactions of chemotherapeutic drugs and tumor necrosis factor-related apoptosis-inducing ligand/Apo-2 ligand on apoptosis and on regression of breast carcinoma in vivo. Cancer Res. 2003;63:5390–400.
226. Asakuma J, Sumitomo M, Asano T, Hayakawa M. Selective Akt inactivation and tumor necrosis actor-related apoptosis-inducing ligand sensitization of renal cancer cells by low concentrations of paclitaxel. Cancer Res. 2003;63:1365–70.
227. Ohtsuka T, et al. Synergistic induction of tumor cell apoptosis by death receptor antibody and chemotherapy agent through JNK/p38 and mitochondrial death pathway. Oncogene. 2003;22:2034–44.
228. Belyanskaya LL, et al. Human agonistic TRAIL receptor antibodies Mapatumumab and Lexatumumab induce apoptosis in malignant mesothelioma and act synergistically with cisplatin. Mol Cancer. 2007;6:66.
229. Rohn TA, et al. CCNU-dependent potentiation of TRAIL/Apo2L-induced apoptosis in human glioma cells is p53-independent but may involve enhanced cytochrome c release. Oncogene. 2001;20:4128–37.
230. Nagane M, Cavenee WK, Shiokawa Y. Synergistic cytotoxicity through the activation of multiple apoptosis pathways in human glioma cells induced by combined treatment with ionizing radiation and tumor necrosis factor-related apoptosis-inducing ligand. J Neurosurg. 2007;106:407–16.
231. Tsurushima H, Yuan X, Dillehay LE, Leong KW. Radiation-inducible caspase-8 gene therapy for malignant brain tumors. Int J Radiat Oncol Biol Phys. 2008;71:517–25.
232. Fiveash JB, et al. Enhancement of glioma radiotherapy and chemotherapy response with targeted antibody therapy against death receptor 5. Int J Radiat Oncol Biol Phys. 2008;71:507–16.
233. Ashkenazi A, Holland P, Eckhardt SG. Ligand-based targeting of apoptosis in cancer: the potential of recombinant human apoptosis ligand 2/Tumor necrosis factor-related apoptosis-inducing ligand (rhApo2L/TRAIL). J Clin Oncol. 2008;26:3621–30.
234. Herbst RS, et al. Phase I dose-escalation study of recombinant human Apo2L/TRAIL, a dual proapoptotic receptor agonist, in patients with advanced cancer. J Clin Oncol. 2010;28:2839–46.
235. Soria JC, et al. Phase 1b study of dulanermin (recombinant human Apo2L/TRAIL) in combination with paclitaxel, carboplatin, and bevacizumab in patients with advanced non-squamous non-small-cell lung cancer. J Clin Oncol. 2010;28:1527–33.

236. Soria JC, et al. Randomized phase II study of dulanermin in combination with paclitaxel, carboplatin, and bevacizumab in advanced non-small-cell lung cancer. J Clin Oncol. 2011;29:4442–51.
237. Yada A, et al. A novel humanized anti-human death receptor 5 antibody CS-1008 induces apoptosis in tumor cells without toxicity in hepatocytes. Ann Oncol. 2008;19:1060–7.
238. Forero-Torres A, et al. Phase I trial of weekly tigatuzumab, an agonistic humanized monoclonal antibody targeting death receptor 5 (DR5). Cancer Biother Radiopharm. 2010;25:13–9.
239. Humphreys RC, Halpern W. Trail receptors: targets for cancer therapy. Adv Exp Med Biol. 2008;615:127–58.
240. Tolcher AW, et al. Phase I pharmacokinetic and biologic correlative study of mapatumumab, a fully human monoclonal antibody with agonist activity to tumor necrosis factor-related apoptosis-inducing ligand receptor-1. J Clin Oncol. 2007;25:1390–5.
241. Hotte SJ, et al. A phase 1 study of mapatumumab (fully human monoclonal antibody to TRAIL-R1) in patients with advanced solid malignancies. Clin Cancer Res. 2008;14:3450–5.
242. Greco FA, et al. Phase 2 study of mapatumumab, a fully human agonistic monoclonal antibody which targets and activates the TRAIL receptor-1, in patients with advanced non-small cell lung cancer. Lung Cancer. 2008;61:82–90.
243. Trarbach T, et al. Phase II trial of mapatumumab, a fully human agonistic monoclonal antibody that targets and activates the tumour necrosis factor apoptosis-inducing ligand receptor-1 (TRAIL-R1), in patients with refractory colorectal cancer. Br J Cancer. 2010;102:506–12.
244. Plummer R, et al. Phase 1 and pharmacokinetic study of lexatumumab in patients with advanced cancers. Clin Cancer Res. 2007;13:6187–94.
245. Wakelee HA, et al. Phase I and pharmacokinetic study of lexatumumab (HGS-ETR2) given every 2 weeks in patients with advanced solid tumors. Ann Oncol. 2010;21:376–81.
246. Motoki K, et al. Enhanced apoptosis and tumor regression induced by a direct agonist antibody to tumor necrosis factor-related apoptosis-inducing ligand receptor 2. Clin Cancer Res. 2005;11:3126–35.
247. Adams C, et al. Structural and functional analysis of the interaction between the agonistic monoclonal antibody Apomab and the proapoptotic receptor DR5. Cell Death Differ. 2008;15:751–61.
248. Doi T, et al. Phase 1 study of conatumumab, a pro-apoptotic death receptor 5 agonist antibody, in Japanese patients with advanced solid tumors. Cancer Chemother Pharmacol. 2011;68:733–41.
249. Herbst RS, et al. A first-in-human study of conatumumab in adult patients with advanced solid tumors. Clin Cancer Res. 2010;16:5883–91.
250. Demetri GD, et al. First-line treatment of metastatic or locally advanced unresectable soft tissue sarcomas with conatumumab in combination with doxorubicin or doxorubicin alone: a phase I/II open-label and double-blind study. Eur J Cancer. 2012;48:547–63.

# The Dark Side of Apoptosis

**Malathy P. V. Shekhar**

**Abstract** Apoptosis is conventionally regarded as an anti-cancer mechanism that eliminates or prevents mutant cell expansion necessary for tumor development and progression. However, evidence for the paradoxical role of apoptosis in tumor progression is accumulating. In this chapter, we describe the mechanisms by which apoptosis serves as a vehicle for accumulating genomic instability to promote malignant progression of tumors, and show a direct association between apoptosis and tumor progression in clinical settings. The negative therapeutic implications of increased apoptosis on clinical outcome and the need to inhibit apoptosis or disable proliferation in apoptotic tumors are discussed.

**Keywords** Genomic instability · Horizontal transfer · Cell–cell fusion · Cancer stem cell · Comedo-ductal carcinoma in situ

## Abbreviations

| | |
|---|---|
| CAD gene | Trifunctional protein with carbamoyl-phosphate synthetase 2, aspartate transcarbamylase, and dihydroorotase activities |
| DCIS | Ductal carcinoma in situ |
| EBV | Epstein-Barr virus |
| REF | Rat embryo fibroblast |
| MEF | Mouse embryo fibroblast |
| IAP | Inhibitors of apoptosis |
| PUMA | P53 upregulated modulator of apoptosis |
| FLIP | FADD-like IL-1$\beta$-converting enzyme)-inhibitory protein |
| TRAIL | TNF-related apoptosis-inducing ligand |
| DR4, DR5 | Death receptors |

M. P. V. Shekhar (✉)
Karmanos Cancer Institute, Department of Oncology, Wayne State University School of Medicine, 110 East Warren Avenue, Detroit, Michigan 48201, US
e-mail: shekharm@karmanos.org

B. Bonavida (ed.), *Molecular Mechanisms of Tumor Cell Resistance to Chemotherapy*, Resistance to Targeted Anti-Cancer Therapeutics 1, DOI: 10.1007/978-1-4614-7070-0_12,

## Introduction

Apoptosis represents a major mechanism by which tissue maintains homeostasis. Apoptosis also plays an important role in protecting tissues against tumorigenesis and malignant conversion. Cell loss occurs once tumors have grown more than a few million cells [1, 2]. Intracellular stresses such as hypoxia, nutrient deprivation, telomere shortening, and cell elimination by the host immune system are thought to contribute to steady attrition of a portion of cancer cells in solid tumors [3–5]. A hypothetical link between apoptosis and tumorigenesis has been described by Hanahan and Weinberg [6]. According to their model, the process of tumorigenesis and malignant conversion forces expression of proapoptotic factors that assist with elimination of 'mutant' cells and protection from transformation. Thus, mechanisms contributing to acquired resistance to apoptosis serve as a potential mechanism for cell survival and neoplastic conversion.

For cells to acquire the ability for limitless replicative potential or resistance to apoptosis, cells need to uncouple their growth/proliferation programs from environmental signals. It is, however, becoming abundantly clear that there are situations in which tumor cells undergo programmed cell death under conditions of optimal growth stimulation and in the absence of environmental stress. In these tumors, the apoptosis program is wired independently of cell–cell and cell-environmental communications. Interestingly, despite the proficiency for spontaneous cell death, the surviving tumor cells exhibit increased ability for malignant progression. This suggests that the very same process used for protecting cells against malignant conversion can contribute to greater vulnerability to malignancy. In metastatic melanoma, spontaneous regression of individual metastatic deposits is seen throughout the life history of the disease, yet the final outcome is usually death [7]. Similarly, spontaneous regression has been reported in metastatic renal carcinoma, neuroblastoma, colorectal carcinoma, and a wide range of solid tumors [8], suggesting that spontaneous cell loss is not an infrequent event. Yet, despite the loss of tumor cells, the final outcome is not often favorable. The fact that cell loss can occur in vitro under optimal growth and normoxic conditions suggests that cell loss seen in vivo is not necessarily due to nutrient deprivation or hypoxia but rather these tumor cells are intrinsically programmed to undergo apoptosis.

**Causes of spontaneous death**. Spontaneous loss of cells occurring in the absence of extracellular factors suggests that the fully transformed cells may be intrinsically unstable due to specific genetic abnormalities, and that in the process of achieving stability they undergo apoptotic cell death. Factors responsible for spontaneous apoptosis in tumors are diverse. In some tumors apoptotic cells are found near foci of confluent necrosis suggesting that mild ischemia may contribute to the initiation of apoptosis. The release of cytokines such as TNF-$\alpha$ by infiltrating macrophages is regarded as one of the contributing factors. The extrinsic or death receptor pathway of apoptosis involves stimulation of external surface receptors by FasL and TRAIL leading to activation of caspase-8. In the intrinsic or mitochondrial pathway, proapoptotic signals meet at the mitochondria with resultant loss of the mitochondrial membrane potential, release of cytochrome c and activation of the caspase-9 signaling cascade [9, 10]. The intrinsic pathway is regulated by members of the Bcl-2 family that have either antiapoptotic (Bcl-2, Bcl-xL, Bcl-W, Mcl-1) or proapoptotic (Bax, Bim, Bak, Bid) function. The p53 tumor suppressor protein functions in apoptosis by transcriptional upregulation of genes (e.g., *Bax*, *PUMA*) that are directly involved in apoptosis [11, 12]. Both apoptotic pathways converge on the same execution pathway resulting in the activation of caspase-3, DNA fragmentation and formation of apoptotic bodies. Tumor cells exhibit resistance to apoptosis through overexpression of FLIP, reduced expression of CD95, TRAIL DR4 or DR5 receptors, or by overexpression of IAPs or altered expression of the Bcl-2 family members [10].

**Paradoxical role of Bcl-2 and Bax in tumor progression**. Upregulation of Bcl-2, Bcl-xL, and Mcl-1 are associated with inhibition of apoptosis; however, their expressions do not often correlate strictly with poor clinical prognosis [13, 14]. In breast cancer, Bcl-2 overexpression is associated with normal ploidy, estrogen receptor positivity, and absence of metastasis; all characteristics associated with better clinical outcome and a more favorable prognosis that is contradictory to its predicted role in apoptosis resistance [15, 16]. In colorectal adenoma, Bcl-2 levels are decreased compared to the adjacent normal tissue [17]. Similar data are reported in cervical, prostate, and endometrial cancers [18]. Overexpression of Mcl-1 is associated with poor prognosis for lung, and head and neck cancers [17–20]. Thus, there are malignancies in which Bcl-2 family member overexpression correlates with favorable prognosis and those in which it is correlative of high tumor grade. Paradoxically, elevated expressions of the proapoptotic proteins Bax and Bak have been associated with poor prognosis in esophageal carcinoma and bladder cancer, respectively [21–23]. In breast tumors, an increase in the proportion of apoptotic cells was observed in recurrent tumors as compared with primary lesions, and patients with tumors with higher apoptotic indices were associated with shorter survival [24]. Several examples of a positive correlation between the apoptotic index and the tumor grade have been reported [25–27]. The role of apoptosis suppression as a vehicle for enhancing genomic instability was tested by measuring the effects of Bcl-2 overexpression on the frequency of CAD gene amplification in cells exposed to the CAD inhibitor PALA. Bcl-2 overexpression failed to increase the frequency of CAD gene amplification. However, similar analysis on cells expressing mutant *p53* showed enhanced frequency of CAD gene amplification. These data suggest that although both p53 inactivation and Bcl-2 overexpression suppress apoptosis, apoptosis inhibition by Bcl-2 overexpression does not make the cells genetically unstable [28]. These data are consistent with low grade tumors produced by Bcl-2 overexpressing cells, further confirming that apoptosis inhibition by Bcl-2 overexpression does not favor selection of tumor cell variants to increase genomic instability [28].

**Apoptosis as a mechanism for driving genomic instability**. Reduced rates of apoptosis correlate with fewer cell divisions and, hence, the possibility of generating fewer mutant clones. On the other hand, higher rates of apoptosis would require increased rates of cell division to compensate for cell loss, and consequently lay the foundation for accruing genomic instability [29, 30]. As discussed above, Bcl-2 overexpression and p53 loss both act by blocking apoptosis; however, only the latter enhances accumulation of genomic instability. In this context, it is interesting to note that Bcl-2 overexpression and mutant p53 are rarely coexpressed in a tumor. Bcl-2 overexpression was found in breast and head and neck cancers expressing wild type p53 [13], [31–34], whereas breast tumors with p53 accumulation (indicative of mutant p53 expression) showed low Bcl-2 expression [31, 35]. p53 is inactivated late in most cancers, suggesting that loss of p53 function in later stages of tumorigenesis may translate into higher rates of apoptosis and consequently higher rates of proliferation during the course of tumor development that would enable expansion and accumulation of mutant clones favoring malignant progression. Clinical data show that noncomedo-ductal carcinoma in situ (DCIS) breast cancers express normal levels of Bcl-2 compared to comedo-DCIS that express weak or negligible Bcl-2, strong Bax and mutant p53. Accordingly, comedo-DCIS lesions are characterized by high apoptotic and mitotic indices [36], and are at greater risk for recurrence and malignant progression compared to noncomedo-DCIS. Evidence for a direct link between increased apoptosis and tumorigenesis was recently demonstrated with a mouse model harboring the loxP-targeted allele of Mcl-1 and albumin promoter-driven Cre-recombinase. Hepatocytes of Mcl-1 fl/fl-AlbCre mice lacking Mcl-1 exhibited increased apoptosis and spontaneously developed hepatocellular carcinoma-like lesions at >50 % higher incidence [37]. These data suggest that apoptosis and proliferation work in concert to drive malignant progression.

**Mechanisms contributing to tumor progression in apoptotic tumors.** Apoptosis allows elimination of unwanted or unstable cells by processing them into an array of smaller bodies called apoptotic bodies by cellular blebbing and fragmentation. Apoptotic bodies have intact plasma membrane and adequate energy supply to maintain their membrane integrity. Cells undergoing apoptosis and apoptotic bodies expose phosphatidylserine on their surface that promote recognition by professional (macrophages, dendritic cells, B-lymphocytes) and nonprofessional (epithelial cells and fibroblasts) phagocytes, resulting in quick ingestion and lysosomal degradation [38]. It is generally believed that apoptosis does not induce inflammation because the apoptotic bodies are rapidly cleared to prevent release of harmful immunogenic materials from the dying cells [39]. We posit that the clearance rates of apoptotic bodies and the presence of associated inflammation will depend upon the extent and rate of apoptosis. Tumor tissues undergoing massive apoptotic cell loss such as comedo-DCIS breast cancers are generally associated with a strong inflammatory response. It is possible that these cells are recruited to mop-up the overwhelming amounts of released apoptotic bodies, or that the slow clearing of apoptotic bodies induces the inflammatory response. Either way, a tumor microenvironment rich in inflammatory cells would promote tumor progression. Additionally, slow or incomplete removal of apoptotic bodies would result in the accumulation of released genetic material into the cellular milieu and, thus, trigger the horizontal transfer or uptake of the DNA by neighboring tumor or stromal cells.

**Horizontal transfer of genetic material via cell–cell fusion**. Horizontal transfer of genes has been reported in bacteria and fungi and plays an important role in the generation of antibiotic resistance and the adaptation to new environments [40]. Thus, horizontal transfer of genes represents a powerful mechanism for bacterial diversification. Cell–cell fusion is considered to play an important role in horizontal transmission of genes and malignant transformation [41–43]. Transfer of genetic information could occur via fusions between tumor and stromal cells or between tumor cells. Human glioblastoma grafted to hamster cheek pouches produced hybrid human/hamster tumors in which human chromosome segregation occurred within the first transplant generation and showed widespread metastasis. At least seven genes from the six human chromosomes were retained of which three genes implicated in oncogenesis (*CD7*, *CXCR4* and *PLAGL2*) showed continued expression [44]. CXCR4 (also called fusin), a G-protein coupled chemokine receptor for SDF-1 has been implicated in proliferation, motility, homing and metastasis of cancer cells and is associated with regions of cell death and angiogenesis. These data suggest that in vivo stability of the resulting hybrids depends upon the selective growth advantage provided by the DNA taken up.

**Horizontal transfer of DNA via apoptotic conversion**. There is increasing evidence that even cell-free cancer DNA can be transferred to induce malignancy. Holmgren et al. [45] demonstrated that DNA can be transferred from apoptotic cells to recipient neighboring cells by phagocytosis. Cocultivation of cell lines containing integrated copies of the Epstein-Barr virus (EBV) resulted in rapid a uptake of EBV DNA to the nucleus of the phagocytosing cell. Once transferred, the expression of EBV encoded genes was detected at both the mRNA and protein levels. Similarly, apoptotic bodies derived from *c-Ha-ras*$^{\mathrm{Val}}$ [12] and *c-myc* transformed rat embryo fibroblasts (REFs) were able to transform p53-/- or p21-/- mouse embryo fibroblasts (MEFs) but not wild type MEFs [46, 47]. FISH analysis confirmed that entire chromosomes are transferred from the apoptotic bodies of REFs and become integrated into the mouse host genome. However, the stability of the integrated DNA can be maintained only if it confers a selective growth advantage to the recipient cell. Following uptake of the apoptotic genetic material, normal recipient cells with the activated Chk2/p53/p21 DNA damage response pathway block replication of the transferred DNA, thereby protecting them from the potentially harmful effects of the apoptotic DNA [48]. Similar uptake of apoptotic bodies with resultant acquisition of and propagation of drug resistance genes has been demonstrated in prostate cancer cells [49]. Since p53 is lost in most cancers, horizontal transfer of genetic material from the dying tumor cells to recipient tumor cells may serve as a driving force for accumulation of genomic instability and high mutability of tumor cells. Apoptotic DNA from the dying tumor cells can also be transferred to recipient stromal cells (fibroblasts, endothelial cells, macrophages). Such transfers could actively modify the structure and behavior of the surrounding microenvironment, and potentially provide an explanation for pro-tumorigenic and pro-metastatic properties of the tumor microenvironment. Apoptotic DNA was detected in the nuclei of ~15 % of the phagocytosing cells [45], suggesting that horizontal DNA transfer is an efficient mode of enhancing genomic diversity of tumor cells that is dictated by the rate and extent of apoptosis in the tumor. In contrast, mutation of specific genes is an inefficient process that requires amplification of the mutated cells and is limited by activities of surveillance mechanisms that monitor and maintain the genomic integrity. It is possible that the fluidity of the cell membrane is increased in cells sensitive to apoptosis, making it more receptive for cell fusion and DNA transfer. Given the manner in which even highly differentiated epithelial cells are stimulated to become phagocytic by the proximity of an apoptotic body, it is one of the less studied but more remarkable components of apoptosis [48, 50]. Horizontal DNA transfer or apoptotic conversion may be clinically important particularly in tumors characterized by high spontaneous apoptosis such as comedo-DCIS breast cancer since these tumor cells also express mutant p53.

Genetic exchanges occurring either via cell–cell fusions or by apoptotic conversion can provide the residual tumor with new attributes for survival, growth, progression and metastasis. It is possible that the phenotypic and genotypic diversity (or heterogeneity) observed in cells within a tumor, between primary and recurrent tumors, primary and metastatic tumors, and/or between metastases of the same tumor arise at least in part by horizontal transmission of genes and gene products by cell–cell fusions or direct incorporation of apoptotic DNA into cancer or stromal cells.

**Cancer stem cell activation**. Recent studies have invoked the role of cancer stem cells to explain the relationship between enhanced apoptosis and tumor progression. It has been proposed that in tumors with high apoptotic rates, the dying cells may free up space for cancer stem cells to proliferate into and populate the tumor [51, 52]. To determine the association between apoptosis and increased tumor growth/progression, Enderling et al. [53] simulated tumor development for different spontaneous cell death rates, and initialized each simulation with one cancer stem cell, stopping the simulation after 35 months or when the tumor reached confluence. When random cell death among tumor cells was increased, an increase in the number of stem cells was observed [53]. While spontaneous cell death can reduce the number of tumor cells in the short run, they can facilitate sufficient symmetric stem cell divisions to enrich the stem cell pool and ultimately promote malignant expansion [54]. These findings have therapeutic implications since conventional anti-cancer therapies are directed towards eradicating apoptosis-sensitive tumor bulk populations while sparing the therapy of resistant cancer stem cells. Thus, the accelerated tumor recurrence following therapy may be explained by the opportunistic proliferation of quiescent tumor cells into the space made available by the initial killing. This treatment recovery cycle could favor the creation of new stem cells by symmetrical division of previously quiescent stem cells as the latter are considered to be more resistant to radiation [55] and chemotherapy [56] compared to their nonstem counterparts [53]. As the tumor returns to its pretreatment size, the tumor could become more refractory to treatment. This notion is supported by observations that high rates of apoptosis in cancer cells correlate with tumor progression [30], whereas upregulation of anti-apoptotic factors suppresses tumor progression and improves prognosis [18, 28]. Alternatively, apoptosis may accelerate tumor generation or progression by preferentially eliminating cells that retain normal apoptosis sensitivity while sparing apoptosis-resistant mutant cells for expansion of mutant clones [57, 58].

**A prototype clinical disease depicting pertinence of apoptosis adverse effects**. The relevance of detrimental effects of apoptosis to clinical settings is best illustrated with comedo-type ductal carcinoma in situ. Among the several ductal carcinoma in situ (DCIS) subtypes of preinvasive breast cancer, comedo type DCIS or comedo-DCIS accounts for $\sim$10 % of all DCIS and confers the greatest risk for progression and post-operative recurrence [59, 60]. Comedo-DCIS tumors are easily distinguished from other DCIS by the characteristic central comedo-necrosis [61] that results from extensive spontaneous apoptosis [62]. Yet despite abundant cell loss, comedo-DCIS often demonstrates microinvasion, chromosome aneuploidy, and higher proliferation and recurrence rates compared to non-comedo DCIS tumors [63–65]. MCF10DCIS.com human breast cancer cells produce tumors that resemble clinical comedo-DCIS and recapitulate the temporal sequence of progression from in situ to invasive cancer [62, 66, 67]. Using the MCF10DCIS.com model, we have demonstrated that spontaneous apoptosis contributes to the etiology and progression of comedo-DCIS [62] (Fig. 1). Spontaneous MCF10DCIS.com cell loss is activated by the mitochondrial pathway with upregulation of Bax, decreases in Bcl-2 and loss of p53 [62]. Clinical comedo-DCIS, like MCF10DCIS.com cells, show a significant drop in Bcl-2 expression combined with an increase in mutant p53 levels [36]. MCF10DCIS.com cells undergo spontaneous apoptosis in vitro under optimal growth conditions indicating that cell loss in vivo is not due to extraneous factors, but rather they are preprogrammed to undergo apoptosis [62]. The high rates of apoptosis in comedo-DCIS are accompanied by compensatory increases in PCNA-positive cells that are enriched for the $CD44^{+}/CD24^{-}$ phenotype (Fig. 2). $CD44^{+}/CD24^{-}$ cells have tumor-initiating properties in breast cancer [68]. The $CD44^{+}/CD24^{-}$ phenotype is associated with stem cell-like characteristics [69], enhanced potential for invasion [70], radiation resistance [71] and with distinct genetic profiles that correlate with adverse prognosis [72]. These data corroborate that abundant cell loss in comedo-DCIS provides increased opportunities for selecting cell variants with greater malignancy potential. The presence of an intact myoepithelial layer is indicative of DCIS lacking invasive potential. Interestingly, apoptosis is also implicated in the initiation of progression of clinical and MCF10DCIS.com-derived comedo-DCIS tumors as both luminal epithelial and myoepithelial cells are concurrently eliminated by apoptosis [62] (Fig. 2). These data provide clinical support for the adverse effects of apoptosis in neoplastic progression and show that apoptosis is tightly linked with the coordinated and concerted events that lead up to malignant progression (summarized in Fig. 3).

**Therapeutic implications**. The majority of anti-cancer therapies work by inducing apoptosis in tumors. In tumors undergoing high rates of natural (spontaneous) or therapy-induced apoptosis, increased horizontal transfer of genetic

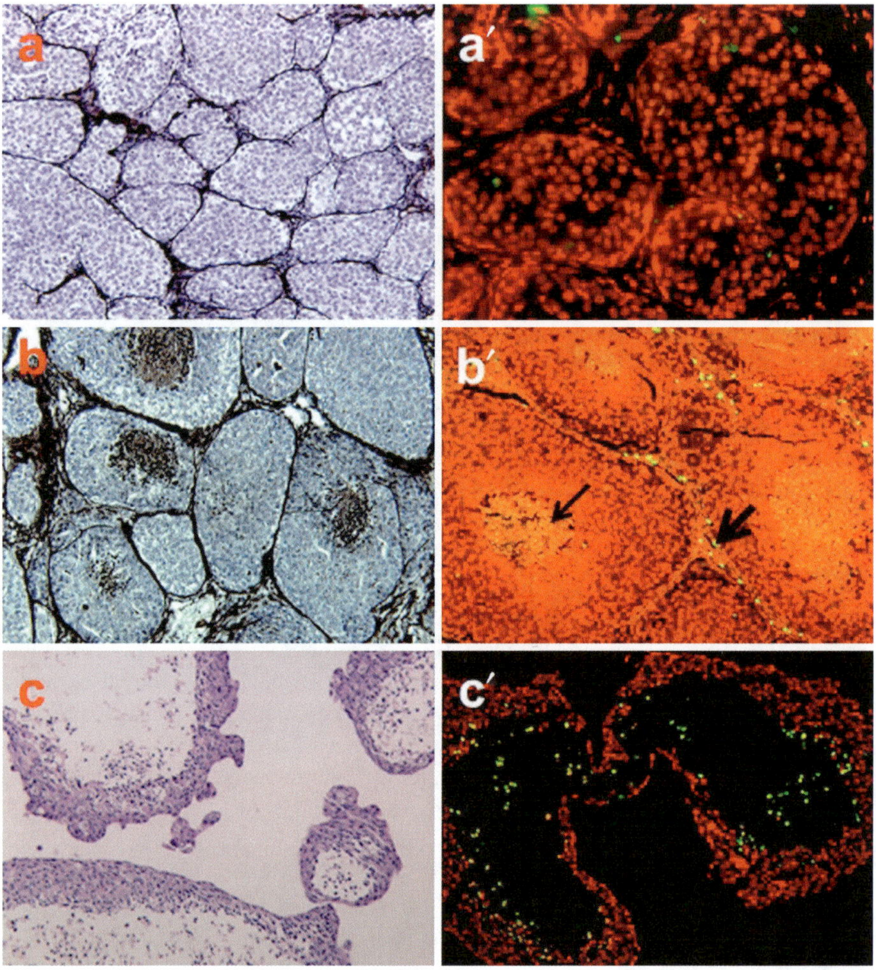


**Fig. 1** Progression of MCF10DCIS.com to comedo-DCIS occurs by high rates of spontaneous apoptosis of luminal epithelial and myoepithelial cells. Panels **a**, **a′**: early DCIS lesions; **b**, **b′**: comedo-DCIS lesions; **c**, **c′**: MCF10DCIS.com multicellular tumor spheroids undergoing spontaneous apoptosis in vitro. Panels **a** and **c**, H&E staining; **b**, silver staining; **a′**–**c′**, TUNEL staining. *Thin* and *thick arrows* in **b′** show apoptotic luminal epithelial and myoepithelial cells, respectively

material by cell–cell fusions or apoptotic conversion, cancer stem cell activation, and/or associated inflammatory response may all contribute to the adverse effects of chemo- and radio-therapy, viz., initial tumor regression that is accompanied by quicker relapse, a greater tumor burden and therapy resistance [73]. Thus, it would seem that using therapies that induce apoptosis in tumors experiencing high rates of spontaneous apoptosis is counterintuitive. Inhibiting apoptosis in these tumor tissues may be beneficial as it could decrease compensatory proliferation of

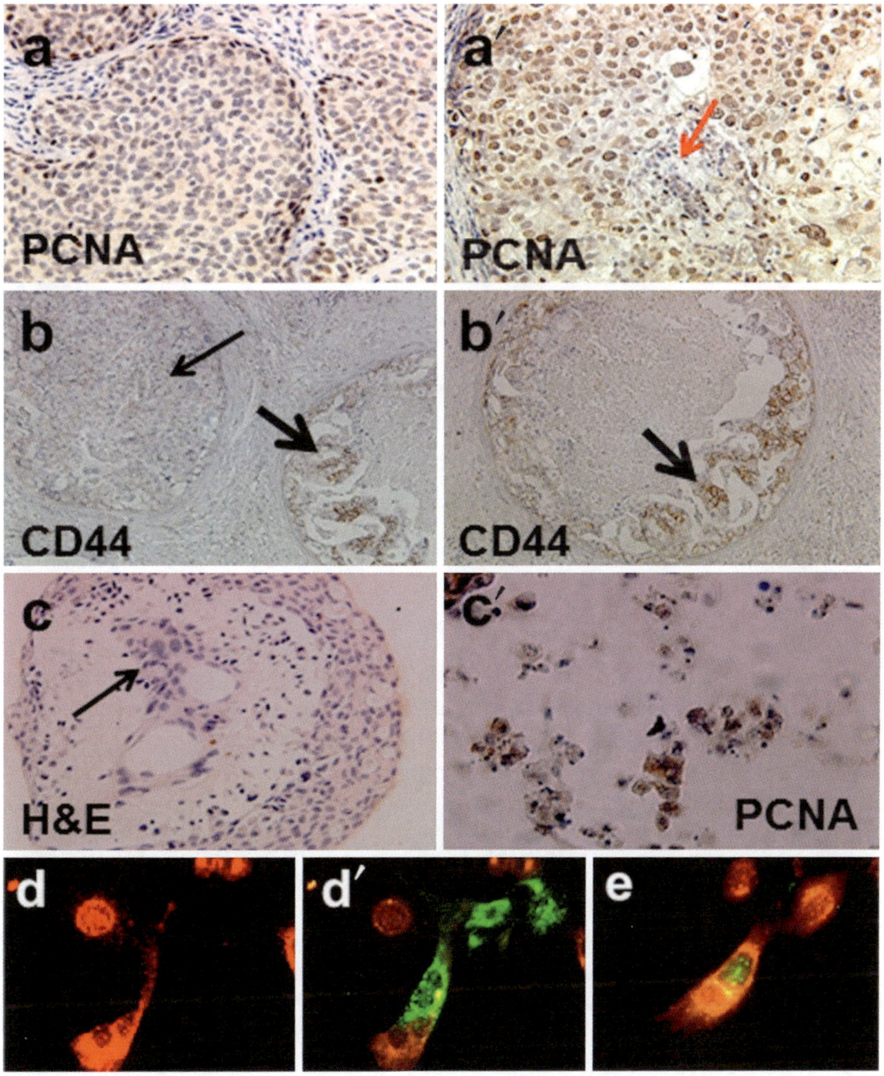


**Fig. 2** Progression of MCF10DCIS.com-derived comedo-DCIS lesions is accompanied by increase in PCNA-positive and $CD44^{+}$ ($CD24^{-}$negative, not shown) cells. Panels **a**, **a′** PCNA expression in early and comedo-DCIS lesions, respectively. Note the presence of cells with uniform nuclei in early lesions (*panel* **a**) compared to cells with larger or pleiomorphic nuclei in the vicinity of the apoptotic core (indicated by *arrow*) of advanced lesions (*panel* **a′**). Panels **b**, **b′** CD44 expression. Note low CD44 expression in early lesion (*thin arrow* in **b**) and strong CD44 expression in advanced comedo-DCIS (*thick arrow* in **b** and **b′**). Panels **c**, **c′** H&E and PCNA staining, respectively, of MCF10DCIS.com multicellular tumor spheroids generated in vitro. Note the appearance of large nucleated cells in the apoptotic core (*arrow* in **c**) with PCNA expression (**c′**). Panels **d–e** cell–cell fusion propensity. MCF10DCIS.com cells were prelabeled with DiI (*red*) or DiO (*green*) and cocultured. Note the formation of fused cells in panel **d′** and **e**

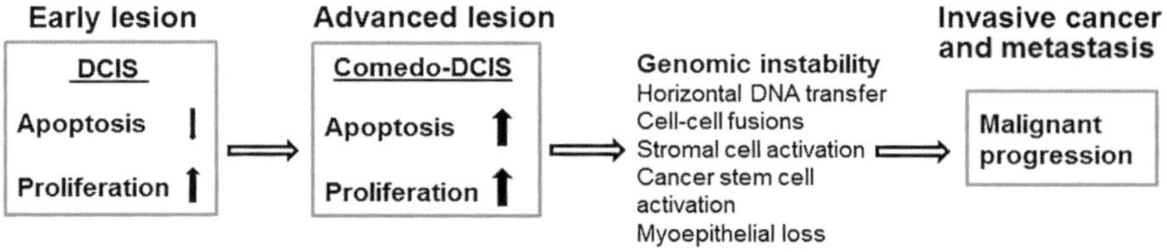


**Fig. 3** Schematic model depicting the role of high apoptosis in malignant progression using the comedo-DCIS model of breast cancer

surviving tumor cells, decrease activation and expansion of initiated or cancer stem cells, decrease acquisition of genetic instability by inhibiting horizontal DNA transfer and reduce inflammation. Taking into account the adverse effects of apoptosis in malignant transformation and progression, it would seem that disabling proliferation of stem and nonstem cells in "apoptotic tumors" may be necessary to achieve a longer lasting clinical response.

## Conclusion

Contrary to the popular belief that apoptosis acts to safeguard cells from neoplastic conversion and progression, apoptosis must also be viewed as a "not so innocent" participant that actively promotes tumor progression. The latter paradoxical role becomes relevant to tumors experiencing high rates of natural (spontaneous) apoptosis or therapy-induced apoptosis, and must be taken into account when therapy decisions are made.

**Acknowledgments** Supported by U.S. Army Medical Research and Materiel Command DAMD17-02-1-0618.

**Conflicts of Interest** No potential conflicts of interest were disclosed.

## References

1. Steel GG. Cell loss from experimental tumors. Cell Tissue Kinet. 1968;1:193–207.
2. Bertuzzi A, Gandolfi A, Sinisgalli C, Starce G, Ubezio P. Cell loss and the concept of potential doubling time. Cytometry. 1997;29:34–40.
3. Kerr J, Winterford C, Harmon B. Apoptosis, Its significance in cancer and cancer therapy. Cancer. 1994;73:2013–23.
4. Elston D. Mechanisms of regression. Clin Med Res. 2004;2:85–8.
5. Smyth M, Kershaw M. Discovery of an innate cancer resistance gene. Mol Intervent. 2003;3:186–9.
6. Hanahan D, Weinberg RA. The hallmarks of cancer. Cell. 2000;100:57–70.

7. McGovern V. Spontaneous regression of melanoma. Pathology. 1975;2:91–9.
8. Chang W. Complete spontaneous regression of cancer: four case reports, review of literature and discussion of possible mechanisms involved. Haw Med J. 2000;59:379–87.
9. Green DR, Reed JC. Mitochondria and apoptosis. Science. 1998;281:1309–12.
10. Fulda S. Inhibitor of apoptosis proteins as targets for anticancer therapy. Expert Rev Anticancer Ther. 2007;7:1255–64.
11. Miyashita T, Reed JC. Tumor suppressor p53 is a direct transcriptional activator of the bax gene. Cell. 1995;80:293–9.
12. Nakano K, Vousden KH. PUMA, a novel proapoptotic gene, is induced by p53. Mol Cell. 2001;7:683–94.
13. Joensuu H, Pylkkanen L, Toikkanen S. Bcl-2 protein expression and long term survival in breast cancer. Am J Pathol. 1994;145:1191–8.
14. Hamilton A, Piccart M. The contribution of molecular markers to the prediction of response in the treatment of breast cancer: a review of the literature on Her-2, p53, and BCL-2. Ann Oncol. 2000;11:647–63.
15. Kobayashi S, Iwase H, Ito Y, Yamashita H, Iwata H, Yamashita T, Ito K, Toyama T, Nakamura T, Masaoka A. Clinical significance of bcl-2 gene expression in human breast cancer tissues. Breast Cancer Res Treat. 1997;42:173–81.
16. Inada T, Kikuyama S, Ichikawa A, Igarashi S, Ogata Y. Bcl-2 expression as a prognostic factor of survival of gastric carcinoma. Anticancer Res. 1998;18:2003–10.
17. Krajewska M, Moss SF, Krajevski S, Song K, Holt PR, Reed JC. Elevated expression of Bcl-X and reduced Bak in primary colorectal adenocarcinomas. Cancer Res. 1996;56:2422–7.
18. Gurova KV, Gudkov AV. Paradoxical role of apoptosis in tumor progression. J Cell Biochem. 2003;88:128–37.
19. Hotz MA, Bosq J, Zbaeren P, Reed J, Schwab G, Krajewski S, Brousset P, Borner MM. Spontaneous apoptosis and the expression of p53 and Bcl-2 family proteins in locally advanced head and neck cancer. Arch Otolaryngol Head Neck Surg. 1999;125:417–22.
20. Eerola AK, Ruokolainen H, Soini Y, Raunio H, Paakko P. Accelerated apoptosis and low bcl-2 expression associated with neuroendocrine differentiation predict shortened survival in operated large cell carcinoma of the lung. Pathol Oncol Res. 1999;5:179–86.
21. Haitek A, Posch B, El-Baz M, Mokhtar AA, Susani M, Ghoneim MA, Marberger M. Bilharzial related, organ confined, muscle invasive bladder cancer: Prognostic value of apoptotic markers, p53, E-cadherin, epidermal growth factor receptor and c-erbB-2. J Urol. 2001;165:1481–7.
22. Kurabayashi A, Furihata M, Matsumoto M, Ohtsuki Y, Sasaguri S, Ogoshi S. Expression of Bax and apoptosis-related proteins in esophageal squamous cell carcinoma including dysplasia. Mod Pathol. 2001;14:741–7.
23. Takayama T, Nagao M, Sawada H, Yamada Y, Emoto K, Fujimoto H, Ueno M, Hirao S, Nakajima Y. Bcl-X expression in esophageal squamous cell carcinoma: association with tumor progression and prognosis. J Surg Oncol. 2001;78:116–23.
24. Vakkala M, Lahteenmaki K, Raunio H, Paakko P, Soini Y. Apoptosis during breast carcinoma progression. Clin Cancer Res. 1999;5:319–24.
25. Tanji N, Yokoyama M, Sugamoto T, Takeuchi M, Terada N. Apoptosis in prostatic adenocarcinomas: a study of relationship to Ki-67 and Bcl-2 protein expression. Anticancer Res. 1998;18:1111–6.
26. Li L, Yan L, Wang Z, Liu Z, Wei Y, Huang G. Change of apoptotic status in human colorectal adenoma-carcinoma stage sequences and its correlation with carcinogenesis and prognosis. Clin Med J. 2000;113:886–8.
27. Sjostrom J, Bergh J. How apoptosis is regulated, and what goes wrong in cancer. BMJ. 2001;322:1538–9.
28. Gurova KV, Kwek SS, Koman IE, Kamarov AP, Kandel E, Nikiforov MA, Gudkov AV. Apoptosis inhibitor as a suppressor of tumor progression: expression of Bcl-2 eliminates selective advantages for p53-deficient cells in the tumor. Cancer Biol Ther. 2002;1:39–44.
29. Nowell PC. The clonal evolution of tumor cell populations. Science. 1976;194:23–8.

30. Wodarz D, Komarova N. Can loss of apoptosis protect against cancer? Trends Genet. 2007;23:232–7.
31. Berardo MD, Elledge RM, de Moor C, Clark GM, Osborn CK, Allred C. Bcl-2 and apoptosis in lymph node positive breast carcinoma. Cancer. 1998;82:1296–302.
32. Castiglione F, Sarotto I, Fontana V, Destifanis M, Venturino A, Ferro S, Cardaropoli S, Orengo MA, Porcile G. Bcl-2, p53 and clinical outcome in series of 138 operable breast cancer patients. Anticancer Res. 1999;19:4555–63.
33. Tete S, Pappalardo S, Fioroni M, Salini L, Imperatrice AM, Perfetti G. Bcl-2, p53, Ki-67 and apoptotic index in cancerous and precancerous lesions of the oral mucosa. Minerva Stomatol. 1999;48:419–25.
34. Lazaris AC, Lenardi I, Kavantzas N, Kandiloros D, Adamapoulos G, Davaris P. Correlation of tumor markers p53, bcl-2 and cathepsin-D with clinicopathological features and disease-free survival in laryngeal squamous cell carcinoma. Pathol Int. 2000;50:717–24.
35. van Slooten HJ, van de Vijver MJ, van de Velde CJ, van Dierendonck JH. Loss of bcl-2 in invasive breast cancer is associated with high rates of cell death, but also with increased proliferative activity. Br J Cancer. 1998;77:789–96.
36. Megha T, Ferrari F, Arcuri F, Lalinga AV, Lazzi S, Cardone C, Cevenini G, Leoncini L, Tosi P. Cellular kinetics and expression of bcl-2 and p53 in ductal carcinoma of the breast. Oncol Rep. 2000;7:473–8.
37. Weber A, Boger R, Vick B, Urbanik T, Haybaeck J, Zoller S, Teufel A, Krammer PH, Opferman JT, Galle PR, Schuchmann M, Heikenwalder M, Schulze-Bergkamen H. Hepatocyte-specific deletion of the anti-apoptotic protein Mcl-1 triggers proliferation and hepatocarcinogenesis in mice. Hepatology. 2010;51:1226–36.
38. Grimsley C, Ravichandran KS. Cues for apoptotic cell engulfment: eat-me, don't eat-me and come-get-me signals. Trends Cell Biol. 2003;13:648–56.
39. Ren Y, Savill J. Apoptosis: the importance of being eaten. Cell Death Differ. 1998;5:563–8.
40. Akiba T, Koyama K, Ishiki Y, Kimura S, Fukushima T. On the mechanism of the development of multiple-drug-resistant clones of Shigella. Jpn J Microbiol. 1960;4:219–27.
41. Goldenberg DM. Ü ber die Progression der Malignita ät: Eine Hypothese [on the progression of malignancy: a hypothesis]. Klin Wochenschr. 1968;46:898–9.
42. Goldenberg DM, Pavia RA, Tsao MC. In vivo hybridization of human tumour and normal hamster cells. Nature. 2012;250:649–51.
43. Goldenberg DM, Pavia RA. Horizontal transmission of malignant conditions rediscovered. New Engl J Med. 1981;305:283–4.
44. Goldenberg DM, Zagzag D, Heselmeyer-Haddad KM, Berroa Garcia LY, Ried T, Loo M, Chang CH, Gold DV. Horizontal transmission and retention of malignancy, as well as functional human genes, after spontaneous fusion of human glioblastoma and hamster host cells in vivo. Int J Cancer. 2012;131:49–58.
45. Holmgren L, Szeles A, Rajnavölgyi E, Folkman J, Klein G, Ernberg I, Falk KI. Horizontal transfer of DNA by the uptake of apoptotic bodies. Blood. 1999;93:3956–63.
46. Bergsmedh A, Szeles A, Henriksson M, Bratt A, Folkman MJ, Spetz AL, Holmgren L. Horizontal transfer of oncogenes by uptake of apoptotic bodies. Proc Natl Acad Sci. 2001;98:6407–11.
47. Bergsmedh A, Szeles A, Spetz AL, Holmgren L. Loss of the p21(Cip1/Waf1)cyclin kinase inhibitor results in propagation of horizontally transferred DNA. Cancer Res. 2002;62:575–9.
48. Bergsmedh A, Ehnfors J, Kawane K, Motoyama N, Nagata S, Holmgren L. DNase II and the chk2 DNA damage pathway form a genetic barrier blocking replication of horizontally transferred DNA. Mol Cancer Res. 2006;4:187–95.
49. de la Taille A, Chen MW, Burchardt M, Chopin DK, Buttyan R. Apoptotic conversion: evidence for exchange of genetic information between prostate cancer cells mediated by apoptosis. Cancer Res. 1999;59:5461–3.
50. Chakraborty AK, Sodi S, Rachkovsky M, Kolesnikova N, Platt JT, Bolognia JL, Pawelek JM. A spontaneous murine melanoma lung metastasis comprised of host x tumor hybrids. Cancer Res. 2000;60:2512–9.

51. Ehemann V, Sykora J, Vera-Delgado J, Lange A, Otto HF. Flow cytometric detection of spontaneous apoptosis in human breast cancer using the TUNEL-technique. Cancer Lett. 2003;194:125–31.
52. Meggiato T, Calabrese F, Valente M, Favaretto E, Baliello E, Del Favero G. Spontaneous apoptosis and proliferation in human pancreatic cancer. Pancreas. 2000;20:117–22.
53. Enderling H, Anderson ARA, Chaplain MAJ, Behesti A, Hlatky L, Hahnfeldt P. Paradoxical dependencies of tumor dormancy and progression on basic cell kinetics. Cancer Res. 2009;69:8814–21.
54. Lynch MD. The role of cellular senescence may be to prevent proliferation of neighboring cells within stem cell niches. Ann NY Acad Sci. 2004;1019:191–4.
55. Diehn M, Cho RW, Lobo NA, Kalisky T, Dorie MJ, Kulp AN, Qian D, Lam JS, Ailles LE, Wong M, Joshua B, Kaplan MJ, Wapnir I, Dirbas F, Somlo G, Garberoglio C, Paz B, Shen J, Lau SK, Quake SR, Brown JM, Weissman IL, Clarke MF. Association of reactive oxygen species levels and radioresistance in cancer stem cells. Nature. 2009;458:780–3.
56. Gangemi R, Paleari L, Orengo AM, Cesario A, Chessa L, Ferrini S, Russo P. Cancer stem cells: a new paradigm for understanding tumor growth and progression and drug resistance. Curr Med Chem. 2009;16:1688–703.
57. Ziegler A, Jonason AS, Leffell DJ, Simon JA, Sharma HW, Kimmelman J, Remington L, Jacks T, Brash DE. Sunburn and p53 in the onset of skin cancer. Nature. 1994;372:773–6.
58. Graeber TG, Osmanian C, Jacks T, Housman DE, Koch CJ, Lowe SW, Giaccia AJ. Hypoxia-mediated selection of cells with diminished apoptotic potential in solid tumors. Nature. 1996;379:88–91.
59. Fisher ER, Land SR, Saad RS, Fisher B, Wickerham DL, Wang M, et al. Pathologic variables predictive of breast events in patients with ductal carcinoma in situ. Am J Clin Pathol. 2007;128:86.
60. Yang M, Moriya T, Oguma M, Cruz CDL, Endoh M, Ishida T, Hirakawa H, Orita Y, Ohuchi N, Sasano H. Microinvasive ductal carcinoma (T1mic) of the breast. The clinicopathological profile and immunohistochemical features of 28 cases. Pathol Int. 2003;53:422–8.
61. Moinfar F, Mannion C, Man YG, Tavassoli FA. Mammary "comedo"-DCIS: apoptosis, oncosis, and necrosis: an electron microscopic examination of 8 cases. Ultrastruct Pathol. 2000;24:135.
62. Shekhar MP, Tait L, Pauley RJ, Wu GS, Santner SJ, Nangia-Makker P, Shekhar V, Nassar H, Visscher DW, Heppner GH, Miller FR. Comedo-ductal carcinoma in situ: a paradoxical role for programmed cell death. Cancer Biol Ther. 2008;7:1774.
63. Jaffer S, Bleiweiss IJ. Histologic classification of ductal carcinoma in situ. Microsc Res Tech. 2002;59:92.
64. Aasmundstad T, Haugen O. DNA ploidy in intraductal breast carcinomas. Eur J Cancer. 1992;26:956.
65. Meyer J. Cell kinetics of histologic variants of in situ breast carcinoma. Breast Cancer Res Treat. 1986;7:171.
66. Miller FR, Santner SJ, Tait L, Dawson PJ. MCF10DCIS. com xenograft model of human comedo ductal carcinoma in situ. J Nat Cancer Inst. 2000;92:1185.
67. Tait LR, Pauley RJ, Santner SJ, Heppner GH, Heng HH, Rak JW, et al. Dynamic stromal-epithelial interactions during progression of MCF10DCIS.com xenografts. Int J Cancer. 2007;120:2127.
68. Al-Hajj M, Wicha MS, Benito-Hernandez A, Morrison SJ, Clarke MF. Prospective identification of tumorigenic breast cancer cells. Proc Natl Acad Sci USA. 2003;100:3983–8.
69. Ponti D, Costa A, Zaffaroni N, Pratesi G, Petrangolini G, Coradini D, Pilotti S, Pierotti MA, Daidone MG. Isolation and in vitro propagation of tumorigenic breast cancer cells with stem/progenitor cell properties. Cancer Res. 2005;65:5506–11.
70. Sheridan C, Kishimoto H, Fuchs RK, Mehrotra S, Bhat-Nakshatri P, Turner CH, Goulet R Jr, Turner CH, Badve S, Nakshatri H. CD44+/CD24− breast cancer cells exhibit enhanced invasive properties: an early step necessary for metastasis. Breast Cancer Res. 2006;8:R59.

71. Phillips TM, McBride WH, Pajonk F. The response of CD24(-/low)/CD44+ breast cancer-initiating cells to radiation. J Natl Cancer Inst. 2006;98:1777–85.
72. Shipitsin M, Campbell LL, Argani P, Weremowicz S, Bloushtain-Qimron N, Yao J, Nikolskaya T, Serebryiskaya T, Beroukhim R, Hu M, Halushka MK, Sukumar S, Parker LM, Anderson KS, Harris LN, Garber JE, Richardson AL, Schnitt SJ, Nikolsky Y, Gelman RS, Polyak K. Molecular definition of breast tumor heterogeneity. Cancer Cell. 2007;11:259–73.
73. Michor F, Hughes TP, Iwasa Y, Branford S, Shah NP, Sawyers CL, Nowak MA. Dynamics of chronic myeloid leukemia. Nature. 2005;435:1267–70.

# Index



B. Bonavida (ed.), *Molecular Mechanisms of Tumor Cell Resistance to Chemotherapy*, Resistance to Targeted Anti-Cancer Therapeutics 1, DOI: 10.1007/978-1-4614-7070-0,

Printed by Printforce, the Netherlands